Electronic Circuits and Systems

Electronic Circuits and Systems

J. D. Ryder

Formerly Professor of Electrical Engineering
Michigan State University
East Lansing, Michigan

Charles M. Thomson

Formerly Senior Dean of Instruction
Wentworth Institute and
Wentworth College of Technology
Boston, Massachusetts

PRENTICE-HALL, INC., Englewood Cliffs, New Jersey

Library of Congress Cataloging in Publication Data

Ryder, John Douglas (date)
    Electronic circuits and systems.

    Includes index.
    1. Electronics.   2. Semiconductors.   3. Electronic
circuits.   I. Thomson, Charles M., joint author.
II. Title.
TK7815.R89
ISBN   0-13-250407-3

PRENTICE-HALL INTERNATIONAL, INC., *London*
PRENTICE-HALL OF AUSTRALIA, PTY. LTD., *Sydney*
PRENTICE-HALL OF CANADA, LTD., *Toronto*
PRENTICE-HALL OF INDIA PRIVATE LIMITED, *New Delhi*
PRENTICE-HALL OF JAPAN, INC., *Tokyo*
PRENTICE-HALL OF SOUTHEAST ASIA (PTE.) LTD., *Singapore*

Contents

CHAPTER **5** *DC Bias for the Transistor* *96*

CHAPTER **6** *The Field-Effect Transistor* *120*

CHAPTER **7**   *The Vacuum Tube*   *145*

CHAPTER **8**   *Frequency Response of RC Amplifiers*   *164*

Preface

This book has been written to provide the electronic technician with a broad and basic understanding of the diode and transistor and other devices, the circuits, and the systems in which they are employed in the electronics field. The technician is often most interested in practical electronic applications in the real world. To foster this interest we have included selected applications with the solid-state and transistor theory.

The concept of electrical equivalent circuits or black boxes is developed early in the text and consistently applied. To demonstrate the similarity of the active electronic devices, only one circuit model is used to describe all three amplifying devices, the bipolar transistor, the field-effect transistor, and the vacuum tube. The circuit model employed is the g_m-controlled constant-current source. It is simplified in form for easy understanding of the circuit operation, but is sufficiently accurate for basic circuit design procedures.

The material presented, and the terminology employed, is intended to acquaint the student with practical usage in the field. The mathematics involved is restricted to algebra and simple trigonometry. Derivations are rather thoroughly developed. In this manner the student can discover how algebra is employed in reasoning and leads to results which have electrical meaning. Many worked examples provide the student with a knowledge of the circuit conditions encountered under a variety of operating conditions. It is expected that the student will have completed or is currently taking a course in ac circuit fundamentals.

Fundamental material on the solid-state conduction process is introduced in Chapter 1. This provides curious students with answers to the why and how of diodes and transistors.

Diodes result from the ability to transform silicon and germanium into n and p forms, and diode theory and application are introduced in Chapter 2. Rectifier circuits are presented, along with the function of filter circuits. This material is introduced here since it demonstrates an important application of the diode, as well as furnishing source material for early laboratory experiments.

The idea of a black box as an equivalent model of an electrical circuit or device is introduced in Chapter 3. One-port and two-port forms are discussed, and the equivalent circuit h-parameters are brought forward.

In Chapter 4, the operation of the transistor is developed, and the equivalence between the transistor and the h-parameter two-port circuit is then shown. Thereafter the common-emitter h parameters and the g_m model are used for amplifier analysis throughout the book. The low-frequency performance of the three basic amplifier forms is developed and compared. Techniques used in the manufacturing of several modern forms of the transistor are discussed.

The necessity for choice of an operating point in the mid-region of the transistor output curves to obtain linear and undistorted response is shown in Chapter 5. Bias circuits which establish and maintain this operating point are found. Several circuits are treated as design examples.

The important field-effect transistor is covered in Chapter 6. The g_m current-source model is used again. Bias circuits and complete amplifier designs are developed.

Building on the previous circuit theory, and again employing the g_m model, the vacuum tube is introduced in Chapter 7. Triode and pentode circuits are briefly discussed. The cathode-ray tube is also introduced here.

The response of amplifiers over broad frequency ranges is considered in Chapter 8 and the problems which arise in multistage amplifiers are also introduced. The effect of the internal parasitic capacitances, always present in all devices, on the response of the amplifying devices at high frequencies is discussed.

Feedback is of the utmost importance in most electronic applications, and sufficient background has been developed to permit a discussion of feedback in Chapter 9. The performance improvements obtained with negative feedback are shown, and we also introduce some of the problems of amplifier instability. Feedback operation in multistage amplifiers is given special attention.

The universally used integrated or monolithic differential amplifiers are presented in Chapter 10. A discussion of the basic circuits within the usual amplifier, and the methods of manufacture, follows. The operational amplifier, and a few of its many applications are included.

Chapter 11 provides coverage of the frequency response of transistors in tuned radio-frequency amplifiers. Also included are the rectangular pulse wave form and its use in video amplifiers or in data circuits.

Power amplifiers and their associated heat-removal problems are considered in Chapter 12. General push-pull circuits as well as modern transformerless circuits are covered. The class B linear radio-frequency amplifier is also discussed because of its popularity in increasing power at radio frequencies.

Feedback is used to create the several forms of tuned oscillator circuits, crystal oscillators, and the laboratory form of RC oscillator presented in Chapter 13.

Having established a foundation of various building-block electronic circuits, we proceed to a study of modulation and detection of amplitude-modulated and frequency-modulated signals in Chapter 14. Frequency spectra and bandwidth requirements for the various signals are shown, as well as methods of generation of the modulated signals.

In Chapter 15 the concepts of information content of a signal and the information capacity of a channel are introduced. The previously-developed basic circuits are combined in several complete radio systems, as examples.

Chapter 16 departs from the realm of the continuous or analog signal and introduces the binary code, the binary pulse signal and logic circuits for data manipulation and computation. Modern ECL, TTL, and CMOS forms of logic circuits are included. Logic circuit applications in adding, flip-flop and counting circuits are used as examples.

The text concludes with Chapter 17 which details the characteristics of power-switching devices and their applications in several areas of power control.

Questions to encourage student review and extensive lists of problems are supplied at the end of each chapter.

JOHN D. RYDER

CHARLES M. THOMSON

Electronic Circuits
and Systems

1

The Semiconductor Diode

In recent years we have learned a great deal about electrical conduction in solids and have been able to make semiconductor materials having electrical properties that normally do not occur in nature. From this research into the electrical properties of solids, the *pn junction diode* has been developed. It is widely applied as a rectifier and as a device with a nonlinear voltage-current characteristic. Principles found in the development of the diode are also employed in the transistor.

1.1 Electrons; Photons

Electronic science had its beginning in 1883 when Edison observed that a current would pass between a metal plate and the heated filament in one of his lamps if the plate was made positive in relation to the filament. This "Edison effect" was due to the flow of negative particles, later named *electrons*. The electron is now accepted as the fundamental unit of electric charge, negative in nature and designated as $-e$. It is so small that 6.25×10^{28} electrons must pass per second to represent a current of one ampere.

The physical form of an electron is unknown but we shall assume that it is a spherical particle with a mass measured as 9.11×10^{-31} kilogram.

The *photon* is a bundle of radiant energy appearing as light, heat, X rays, and other electromagnetic radiation. The size of the energy package is inversely related to the wavelength of the radiation and is measured in joules.

1.2 Atoms

The chemical elements are built of *atoms*. The Bohr theory of atomic structure proposes a central nucleus of positive electrical charge and mass, around which electrons move in definite orbits. The positive charge of the nucleus is due to *protons*, each having a positive charge equal to that of the electron. The proton has a mass approximately equal to that of the hydrogen atom. Additional mass is added to the nucleus of heavier elements by the presence of uncharged particles known as *neutrons*.

Each element has a number of protons in its nucleus corresponding to its atomic number, which can be found in the periodic table. A normal atom is electrically neutral since the number of protons in the nucleus is equal to the number of electrons in the surrounding orbits.

There is a fixed amount or *level* of rotative energy associated with each orbit; electrons in the innermost orbits close to the nucleus have the least energy and those at greater distance from the nucleus have greater energy. Orbits with common energy levels are called *shells* or *rings*. Each shell or ring has positions for a definite number of electrons. When these positions are occupied by electrons, the shell is said to be *filled*.

The inner shells are normally filled and their electrons are shielded from external forces by the charges of the outer shell electrons. The electrons in the outermost shell are called the *valence electrons* since it is these outer electrons that contact the neighboring atoms and give the element its expected properties in forming chemical compounds.

Hydrogen has only one proton and one electron, and possession of a single electron gives hydrogen a valence of one and places it in Group I of the periodic table. Copper, another Group I element with one valence electron in its outer shell, has its 29 electrons grouped as

Shell	K	L	M	N
No. of electrons	2	8	18	1

The elements found in Group I have many unfilled positions in the outer shell and easily join with other elements having greater numbers of valence electrons. Thus Group I elements are considered chemically active.

Elements of Group VIII such as helium, neon, and argon have filled outer shells with eight electrons. There are no unfilled energy levels available to valence electrons of other atoms and Group VIII elements are chemically inactive.

The Group IV elements include our important semiconductors, silicon and germanium, with the following electron arrangement in the shells:

	K	L	M	N
Carbon	2	4		
Silicon	2	8	4	
Germanium	2	8	18	4

The four electrons in the valence shell provide desirable semiconductor properties. Even carbon becomes a semiconductor when in diamond form and heated sufficiently.

An atom bombarded by a high velocity electron or other particle carrying sufficient energy may have an electron knocked out of the valence shell. This leaves the atom with a net positive charge of $+e$, and the positively charged atom is known as a *positive ion*. The energy supplied by the bombarding particle is known as the *ionizing energy*.

1.3 Crystals

In a liquid there is a random location of the atoms but as an element freezes from the liquid state, the distance between atoms decreases and the binding forces increase. This gives the material its strength as a solid. In many solids the atoms assume regular, lattice-like arrangements much as spheres pack in a box in layered order. The resultant structural arrangement of atoms is called a *crystal*. Crystals repeat the unit structure, shown in Fig. 1.1, as well as other forms. The form for a given element is determined by the nature of the forces that bind the atoms in the solid.

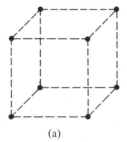 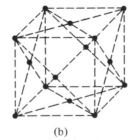 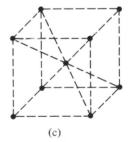

(a)   (b)   (c)

Figure 1.1 Unit crystal cells: (a) simple cubic lattice; (b) face-centered cubic lattice; (c) body-centered cubic lattice.

Mechanical working of a material, such as the stretching of a wire, tends to create a structure of many small crystals. The random boundaries of these individual crystals cause erratic electrical conduction properties. Semiconductor materials are carefully grown in single crystal form, however, and this ensures that crystal boundaries do not disturb the desired conduction properties.

1.4 Metals, Insulators, and Semiconductors

Our good electrical conductors are metals, such as silver, copper, or gold. In metals the interatomic spacing is so close that the orbits of the outer-most electrons of neighboring atoms overlap and a valence electron will be in the attractive fields of several nuclei. The fields essentially cancel, leaving these outer electrons very loosely held. They are freed from their nuclei by the random thermal vibrations of the atoms at room temperature. Copper and silver have one valence electron and thus we have one electron per atom available to move as a free charge. These metals have about 10^{23} atoms per cm³ and so we have 10^{23} electrons per cm³ as free charges available to move in the conduction process.[1]

When we apply a voltage to a wire, individual charges move in a random manner toward the positive terminal at low velocity. With the large number of free charges we find that the electrical resistivity of metals is low (see Table 1.1).

TABLE 1.1   **Electrical Properties of Materials**

Material	Property	Resistivity (Ω-m)
Silver	Conductor	1.6×10^{-8}
Germanium	Semiconductor	0.6
Silicon	Semiconductor	1500
Polystyrene	Insulator	10^{18}

There are other materials that bind their valence electrons so tightly that there are few charges able to move at the usual ambient temperatures. With few free charges these materials cannot carry an appreciable current and they are classed as *insulators* (quartz, porcelain, and polystyrene are examples).

[1]We shall use the scientific method of notation, stating quantities as small numbers times powers of 10, to avoid writing many zeros to the right or left of the decimal point. We shall also designate major magnitudes by the following prefixes, used with the unit name, as in *centi*meter = 0.01 meter:

Multiple	Prefix	Symbol	Multiple	Prefix	Symbol
10^{12}	tera	T	10^{-2}	centi	c
10^{9}	giga	G	10^{-3}	milli	m
10^{6}	mega	M	10^{-6}	micro	μ (lowercase Greek mu)
10^{3}	kilo	k	10^{-9}	nano	n
			10^{-12}	pico	p

In pure form the *semiconductors* have conduction properties intermediate between the good conductors and the insulators. A pure semiconductor is an insulator at temperatures near absolute zero ($-460°F$ or $-273°C$) but the resistance falls as the temperature rises. Such materials, whose resistance decreases with rising temperature, are said to have negative *temperature coefficients of resistance*.

Semiconductors of major importance are *silicon* and *germanium*; their properties are compared to a good conductor and to an insulator in Table 1.1. A number of metallic sulfides and oxides and compounds such as gallium arsenide and gallium phosphide are also useful in semiconductor applications.

1.5 Conduction in Silicon and Germanium

Our most important semiconductors, silicon and germanium, are from Group IV and they have four valence electrons. They form crystals having *covalent bonds* between atoms in which the valence electrons are shared in pairs with four adjacent atoms. This is diagrammed in Fig. 1.2. The electrons in the covalent bonds are spinning on their axes but in opposite directions. The spin creates a magnetic field and the interlocking of these magnetic fields

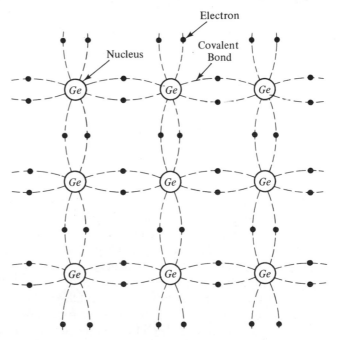

Figure 1.2 Schematic drawing of covalent bonds in germanium at absolute zero temperature.

provides the binding force of the covalent bond that holds the crystal together.

At absolute zero temperature ($-460°F$ or $-273°C$) all the valence electrons of the semiconductor are tightly held in the covalent bonds with neighboring atoms. There are no charges free to move if a voltage is applied. Electrical conduction is not possible and the materials behave as insulators. As the temperature rises to the ambient range ($80°F$ or $27°C$), thermal energy is absorbed by the atoms and the electrons of the solid; this energy appears as random vibration or agitation of these particles about their lattice locations. Some electrons acquire sufficient energy to break the covalent bond and the electrons become free and mobile. Energy for breakage of a covalent bond may also be supplied by a high voltage across the material or by radiation with photons of appropriate wavelength.

When a bond is broken and an electron freed, an electron vacancy is left in the covalent bond. The vacant electron site is called a *hole*, indicated in Fig. 1.3. The hole represents the absence of a negative charge and is attractive to electrons; therefore the hole appears to have a positive charge of $+e$ value.

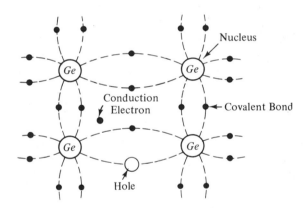

Figure 1.3 Formation of a hole.

An atom that has lost an electron is a positive ion fixed in the crystal lattice. The total material remains electrically neutral, however, since the number of positive ions is still equal to the number of free electrons.

The hole is also considered to be mobile. Examine Fig. 1.4, where a broken covalent bond has left a hole at A. Another electron may move from a bond at B to the hole at A, under the force of a small voltage applied across the material as shown. When the electron fills the hole at A, there is now a hole at B; then another electron may move from a bond at C to the hole at B and the hole has moved to C. In a similar manner an electron may move from a bond at D to the hole at C and so the hole has moved from A to D in stepwise fashion.

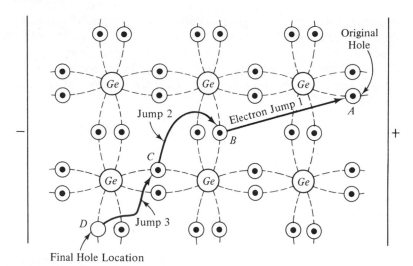

Figure 1.4 Conduction by a hole.

The hole and the electron are both mobile charges that can take part in electrical conduction in these materials. The positive hole progresses toward the negative terminal of an applied voltage, however, while the electron moves toward the positive terminal. In this text we define *conventional electric current* as a flow of positive charges so that the movement of holes corresponds to a conventional current.

Since holes and free electrons are simultaneously produced when a covalent bond is broken, we speak of the process as the generation of *electron-hole pairs*. This is a natural and inherent process in semiconductors and conduction with charges produced by pair generation is called *intrinsic conduction*.

The number of pairs produced and the intrinsic conductivity are both low at room temperature since intrinsic conduction by holes and electrons is dependent on the thermal agitation of the atoms to provide the energy to break the valence bonds. The process is so sensitive to the temperature that intrinsic current levels may be expected to double for every increase of 10°C or 18°F, limiting the use of silicon devices to 390°F or 200°C, and germanium devices to 200°F or 90°C.

Intrinsic conduction is not desired in transistor operation and the effect is suppressed by operation of those devices below these maximum temperature limits.

1.6 The Forbidden Energy

To become free and mobile, a valence electron in a semiconductor must acquire sufficient vibrational energy to break the covalent bond. The breakage energy of a bond in a given material is a fixed quantity, called E_G.[2] We cannot have a partially broken bond and so energies less than E_G are unacceptable or forbidden to the valence electrons. As a result, E_G is called the *forbidden energy* or sometimes the *gap energy* in semiconductors.

Measured in electron volts, the forbidden energies of the major semiconductors are

Silicon	1.15 eV
Germanium	0.75 eV

Since the forbidden energy of silicon is greater than that of germanium, more energy is needed to break a covalent bond in silicon and there will be fewer intrinsic charges. Thus silicon has lower intrinsic conductivity at ambient temperatures.

1.7 n and p Semiconductors

Germanium is obtained as a by-product in zinc refining and silicon is derived from the manufacture of silicones. Both materials are highly purified to remove the random impurities of manufacture.

To obtain the desired conduction properties, the pure silicon or germanium is "doped" with minute amounts of selected elements as controlled impurities. Desirable doping elements include boron, aluminum, gallium, and indium; these have three valence electrons and are said to be *trivalent* impurities. Other doping elements are chosen from arsenic, antimony, and phosphorus; these have five valence electrons and are referred to as *pentavalent* impurities.

The impurities are added at rates of one atom of doping element to 10^5 to 10^8 atoms of semiconductor. The small density of doping atoms causes them to be separated by thousands of semiconductor atoms in every direction and the form of the crystal is not altered by the presence of the impurity atoms. An impurity atom can only find a place in the crystal by substituting for one of the germanium or silicon atoms; it cannot alter the crystal form.

When a pentavalent atom, such as arsenic, substitutes for a silicon atom, there are places in the covalent bonds for only four of its five valence electrons and one electron is left over. This is indicated in Fig. 1.5(a). The binding force of this electron to its nucleus is slight and the thermal agitation at room temperature is sufficient to break this electron free.

---

[2]We measure E_G and other energies associated with atomic particles in *electron volts* (eV). An electron receives 1 eV of energy when accelerated through 1 V; it receives 500 eV of energy when accelerated through a potential of 500 V. This numerical equivalence of potential and energy gained makes the electron volt a convenient measure of energy.

The electron volt represents an energy of 1.60×10^{-19} joule.

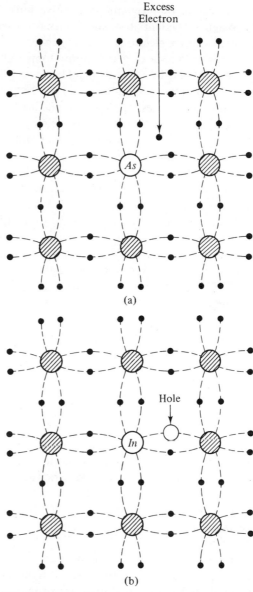

Figure 1.5 Semiconductor with (a) a pentavalent arsenic impurity; (b) a trivalent indium impurity.

By addition of a pentavalent impurity we have created a material having only electrons as free and mobile charges; such a semiconductor is said to be *n material*. The pentavalent atoms have donated an electron for conduction and are called *donor atoms*. Since an electron has been removed from the donor atoms, they remain as positive ions fixed in their crystal positions.

When a trivalent atom substitutes for a semiconductor atom in a crystal, its three valence electrons enter the covalent bonds with electrons from three neighboring silicon or germanium atoms but an electron vacancy or hole is left in the bond to a fourth neighbor atom, as illustrated in Fig. 1.5(b). At room temperature the built-in hole can be filled by an electron from a nearby bond and this process creates a mobile hole in the adjacent area.

The use of a trivalent impurity element has created a material with free holes so that conduction will occur by hole transfer and the impurity semiconductor is said to be *p material*. The trivalent impurity atoms accept electrons to fill their bond vacancies and are called *acceptor atoms*. Since these atoms have acquired an extra electron, they remain fixed in the crystal as negative ions.

In an *n* semiconductor the conduction will be predominantly by electrons and the electrons will be called the *majority carriers*. There will be a few holes present in an *n* material due to electron-hole generation at usual ambient temperatures. The holes present in an *n* material are known as the *minority carriers*.

The conduction in a *p* material will be primarily due to holes and they are known as the *majority carriers*. There will be a few electrons present in a *p* material due to thermal-pair generation; these electrons in a *p* material are known as the *minority carriers*.

1.8 Purification of a Semiconductor

The purification of germanium or silicon to semiconductor standards is usually done by zone refining in an inert gas atmosphere. If the temperature of a short section of impure rod is raised to the molten state and the molten zone moved slowly, the impurities tend to remain in the molten zone and travel with it to the end of the rod. By multiple repetition of the process, the impurities can be concentrated at one end of the rod. This section is then removed and discarded.

To grow single crystal *n* or *p* material, the purified semiconductor is melted in a crucible and the desired trivalent or pentavalent impurity added. A small seed crystal is dipped into the molten semiconductor and as it is slowly rotated and withdrawn, a large crystal grows on the seed with the same lattice orientation. This is illustrated in Fig. 1.6(a).

The large crystals may be 2 or 3 cm in diameter. Properly oriented, they are sliced by diamond saws into thin wafers. The surfaces are then etched and polished to remove crystal dislocations caused by the sawing process.

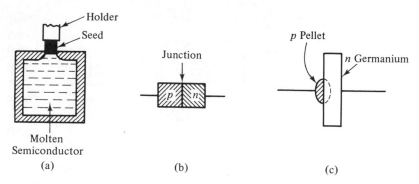

Figure 1.6 (a) Growing the single crystal; (b) a grown *pn* junction; (c) an alloyed junction.

1.9 Manufacture of a pn Junction

The semiconductor diode employs the properties of a *junction* between *p* and *n* layers of semiconductor material. In the *grown junction* process, a crystal is grown from an *n* melt and at a proper time a *p* impurity is added. The *p* material overrides the *n* doping and yields a *p* material. The junction occurs where the *n* material changes to *p*, as indicated in Fig. 1.6(b).

Alloyed junctions are produced by fusing a pellet of a *p* impurity, such as indium, onto the *n*-silicon or *n*-germanium wafer. The indium alloys with the semiconductor and creates a thin *p*-silicon or *p*-germanium layer. The remainder of the pellet serves as a contact to the *p* side of the junction, as shown in Fig. 1.6(c).

In the *diffusion process*, a semiconductor wafer of *n*-impurity type is heated in a furnace with a gaseous atmosphere containing the desired *p* impurity. At elevated temperature the impurity atoms migrate into the semiconductor and create a *p* layer in the *n* wafer. Numbers of large wafers can be processed in one operation, resulting in uniformity of characteristics.

There is no crystalline discontinuity at the junction formed between *n* and *p* layers in a single crystal. Such a junction cannot be created mechanically by forcing two individual crystals together since the junction must be atomic in nature.

1.10 The Junction Diode

Our ability to make *n* and *p* materials by controlled impurity additions to a pure semiconductor results in the *pn* junction, which has the property of one-way conduction. Consider the junction diode of Fig. 1.7 with positive voltage connected to the *p* region and the negative battery terminal to the

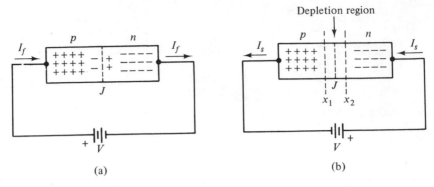

Figure 1.7 (a) Forward voltage across a *pn* junction; (b) junction with a reverse voltage applied.

n region. This is the *forward current* condition, with the forward current into *p* shown as I_f.

There are supplies of electrons in the *n* region supported by electron flow or *injection* into *n* from the negative battery terminal. There are holes in the *p* region supported by hole injection from the positive terminal. With the voltage connected, holes move across the junction from *p* and into the electron-rich *n* region, where they meet and combine with electrons. But this positive charge has disturbed the neutrality of the *n* region and so electrons flow from the negative terminal toward the junction to restore the charge balance. This is indicated by the several flows of Fig. 1.8.

Likewise, free electrons move from the electron-rich *n* region across the junction into the *p* region. This flow has made the *p* region more negative.

We have a current composed of holes in the *p* region and a negative current due to electrons in the *n* region. The holes and electrons are continuously replenished from the battery and the movement of these charges to the terminals constitutes the forward current I_f into the *p* and out of the *n* regions.

Figure 1.8 shows how the number of charges varies with distance from the junction. The currents far from the junction are carried by the majority carriers of each region. Of course, in the metallic wires between diode and battery the current is due to motion of electrons. Electrons are supplied to the *n* region from the negative battery terminal and electrons are extracted from the covalent bonds at the *p*-region terminal by the positive battery potential, in effect injecting holes into the *p* region.

With *reverse polarity*, the diode has a positive voltage connected to the *n* region and negative voltage to the *p* region, as shown in Fig. 1.7(b). With the *n* region positive, free electrons are attracted away from the *n* junction. With the *p* region negative, the free holes are also pulled away from the junction and into the p region. As a result, there is a region extending between x_1 and x_2 on both sides of the junction that has been depleted of mobile charges. Accordingly this is called the *depletion region*. It may be only a small fraction of a millimeter thick but without mobile charges it cannot conduct. Thus with reverse voltage across the diode we have an effective insulating layer at the junction.

The depletion region is still subject to electron-hole pair generation, however, due to thermal energy. All such intrinsic charges are swept out of the region by the applied voltage and this flow is known as the *reverse saturation current I_s*. This reverse current depends on the number of broken covalent bonds at the operating temperature. The current is small in magnitude but it is strongly dependent on temperature.

The *pn* diode easily passes current in the forward direction and has only a small current in the reverse direction. This is the property of a *rectifier diode*.

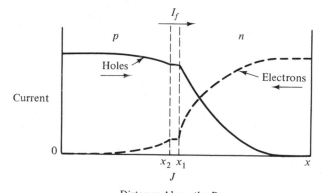

Distance Along the Bar

Figure 1.8 Electron and hole currents in a forward-biased junction.

1.11 The Diode Voltage-Current Equation

The current through a *pn* junction can be predicted by the equation

$$I = I_s(\epsilon^{39V} - 1) \tag{1.1}$$

at 80°F or 27°C. The voltage V is positive for forward current and negative for reverse current.

With forward voltages greater than 0.1 V we can use a simpler form of equation as

Forward Current

$$I = I_s \epsilon^{39V} \tag{1.2}$$

and with reverse voltage more negative than -0.1 V we can use the relation

Reverse Current

$$I = -I_s \tag{1.3}$$

This identifies the reverse current as the reverse saturation current of the junction. It is dependent on the rate of electron-hole pair generation in the depletion region and is not dependent on the reverse voltage up to the Zener (see Sec. 1.12) breakdown point, which occurs at about -40 V for the typical diode shown in Fig. 1.9. The current I_s, being dependent on the rate of pair generation, is variable with temperature.

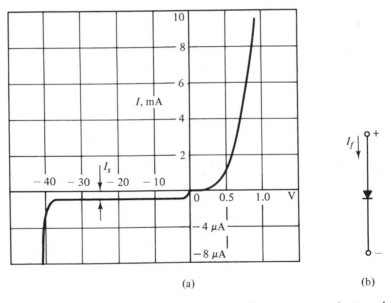

(a)   (b)

Figure 1.9 (a) Typical semiconductor diode volt-ampere curve (note scale changes); (b) diode circuit symbol.

Table 1.2 is calculated by use of Eq. 1.1 and presents the data in terms of the current ratio I_f/I_s. It shows that the diode is a one-way electrical device.

The volt-ampere curve of Fig. 1.9 is from actual measurement of a sample diode and the voltage includes resistance drops in the semiconductor material outside the junction region. This accounts for greater voltages than are given in Table 1.2.

The diode circuit symbol is shown in Fig. 1.9(b).

TABLE 1.2 Diode Forward Current

$+V$	I_f/I_s	$+V$	I_f/I_s
0.04	3.76	0.22	5.32×10^3
0.06	9.38	0.24	11.6×10^3
0.08	21.6	0.26	25.3×10^3
0.10	49.0	0.28	55.3×10^3
0.12	108	0.30	120×10^3
0.14	235	0.35	8.47×10^5
0.16	513	0.40	5.96×10^6
0.18	1119	0.45	4.19×10^7
0.20	2440	0.50	2.94×10^8

1.12 Zener Diodes

In theory, the reverse current of the *pn* diode is a constant, I_s, and independent of reverse voltage. Actually the diode has a *breakdown* at some value of reverse voltage. This occurs at -40 V for the diode of Fig. 1.9. The breakdown voltage is stable for a given diode. This breakdown is not harmful if the power dissipation in the diode is limited to the rated value and a safe temperature is maintained.

The Zener diode circuit symbol is shown in Fig. 1.10(a). The diode voltage cannot exceed the breakdown value and V_z at the output terminals will not rise above that level in the circuit of Fig. 1.10(b), regardless of increases in V_i. The diode current increases and the excess of voltage V_i over V_z appears as an increased voltage drop in R. Therefore the Zener diode is useful to maintain a fixed voltage across its terminals as a *voltage regulator*. Its effect is much as if a battery V_z was connected across the output terminals whenever $V_i > V_z$, as indicated in Fig. 1.10(c).

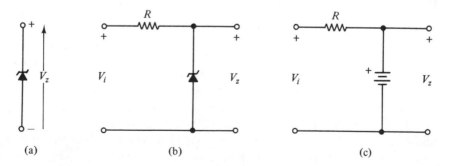

Figure 1.10 (a) Zener diode symbol; (b) regulating circuit; (c) illustrating Zener circuit action.

The reverse-voltage region of operation was investigated by Zener and the diode bears his name. The reverse voltage appears across the depletion region and at some value is sufficient to accelerate the thermally generated electrons in the region to a velocity at which they are able to free more electron-hole pairs by collision with covalent bonds. After each collision there is an additional pair of charges and in turn these are accelerated and collide with atoms, creating still more charges; the current builds up like an *avalanche*.

The breakdown voltage can be lowered by increasing the impurity doping. Diodes are available with Zener voltages ranging from about 3 to 200 V and with power ratings from $\frac{1}{4}$ to 200 W.

1.13 Light-Emitting Diodes

The gap energy E_G is acquired by the electron when it is broken free from a covalent bond in a semiconductor (see Section 1.6). Consequently, when a free electron falls back into a hole in a covalent bond, the electron must give up the same amount of energy. This energy is radiated from the electron as a single photon of light, with the color of the light related to the gap energy of the material.

For a gallium arsenide *pn* diode with $E_G = 1.5$ eV, the emitted light is a dark red. Green is obtained from a gallium phosphide diode. Because the energies of the electrons vary slightly, the light emitted is not of one wavelength but spreads over a narrow wavelength band. The intensity of the light can be varied with the applied voltage, changing the number of recombinations per second.

The light is entirely electronic in nature and very fast on-off switching is possible. These *light-emitting diodes* (LEDs) are being used as point sources for numerical readouts for calculators and in alphabetic display devices.

1.14 Photodiodes

Reverse-voltage diodes in which the *p* layer is made so thin as to be transparent will conduct if light of proper wavelength falls on the junction layer. The photons received must carry sufficient energy to supply the gap energy and break a covalent bond; we then have a diode in which electron-hole pairs are generated in the depletion region by radiant luminous energy instead of thermal energy. The needed energy for silicon can be obtained with photons in the red and near infrared. An interesting application is made with the gallium arsenide LED, emitting in the red, and the silicon photodiode, sensitive in the red region. The two devices form a *photoisolator*,

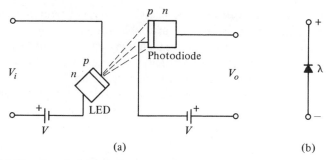

Figure 1.11 (a) The photoisolator, using an LED and a photodiode; (b) photodiode symbol.

as shown in Fig. 1.11. The two circuits are isolated from each other except for the light transmitted from the LED to the photodiode. There is no reverse transfer of energy.

The reverse saturation current of the junction will still be present when the cell has no illumination and is known as the *dark current*. Its magnitude should be small compared to the current obtained with useful illumination levels.

1.15 Capacitance Effects in the pn Diode

The *pn* junction with reverse voltage has a depletion region, acting as an insulator, between the *n*- and *p*-conductive regions. The combination acts as a capacitor. The depletion layer thickens with increased voltage as more charges are pulled away from the junction so that the capacitance has an inverse relation with applied voltage.

Variation of the reverse voltage can be used to change the capacitance and ranges of 2 to 50 pF are possible. These diodes are being used as tuning capacitors in high-frequency and TV receiving systems. They are known as *varactors* or *voltage-variable capacitors* (VVCs).

1.16 General Comments

Our modern era of solid-state electronics relies on an understanding of atomic structure and electrical conduction processes. We employ the Bohr atom model with a positive nucleus surrounded by electrons in orbits, each having a fixed level of energy. Metals, as good conductors, have a great many free valence electrons to take part in electrical conduction. Conversely, insulators hold their charges very tightly and have very few free charges at usual temperatures; therefore, they do not readily conduct. The semiconductors in the pure or intrinsic state have strong covalent bonding and as a result there are few free charges available for conduction at room temperature. Impurity additions add free charges in controlled numbers as holes or electrons; with these tailor-made materials we are able to develop the one-way conduction properties of the *pn* junction.

With forward bias the free majority charges are urged across the junction and we have a condition of easy conduction. With reverse bias the mobile charges are withdrawn from the junction region leaving a high-resistance depletion zone. This acts as an insulator.

The *pn* diode is a polarity-controlled switch, with remarkably low voltage required for forward conduction and a high resistance to conduction in the reverse direction.

REVIEW QUESTIONS

1.1 What are the parts of an atom?

1.2 How does a germanium atom differ from an atom of copper or an atom of argon?

1.3 How many electrons does an atom of potassium have in its valence shell?

1.4 Why are silver and copper considered good conductors of electricity?

1.5 What is meant by a free electron in a metal?

1.6 Describe covalent bonding.

1.7 Describe the mechanism of conduction in intrinsic silicon at room temperature.

1.8 At what temperature does intrinsic conduction become limiting in silicon electronic devices? What is the equivalent temperature in germanium devices?

1.9 What is a hole in a semiconductor? Why are there no holes in copper?

1.10 Why is a single crystal structure desirable in semiconductor devices?

1.11 What is accomplished in zone refining of semiconductors?

1.12 What is the reason for adding a donor element to a semiconductor? Choose an element that might constitute a donor impurity.

1.13 What is the reason for adding an acceptor element to a semiconductor? Choose a suitable element for an acceptor impurity.

1.14 What impurities might be used in the two sides of a *pn* junction?

1.15 Contrast *n*- and *p*-silicon materials in several ways.

1.16 Why is the permissible operating temperature of *n* silicon higher than it is for *n* germanium?

1.17 Explain the development of the depletion region of a junction.

1.18 What is meant by majority and minority charges in *n* silicon? What is the minority charge in *p* germanium?

1.19 Explain how a hole moves through a *p* semiconductor, remembering that the impurity atoms are very far apart.

1.20 Where are the minority carriers, and what are they, in a *pn* junction?

1.21 At a given temperature, I_s in germanium is larger than it is in silicon. Why?

1.22 What kind of charges enter each end of the *p* region in a forward-biased *pn* junction?

1.23 What is the effect of a forward bias on the depletion region in a *pn* junction?

1.24 What is the effect of increasing the reverse voltage on the depletion region of a *pn* junction?

1.25 What is the cause of the reverse saturation current?

1.26 Why does the reverse saturation current increase with temperature? How fast does it increase in silicon?

1.27 Why does the reverse saturation current not vary with reverse voltage?

1.28 What is a photon?

1.29 What is the relation between the energies carried by a photon of red light with wavelength of 7.5×10^{-7} m and a photon of ultraviolet light with wavelength of 3.75×10^{-7} m?

1.30 Explain the phenomenon of avalanching in a Zener diode.

1.31 Explain how a varactor diode acts as a variable capacitance.

1.32 Why is a *pn* junction sometimes said to be a nonlinear circuit element?

1.33 Why is a diode called a polarity switch?

1.34 The brightness of the LED is dependent on what factor?

1.35 The color of an LED is determined by what factor?

PROBLEMS

1.1 Simplify the following quantities with use of the appropriate prefixes on the units:

0.0001 m	9534×10^{-7} F
0.0000075 g	12,473 Ω
1.75×10^{-5} mm	0.00475 V
0.0176×10^{-4} s	0.0146 A

1.2 What is the energy in electron volts of an electron that has risen through a potential of 3.25 V?

1.3 For a silicon diode with $I_s = 25$ μA, find the forward current with a voltage of 0.2 V.

1.4 A given diode has $I_s = 5 \times 10^{-6}$ A; find forward and reverse currents for $V = 0.25$ V.

1.5 A diode has $I_s = 5 \times 10^{-6}$ A; what value of $+V$ is required to obtain a 75-mA forward current?

1.6 A certain *pn* diode requires 0.28 V to cause a current of 180 mA in the forward direction. What is the reverse current at -35 V?

1.7 A Zener diode has a breakdown voltage of 20 V. The input to the circuit of Fig. 1.10(b) is supplied from a 45-V battery and the series resistor $R = 2500\ \Omega$. What should be the current and power rating of the Zener diode?

1.8 A Zener diode is rated at $V_z = 12$ V and 6 W power. With $V_i = 15$ V, find the needed series resistance R of Fig. 1.10(b) to keep the diode within power rating.

1.9 A diode conducts 2 μA with 10 V applied and the same current with 150 V applied; is it forward- or reverse-biased in each case? What current will it carry when 0.2 V is applied in the forward direction?

1.10 A *pn* diode carries 50 mA when 0.25 V is applied in the forward direction. What voltage is needed to raise the current to 150 mA?

2

The Diode as Rectifier and Switch

The *pn*-semiconductor diode acts as a switch whose on-off action is controlled by the polarity of the applied voltage. The switch has a low resistance and very small voltage drop under forward voltage and a very high resistance under reverse-voltage conditions. Numerous applications depend on this switching property, a major use being the rectification of alternating current for dc power supply.

2.1 The Ideal Diode Model

The semiconductor diode has a voltage-current curve as in Fig. 2.1(a), previously discussed in Sec. 1.11. The straight vertical line at Fig. 2.1(b) is an approximation to the actual curve and represents the *ideal diode*, assumed to be a closed circuit for a forward voltage and an open circuit for any reverse voltage. This seems a reasonable assumption since the usual forward voltage is less than 1 V and the reverse current is negligibly small at reverse voltages below the Zener level. In effect, we have said that a small forward voltage is required to make a diode conduct but we are going to call that small voltage *zero* for ease in calculation.

The diode electrodes are called the *anode* and the *cathode*. The anode is represented by the arrow of the diode symbol and is internally connected to the *p* region, while the cathode is the bar of the symbol and is connected to the *n* region. For forward conduction the anode should be positive and the cathode negative.

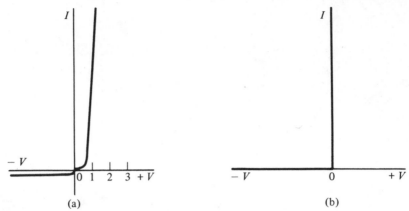

Figure 2.1 (a) *pn* diode volt-ampere curve; (b) ideal diode curve.

2.2 The Half-Wave Rectifier Circuit

On the half cycle of transformer voltage in which the diode anode is positive in Fig. 2.2(a), the diode acts as a closed switch. Voltage v of the transformer, Fig. 2.2(b), is then directly connected to the load and a current i is present in the circuit as a pulse of half-sine form shown in Fig. 2.2(c). On the next half cycle of transformer voltage the diode anode is negative; the diode is in its reverse condition and acts as an open circuit. No current exists in the load.

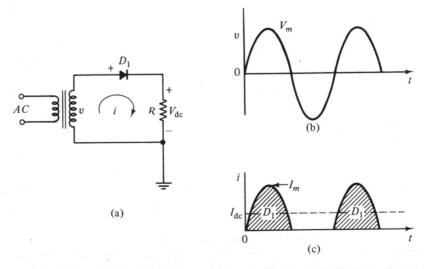

Figure 2.2 (a) Half-wave rectifier circuit; (b) transformer secondary voltage; (c) load-current pulses.

The successive half-sine pulses of Fig. 2.2(c) all lie above the zero axis and the waves have an average positive value, shown as I_{dc}. The peak of the ac voltage from the transformer is V_m and on the conducting half cycle this voltage appears across R. Then the peak of the current pulse is

$$I_m = \frac{V_m}{R} = \frac{1.41V_{rms}}{R} \tag{2.1}$$

With this definition we can determine the dc or average load current as

$$I_{dc} = \frac{I_m}{\pi} = \frac{V_m}{\pi R} = \frac{0.318V_m}{R} \tag{2.2}$$

and the dc voltage across the load is

$$V_{dc} = I_{dc}R = 0.318V_m \tag{2.3}$$

On the reverse half cycle the diode is an open circuit and the entire transformer voltage appears across the diode. At the peak of the wave this is V_m. The insulating depletion layer in the diode has to withstand the *peak reverse voltage* (PRV) in order to prevent a current. The PRV is an important rating of a diode. For the half-wave circuit

$$\text{PRV} = V_m = 1.41V_{rms} \tag{2.4}$$

The *average current rating* of the diode selected for a particular application must be greater than the expected I_{dc} of Eq. 2.2 and the *peak current rating* of the diode must be greater than the expected peak current I_m of Eq. 2.1.

Example: A transformer has a rated secondary voltage of $24V_{rms}$. Find the average current, the peak current, the dc voltage across the load resistance of 50 Ω, and the needed PRV rating.

From our knowledge of sine waves we know that

$$V_m = 1.41V_{rms} = 1.41 \times 24 = 33.8V \text{ peak}$$

Then by Eq. 2.2

$$I_{dc} = \frac{V_m}{\pi R} = \frac{33.8}{\pi \times 50} = 0.216A$$

By Eq. 2.1

$$I_m = \frac{V_m}{R} = \frac{33.8}{50} = 0.676A \text{ peak}$$

By Eq. 2.3

$$V_{dc} = I_{dc}R = 0.216 \times 50 = 10.8V$$

Using Eq. 2.4

$$\text{PRV} = V_m = 33.8V \text{ peak}$$

2.3 The Full-Wave Rectifier Circuit

The *full-wave rectifier circuit* of Fig. 2.3(a) operates in a more efficient manner by supplying current to the load on both half cycles. The transformer is center-tapped and each half develops a peak voltage V_m. These windings are oppositely connected to two diodes so that positive voltages are supplied to the diode anodes on alternate half cycles, Fig. 2.3(b), and they transmit current pulses to the load as shown in Fig. 2.3(c).

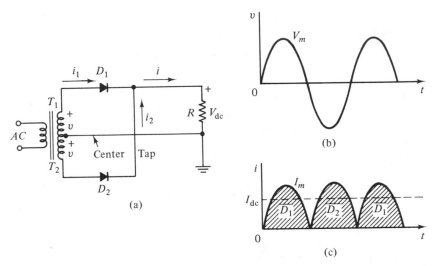

Figure 2.3 (a) Full-wave rectifier circuit; (b) transformer voltage; (c) load-current pulses.

In the first half cycle, diode D_1 is forward-biased from T_1 and conducts a current pulse to the load; diode D_2 has a reverse voltage from T_2 and appears as an open circuit. In the second half cycle, the transformer polarity reverses and diode D_2 is forward-biased from T_2; it transfers a current pulse to the load. At this time diode D_1 is reverse-biased and open.

With the transformer windings and diodes operating independently as two half-wave rectifiers, V_m is the peak voltage of each half of the transformer to center tap. The peak current of a diode remains as

$$I_m = \frac{V_m}{R} \tag{2.5}$$

The load dc current is the average of two half-sine pulses, however, and is therefore twice the dc value obtained for the half-wave circuit, giving

$$I_{dc} = \frac{2(0.318V_m)}{R} = \frac{0.636V_m}{R} \tag{2.6}$$

The dc load voltage is

$$V_{dc} = I_{dc}R = 0.636V_m \qquad (2.7)$$

The two diodes can be seen as in series across the full secondary voltage of the transformer with a peak voltage of $2V_m$. One diode is a closed path and the other diode is open at any time. The voltage across the open diode at the peak is the PRV, which is

$$PRV = 2V_m \qquad (2.8)$$

The full-wave rectifier circuit requires two diodes and a larger center-tapped transformer secondary, from which we obtain twice the dc voltage obtained from the half-wave circuit. However, the required PRV rating for the diodes also doubles.

More importantly, the "bumps" of the load current pulses of Fig. 2.3(c) are a better approximation to a "smooth" and steady dc current and are easier to filter into smooth dc current. The use of the full-wave circuit is strongly favored over the half-wave circuit.

2.4 The Bridge Rectifier Circuit

At the small cost of two more diodes we can eliminate the center-tapped transformer and obtain full-wave rectification in the *bridge rectifier circuit* of Fig. 2.4(a).

The transformer has one winding with a peak voltage of V_m. On the first half cycle, with terminal A positive, diodes D_1 and D_3 have forward-voltage conditions and conduct in series as shown in the equivalent circuit at Fig. 2.4(b). Diodes D_2 and D_4 are reverse-biased by the transformer voltage in this half cycle and are open circuits. A current pulse having $I_m = V_m/R$ passes downward in the load R.

On the next half cycle the transformer voltage reverses and terminal B is positive; diodes D_2 and D_4 conduct, with the current path as in Fig. 2.4(c). Diodes D_1 and D_3 are reverse-biased by the transformer voltage and are open. Another current pulse passes downward in the load R.

The current pulses in the load are in the same direction and the load current waveform is that of Fig. 2.3(c) as a full-wave output. The equations for dc current and voltage of Sec. 2.3 apply.

Now consider point C in the circuit at Fig. 2.4(c). Point C is effectively connected to A of the transformer because the voltage drop across D_2 is negligible during conduction. With C at the voltage of A and point E connected to B, the open diode D_3 appears connected across the full transformer and has to withstand a peak voltage V_m. The diode PRV rating therefore should be V_m. Similar reasoning could be applied to each of the other diodes during the appropriate half cycles.

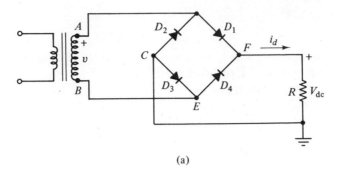

(a)

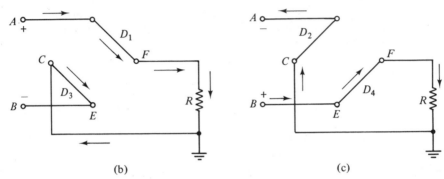

(b)   (c)

Figure 2.4 (a) The bridge rectifier circuit; (b) with terminal *A* positive; (c) with terminal *B* positive.

The bridge rectifier is extensively used because of the lower cost of the transformer and the lower PRV rating for the diodes compared to the center-tapped transformer circuit.

2.5 Measurement of the Ripple in the Rectifier Circuit

A battery provides a constant output voltage and is considered an ideal dc voltage supply. However, it is usually more economical to supply dc to electronic equipment from the ac line through rectifier circuits, with batteries used only for portable equipment. As has been shown, the output voltage from rectifier circuits is not steady, and has varying components or *ripple* superimposed on the average value of the dc voltage, as represented in Fig. 2.5(a). Such variations, remaining from the ac supply as 60- or 120-Hz frequencies, will appear in an amplifier output as "hum" or noise.

The amount of ripple in the rectifier circuit output is measured by the *ripple factor*, γ (gamma), defined as

$$\gamma = \frac{\text{ripple voltage, rms}}{\text{dc voltage}} \qquad (2.9)$$

$$\text{Per cent ripple} = 100\,\gamma$$

Smaller values of γ indicate smoother output voltage, or a closer approach to ideal dc.

An ac voltmeter will read the rms value of the ac or ripple component when connected to the rectifier load through a large capacitor to block the dc voltage. The dc voltage can be measured by a dc voltmeter connected across the load, as in Fig. 2.5(b).

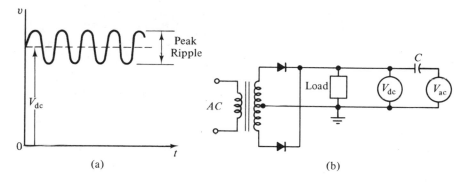

(a)   (b)

Figure 2.5 (a) Ripple on a dc voltage; (b) circuit for measurement of ripple percentage.

If such measurements are made on the output of the half-wave rectifier circuit, shown in Fig. 2.2(c), the ripple factor would be 1.21, or 121 per cent; that is, the ac rms component in the output voltage is 1.21 times as large as the dc voltage. Similarly, the output of the full-wave circuit of Fig. 2.3(c) shows a ripple factor of 0.48 or 48 per cent of the dc voltage.

Neither of these ripple figures is sufficiently low for use with most electronic circuits where the expected ripple factor should be in the range of 0.005 to 0.00005, or 0.5 to 0.005 per cent. The latter figure represents a ripple voltage of 0.5 mV superimposed on a dc voltage of 10 V. These requirements show that further smoothing of the rectifier circuit output is needed. It is best to start with full-wave rectification if filtering is to be added, since the original value of ripple is already lower than with the half-wave circuit.

2.6 The Capacitor Filter

The availability of capacitors of many microfarads makes the simple capacitor filter of Fig. 2.6(a) desirable as a low cost method of smoothing the rectifier output voltage wave.

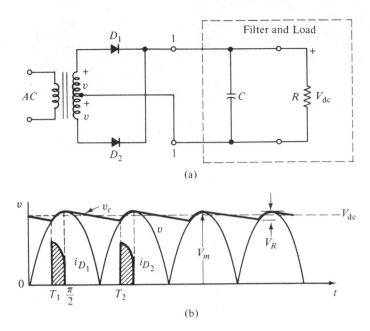

Figure 2.6 (a) Capacitor filter with a full-wave rectifier; (b) load and capacitor voltage.

Figure 2.6(b) shows the capacitor and load voltage as v_C, at the output of a full-wave rectifier circuit. From time T_1 to the peak at $\pi/2$ the diode D_1 is forward-biased by the transformer sine voltage and the capacitor charges through D_1 to the peak of the wave at V_m. When the transformer voltage falls below V_m after $\pi/2$, the diode becomes reverse-biased and conduction stops. If no load resistance were connected to the circuit, the output voltage would remain at a dc level equal to V_m. With a load R connected, the capacitor discharges through R and its voltage falls below V_m until time T_2 is reached. At that time, diode D_2 becomes forward-biased by an excess of transformer voltage over v_C and this diode recharges C to V_m for the second half cycle.

The fall in v_C will be small during the discharge interval from $\pi/2$ to T_2 if C and R are large. This means that the ripple is small. The time given to capacitor charging will be very short, possibly less than 10° of the cycle. We are able to develop an expression for the ripple of the capacitor filter with a full-wave rectifier as

$$\gamma = \frac{0.144}{CRf} \qquad\qquad (2.10)$$

where f is the frequency of the supply line. This result introduces the term CRf as a useful design factor.

What do we measure with the factor CRf? We know that the period of the supply sine wave is $T = 1/f$ and so

$$CRf = \frac{CR}{T}$$

Thus CRf represents the ratio of the capacitor-load resistance time constant to the period of the sine wave of the supply. In more general terms, CRf measures the ability of the capacitor C to store energy and maintain a voltage near V_m until the next diode conduction period occurs to recharge the capacitor.

For low values of ripple factor we want CRf to be large. The result is indicated in Fig. 2.7, plotted for Eq. 2.10. Since the ripple varies inversely with the load resistance, then the *ripple increases with load current* for the capacitor filter.

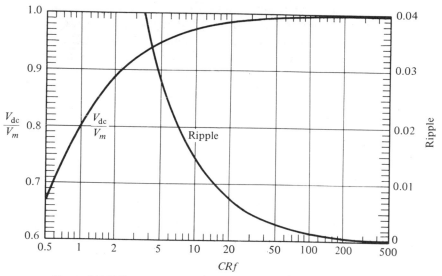

Figure 2.7 Full-wave rectifier: ripple and V_{dc}/V_m versus CRf for a capacitor filter.

Given a specified ripple factor, we have need for

$$C \geq \frac{0.144}{\gamma R f} \quad \text{F} \tag{2.11}$$

The dc load voltage is related to the peak of the ac voltage. By study of the wave form in Fig. 2.6(b), we have

$$V_{dc} = V_m - \frac{V_R}{2}$$

where V_R is the peak-to-peak ripple voltage. The dc voltage is obtainable as

$$\frac{V_{dc}}{V_m} = \frac{1}{1 + \dfrac{1}{4CRf}} \tag{2.12}$$

This ratio again involves the design factor CRf and the equation is plotted in Fig. 2.7 against that factor. It shows that for no load (infinite R) the voltage will be equal to V_m, as predicted.

In Fig. 2.7 we see that $CRf > 5$ is needed for a ripple less than 0.03, or 3 per cent. With $CRf > 5$ the dc voltage will be greater than $0.95V_m$. On this upper plateau of the voltage ratio curve, the load (value of R) may change considerably; yet the dc voltage will remain at 90 per cent or higher of the V_m value. Thus the filter will give nearly constant dc voltage for varying loads as are encountered with some amplifiers.

Example: A full-wave rectifier circuit, operating from a 60-Hz line, supplies a load with 200 mA and 30 V dc at full load. What value of C is needed to limit the ripple factor to 0.01 (1 per cent ripple)? What should V_m be?

The ripple will be maximum at full load current, or minimum load resistance, which is

$$R = \frac{30}{0.200} = 150 \ \Omega$$

Then, using Eq. 2.11 we find C for the filter as

$$C = \frac{0.144}{\gamma R f} = \frac{0.144}{0.01 \times 150 \times 60}$$

$$= 0.00160 \ \text{F} = 1600 \ \mu\text{F}$$

The ripple voltage is $(30 \times \gamma) = 30 \times 0.01 = 0.3 V_{rms}$.

The design factor is

$$CRf = 0.0016 \times 150 \times 60 = 14.4$$

Using Eq. 2.12,

$$\frac{30}{V_m} = \frac{1}{1 + \dfrac{1}{4 \times 14.4}} = \frac{1}{1.017}$$

$$V_m = 30 \times 1.017 = 30.5 V \ \text{peak}$$

$$V_{rms} = \frac{V_m}{\sqrt{2}} = \frac{30.5}{1.41} = 21.6V$$

and this is the needed transformer voltage.

Figure 2.6(b) indicates that the pulse of current through the diode is of short duration but has a large peak value. In this short interval, the diode must pass the total charge which appears in the load as the average current over the half cycle. The peak current requirements of the diode are difficult to calculate because we do not know the length of the diode conduction period. Fortunately we are protected against design inaccuracies by the ability of the diode to withstand large current pulses for a short period of time and by the resistance and reactance introduced into the circuit by the transformer.

At initial turn-on of the circuit, however, the filter capacitor is fully discharged. This capacitor acts as a short circuit on the transformer and conducting diode. The diode will have a surge current rating I_S and the initial surge of charging current must be kept below this rating by use of a resistor in series between transformer and each diode. The value of this surge resistor can be calculated from

$$R_S = \frac{V_m}{I_S}$$

2.7 The π Filter

For applications requiring ripple percentages much less than 1 per cent, the cascaded filtering action of the circuit in Fig. 2.8 is useful. The combination of C_1, L, and C_2 is called a π filter because of its similarity to that Greek letter.

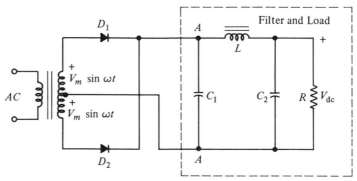

Figure 2.8 Full-wave rectifier with a π filter.

Capacitor C_1 charges to the peak V_m of the applied voltage as does the capacitor C of Sec. 2.6. Therefore we can use what we have already learned about the ripple voltage of the capacitor filter and determine the effect of L and C_2 in reducing the ripple reaching the load.

We shall consider only the full-wave rectifier circuit, as half-wave supply is rarely used with complex filter circuits. Accordingly, the ripple voltage has a frequency twice the line frequency. The effect of the inductor is due to its reactance at double line frequency:

$$X_L = 2\pi(2fL) = 4\pi fL \quad (\Omega) \qquad (2.13)$$

The effect of capacitor C_2 on the alternating ripple at the double frequency is

$$X_{C_2} = \frac{1}{2\pi(2fC_2)} = \frac{1}{4\pi fC_2} \quad (\Omega) \tag{2.14}$$

Now C_2 will be chosen sufficiently large in capacity that its reactance will be very small with respect to the load resistance R ($R = 10X_C$, for example). Let us calculate the parallel impedance of C_2 and R as

$$Z = \frac{RX_{C_2}}{\sqrt{R^2 + X_{C_2}^2}}$$

Using $R = 10X_{C_2}$ we have

$$Z = \frac{10X_{C_2}X_{C_2}}{\sqrt{100X_{C_2}^2 + X_{C_2}^2}} = \frac{10X_{C_2}^2}{\sqrt{101X_{C_2}^2}}$$

But $\sqrt{101} \cong \sqrt{100} = 10$ so

$$Z = \frac{10X_{C_2}^2}{10X_{C_2}} = X_{C_2}$$

The impedance of a parallel combination of a small capacitive reactance and a large resistance is effectively that of the capacitance only. For the ripple components of the current we have changed the actual circuit of Fig. 2.9(a) to the simpler one at (b).

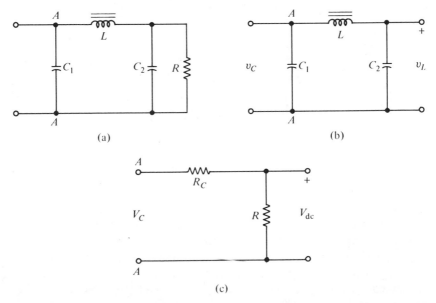

Figure 2.9 (a) The π filter; (b) equivalent circuit for the double-frequency current; (c) dc equivalent circuit.

Equation 2.10 was written for the ripple factor across the capacitor filter and can be used for the ripple factor across C_1 of the π filter in Fig. 2.9, as

$$\gamma_1 = \frac{0.144}{C_1 R f} \tag{2.15}$$

The ripple factor across C_2 may also be calculated. The voltage-divider action of L and C_2 further reduces the ripple voltage and we can write

$$\gamma_2 = \frac{0.144}{C_1 R f} \frac{X_{C_2}}{X_L + X_{C_2}}$$

But we choose L so that $X_L \gg X_{C_2}$ and we can drop X_{C_2} in the denominator, giving

$$\gamma_2 = \frac{0.144}{C_1 R f} \frac{X_{C_2}}{X_L} \tag{2.16}$$

Substituting for the reactances with Eq. 2.13 and 2.14 and using the values of the constants, we have

$$\gamma_2 = \frac{9.1 \times 10^{-4}}{R C_1 C_2 L f^3} \tag{2.17}$$

The ripple factor is inversely proportional to the value of load R. Thus the ripple increases with load current. The inductance and the capacitances have an equal effect on the ripple so that choice of values and their distribution among C_1, C_2, and L is arbitrary.

The dc voltage across C_1 at A, A is given by Eq. 2.12 as

$$V_{A,A} = \frac{V_m}{1 + 1/(4C_1 R f)}$$

The inductor may have a dc resistance R_c, and this resistance and the load R are connected in series across A, A for the dc current, as shown in Fig. 2.9(c), which is the equivalent circuit for dc. Using resistances R_c and R as a voltage divider, we have the dc output voltage of the filter as

$$V_{dc} = \frac{V_m}{1 + \dfrac{1}{4C_1 R f}} \left(\frac{R}{R + R_c} \right) \tag{2.18}$$

For small load currents (large R) the dc output voltage of the filter approximates V_m if C_1 is large. Usually C_1 and C_2 are made large and equal, and L is given a value of 5 or 10 henrys.

Example: Filter components might be chosen as $C_1 = 50 \ \mu F$, $C_2 = 50 \ \mu F$, $L = 5 \ H$, and $R_c = 200 \ \Omega$, with a load taking 200 mA at 125 V dc. The load resistance is

$$R = \frac{125}{0.200} = 625 \ \Omega$$

The ripple factor is

$$\gamma_2 = \frac{0.00091}{625 \times 50 \times 10^{-6} \times 50 \times 10^{-6} \times 5 \times 60^3}$$
$$= 0.00054$$

Per cent ripple $= 100\,\gamma_2 = 0.054$ per cent.

This would be a satisfactorily small ripple factor for electronic amplifiers. It represents an rms ripple voltage of

$$125 \times 0.00054 = 0.067 \text{ V}$$

or 67 mV ripple on the 125-V dc supply.

2.8 The π-R Filter

For filters used with rectifiers having small current output, it is often economically desirable to replace the inductor of a π filter with a resistor, as in Fig. 2.10(a).

The equivalent dc circuit of Fig. 2.10(b) is the same as that for the π filter in Fig. 2.9(c) and so the dc output voltage can be calculated by use of Eq. 2.18, repeated here as

$$V_{dc} = \frac{V_m}{1 + \dfrac{1}{4C_1 R f}}\left(\frac{R}{R + R_f}\right) \qquad (2.19)$$

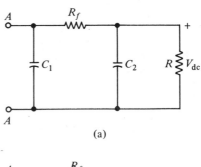

(a)

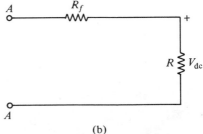

(b)

Figure 2.10 (a) The π-R filter circuit; (b) equivalent dc circuit.

The resistance of the filter inductor R_c usually is in the range of 80 to 400 Ω and similar values are suited to R_f in the π-R filter circuit.

The ripple factor is found by replacing $X_L = 4\pi fL$, the reactance of the filter inductor, with R_f of the filter circuit and so

$$\gamma_2 = \frac{0.0114}{RR_fC_1C_2f^2} \tag{2.20}$$

Since the average power lost in R_f is $P = I_{dc}^2 R_f$, the circuit should be employed only when I_{dc} is small to avoid undue power losses and the resultant heat dissipation in the equipment.

Example: Replacing the 5-H inductor of the example of Sec. 2.7 with its resistance of 100 Ω as R_f, we find the ripple factor becomes

$$\gamma_2 = \frac{0.0114}{625 \times 100 \times 50 \times 10^{-6} \times 50 \times 10^{-6} \times 60^2}$$
$$= 0.0202 \text{ ripple factor}$$

Per cent ripple $= 100\,\gamma_2 = 2.02$ per cent.

This represents a ripple rms voltage of

$$125 \times 0.0202 = 2.53 \text{ V}$$

The ripple has been increased with the removal of the inductor. The dc voltage will be unchanged, however, since we made $R_f = R_c$.

2.9 The Voltage-Doubling Rectifier Circuit

A higher value of dc voltage can be obtained from a given transformer by use of the *voltage-doubling rectifier circuit,* one form of which is shown in Fig. 2.11.

With the upper ac terminal positive we have the equivalent circuit of Fig. 2.11(b). With diode D_1 forward-biased and conducting, diode D_2 is open. Diode D_1 charges C_1 to the peak value of V_m, the applied ac wave. On the next half cycle the circuit becomes that at Fig. 2.11(c), with B positive. Diode D_2 conducts through C_2, charging that capacitor to the peak of the ac wave. The capacitor polarities are additive across the dc output terminals; at no load this voltage is equal to the double-peak value of the ac input wave.

The diodes conduct high peak currents for short intervals and C_1 and C_2 must be of high capacitance to maintain the dc voltage over the nonconducting intervals of the diodes and to give a low ripple percentage.

The doubler circuit is useful for high dc voltages because the transformer has a lower total voltage and the transformer insulation requirements are reduced in comparison to other circuits. Performance data are presented in Fig. 2.12.

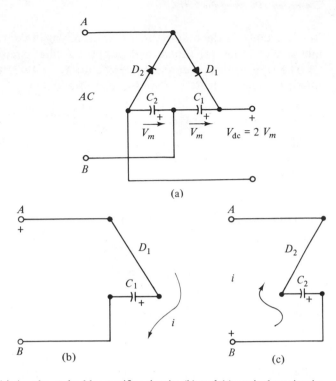

(a)

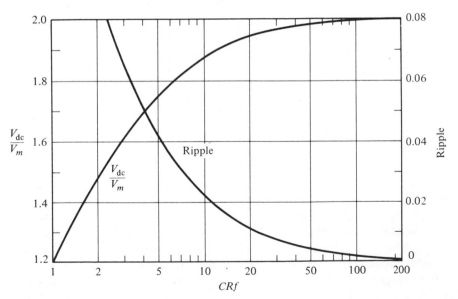

(b)   (c)

Figure 2.11 (a) A voltage-doubler rectifier circuit; (b) and (c) equivalent circuits on alternate half-waves.

Figure 2.12 The voltage-doubler rectifier circuit performance.

The principles used in the voltage-doubling circuit may be extended to obtain still higher voltages. Such voltage multipliers use a ladder arrangement to increase the output dc voltage but applications are limited because of high ripple percentage with small output currents.

2.10 Rectifying AC Voltmeters

Instruments for the measurement of ac cannot employ iron in their field structures because the action of the iron varies with frequency and such variation will introduce errors with varying frequency or waveform. With only air in the magnetic circuits, the ac instruments operate at low field-flux densities and must have large currents in the moving coils to give useful deflections. Permanent magnet-moving coil (PMMC) instruments for dc measurement derive their field flux from permanent magnets, however, and a smaller current can be measured than is possible with ac instruments. Therefore, we often combine diodes with PMMC instruments, rectifying the small ac currents and performing their measurement with low-current dc instruments.

The bridge rectifier circuit of Fig. 2.13(a) is used, leading to ac voltmeters

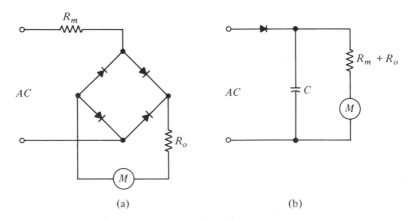

(a)   (b)

Figure 2.13 (a) Full-wave average-reading meter; (b) peak-reading ac voltmeter.

that may have internal resistances of 2000 Ω per V of scale or give full-scale deflection with a current of only 0.5 mA. The dc instrument coil is supplied with full-wave rectified current pulses and the deflection is proportional to the average or dc value of these pulses. From Eq. 2.6 for the full-wave rectifier circuit,

$$I_{dc} = 0.636\frac{V_m}{R} = 0.636\frac{V_m}{R_m + R_0}, \qquad (2.21)$$

where

V_m = peak value of the applied ac voltage
R_m = series multiplying resistance of the instrument
R_0 = the resistance of the instrument movement

We assume that ideal diodes are being used.

Inverting the equation above we have

$$V_m = \frac{R_m + R_0}{0.636} I_{dc}$$

which is the peak value of the ac voltage applied to the instrument for a given dc instrument current. To determine the rms value of the ac voltage at the instrument terminals, we multiply by $1/\sqrt{2} = 0.707$ and have

$$V_{rms} = 0.707 V_m = \frac{0.707}{0.636}(R_m + R_0)I_{dc}$$
$$= 1.11(R_m + R_0)I_{dc} \qquad (2.22)$$

where I_{dc} is the current measured by the deflecting coil of the dc instrument.

The scale introduces the factor 1.11, which is appropriate only for a sine waveform from which we obtained the factors $0.636 = 2/\pi$ and $0.707 = 1/\sqrt{2}$. If the input wave is not a sinusoid, the instrument indication will be in error.

Another form of rectifying voltmeter appears in Fig. 2.12(b) and employs a half-wave diode with a capacitive filter. The voltage applied to R_m and the instrument is equal to the peak of the ac wave. The scale is calibrated to read in rms values of that wave and includes the 0.707 factor. Again, errors arise if the applied waveform is not sinusoidal.

2.11 Diode Wave Clippers

The polarity-switching property of the diode permits its use in simple *clipper circuits* to remove parts of waveforms. The circuits employ a diode, a resistor, and often a battery or voltage source. The output waveform can be clipped at different levels, dependent on the diode connection and the battery polarity and potential. Basic forms include series and parallel diodes and biased and unbiased circuits.

The circuit of Fig. 2.14(a) employs a parallel diode and series resistor R, connected to clip off all positive voltages; essentially the circuit is a half-wave rectifier. When point A is positive to ground, the diode is forward-biased and appears as a short circuit, through use of the ideal diode model. For $v_i > 0$, the state of the circuit is shown in Fig. 2.14(b) and $v_o = 0$. For $v_i < 0$, the diode is reverse-biased and open and the state of the circuit becomes that of Fig. 2.14(c). For a sine-wave input the output wave is that at

Fig. 2.14(e). Resistor R should be large with respect to the diode resistance; 10,000 Ω is suitable for most diodes.

A biased parallel clipper is shown in Fig. 2.15(a), with clipping of all positive voltages above $+10$ V. With $v_i < +10$, point A is negative to the diode

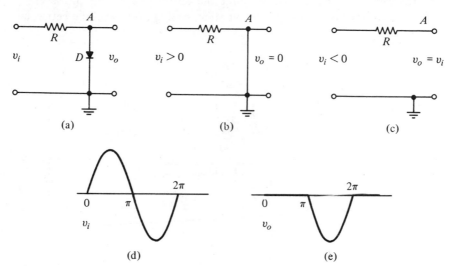

Figure 2.14 (a) Parallel diode positive clipper; (b) circuit of (a) for $v_i > 0$; (c) circuit of (a) for $v_i < 0$; (d) input sine voltage; (e) clipped sine voltage.

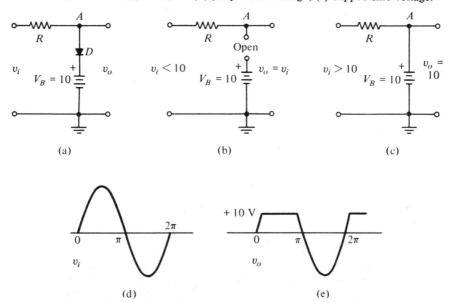

Figure 2.15 (a) Clipping at $+10$ V; (b) for $v_i < 10$; (c) for $v_i > 10$; (d) and (e) waveforms.

cathode at $+10$ V, the diode is open, and the circuit is that of Fig. 2.15(b).
The input voltage is transmitted to the output as v_o. With $v_i > +10$ V, the
diode closes and the circuit becomes that at Fig. 2.15(c), giving $+10$ V across
the output. For sine-wave input the output wave appears at Fig. 2.15(e).

In analysis of the circuit to determine the output waveform, the several
circuits should be drawn for each voltage state, noting that switching between

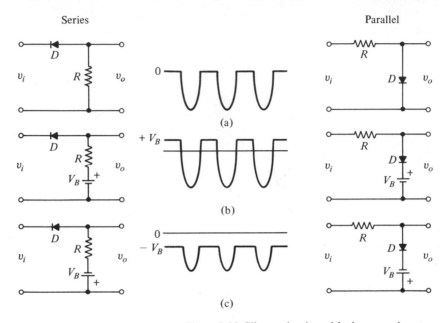

Figure 2.16 Clipper circuits, with sine-wave inputs.

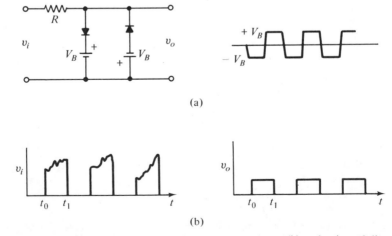

Figure 2.17 (a) Sine wave clipped to form a square wave; (b) reshaping of distorted pulses by clipping.

the circuits occurs at V_B in each case. Series and parallel forms of clipping circuits are compared in Fig. 2.16. Reversal of diode and battery polarities will rotate the waveforms about the zero voltage axis.

Several applications of clipper circuits are shown in Fig. 2.17. At (a) is a *double-diode clipper* that can be used to form approximate square waves from a sine-wave input. When followed by amplification and a second clipping operation the rise time of the pulses can be made quite short. At Fig. 2.17(b) we have the process of restoration of distorted pulses by clipping below the noise level.

2.12 Diode Clampers

A *clamping circuit* employs a diode, a resistor, and a capacitor and will place a wave on a desired dc axis. For instance, Fig. 2.18(a) shows a circuit whereby the unsymmetrical input pulse at (b) is clamped with its positive peaks at zero in (c).

The RC values of the circuit should be chosen so that the time constant is at least five times the period of the input wave; that is, with the period at $1/f$, then $RC > 5/f$. When the wave goes to $+5V$ at t_0, the capacitor charges

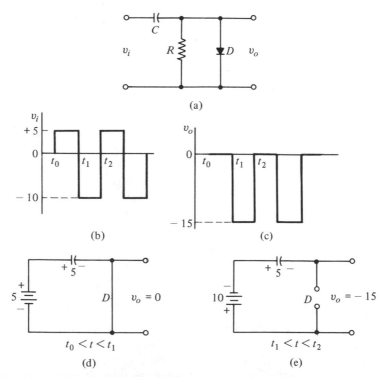

Figure 2.18 (a) Diode clamp; (b) input pulse wave; (c) output clamped at 0 V; (d) and (e) circuits during operation.

quickly through the diode to $+5$ V and we have the circuit condition of Fig. 2.18(c) that exists to t_1. At t_2 the input drops to -10 V, and with the capacitor charged to $+5$ V the diode voltage is -15 V and it opens, leading to the circuit at Fig. 2.18(d) and -15-V output.

The output wave at Fig. 2.18(c) shows that the positive peak is now *clamped* at 0 V and the pulse swings down to -15 V. The positive peaks may be clamped at other dc levels by introducing a voltage V_B in series with the diode.

2.13 Review

The ideal diode model is useful in circuit analysis because of its simplicity; it employs the concept of zero forward-voltage drop when the diode anode is positive to the cathode and zero current with reversed polarity. An actual semiconductor diode provides a good approximation to the performance of the ideal diode in most applications.

The full-wave rectifier circuit is an important application of the polarity-switching diode. The dc quantities in a resistance load are

$$I_{dc} = \frac{2V_m}{\pi R}; \qquad V_{dc} = \frac{2V_m}{\pi}$$

The ripple, at 48 per cent of the dc voltage, is too large for a dc supply in most electronic circuits.

The bridge rectifier circuit reduces the needed transformer voltage by one-half. This eliminates the relatively high cost of a center-tapped winding at the expense of adding two additional diodes. It is now a much-used circuit form.

Ripple can be reduced by the addition of a filter capacitor across the load. The dc voltage will approach V_m of the ac wave applied. The parameter CRf, with f as the supply frequency, is useful as a performance factor. The ripple reduction improves with increasing CRf values.

Further reduction of ripple voltage is possible by use of the π filter but this adds an expensive and bulky inductor. Replacement of the inductor with a resistance saves cost, space, and weight but sacrifices dc voltage if the ripple is to be small.

Filter circuits are compared in Table 2.1, several employ an input capacitor of many microfarads. This seems a better choice for modern small equipment than filter circuits using inductor input since the capacitor is lighter, cheaper, and smaller per unit of filtering effect than is the iron-cored inductor. A capacitor input filter also provides output voltages near V_m and gives good control of that voltage if sufficient filter capacitance is used (large CRf). Thus these circuits make efficient use of the transformer.

TABLE 2.1 **Filter-Circuit Comparison (Full-Wave Rectifier)**

	V_{dc}	*Ripple*
No filter	$\dfrac{2V_m}{\pi}$	0.485
Capacitor filter	$\dfrac{V_m}{1 + \dfrac{1}{4CRf}}$	$\dfrac{0.144}{CRf}$
π filter	$\dfrac{V_m}{1 + \dfrac{1}{4C_1 Rf}}\left(\dfrac{R}{R + R_c}\right)$	$\dfrac{0.00091}{RC_1 C_2 Lf^3}$
π-R filter	$\dfrac{V_m}{1 + \dfrac{1}{4C_1 Rf}}\left(\dfrac{R}{R + R_f}\right)$	$\dfrac{0.0114}{RR_f C_1 C_2 f^2}$

REVIEW QUESTIONS

2.1 What is an ideal diode?

2.2 A sine wave of voltage is applied to an ideal diode. At what angles in the wave does the diode switch on and off?

2.3 The secondary voltage of a transformer for a half-wave circuit is $35V_{rms}$. What should the voltage rating of a transformer for a full-wave center-tap rectifier be, assuming equal dc voltage output from the rectifier?

2.4 Define
(a) Peak reverse voltage
(b) Average current
(c) Maximum current
(d) Ripple

2.5 Name several advantages of full-wave rectification over half-wave rectification.

2.6 For rectifiers with resistance loads and 60-Hz supply, what is the frequency of the ripple component in the load voltage of
(a) Half-wave rectifier
(b) Center-tap circuit
(c) A bridge rectifier

2.7 What is the effect of a defective diode that is continuously open on the output voltage and the ripple of a full-wave rectifier?

2.8 What are several advantages of the bridge rectifier circuit over the full-wave center-tap circuit?

2.9 Name at least one disadvantage of the bridge rectifier over the full-wave center-tap circuit.

2.10 What is the purpose of a filter circuit? How does it accomplish this purpose?

2.11 Why do diodes carry short current pulses when a capacitor filter is used?

2.12 What components of the diode current pass through the load in a capacitor filter circuit?

2.13 Why does the voltage output of a capacitor filter approach V_m in magnitude?

2.14 Name an advantage of a π filter over a capacitor filter. Name a disadvantage.

2.15 What is the effect of the resistance of the inductor winding in a π filter?

2.16 Name several disadvantages of filters employing inductors.

2.17 Under what conditions do we use a π-R filter?

2.18 What is the relation of the maximum voltage rating for the filter capacitors to the rms voltage of the transformer with a bridge rectifier circuit?

2.19 What is an advantage of a voltage-doubler circuit?

2.20 Is the voltage-doubler circuit of Fig. 2.11 a half-wave or a full-wave circuit?

2.21 If a PMMC instrument measures the load voltage of a half-wave rectifier, what is the component measured?

2.22 The diodes of the bridge rectifier of Fig. 2.13 are not ideal. How would you alter Eq. 2.22 to include the effective diode resistance?

2.23 A dc voltmeter reads 680 V across the output of a voltage-doubler rectifier, at no load. What would an rms ac voltmeter read across the transformer secondary that supplies the rectifier?

PROBLEMS

2.1 Determine the dc voltage obtained by half-wave rectification of an ac wave of 180-V rms.

2.2 What PRV rating should the diode of Problem 2.1 have?

2.3 A half-wave rectifier circuit supplies 100 mA dc to a 500 Ω load. Find the dc load voltage, the PRV rating for the diode, and the rms rated voltage of the transformer supplying the rectifier.

2.4 Find the dc load voltage obtained from a full-wave rectifier in which the transformer supplies a peak voltage of 150 V on each side of the center tap.

2.5 The dc output voltage of a full-wave rectifier is 120 V. What is the needed rms voltage rating for each half of the center-tapped transformer? What is the peak ac voltage rating for the full transformer secondary?

2.6 A center-tapped transformer supplies the diodes of a full-wave rectifier giving a dc output across the load of 180 V. What PRV rating should be used for the diodes?

2.7 Determine the needed PRV rating for a diode in a bridge rectifier supplying 80 V dc to a load.

2.8 A full-wave rectifier has 125 V dc output to a resistance load. What is the rms value of the ripple voltage?

2.9 Calculate the needed transformer voltages to supply a load at 50 V dc, from
(a) A half-wave rectifier circuit

(b) A center-tapped full-wave circuit

(c) A bridge rectifier circuit

(d) Determine the diode PRV ratings for each case.

2.10 A rectifier output following a filter has an rms ripple voltage of 1.72 V, with a dc load voltage of 90 V. What is the ripple factor?

2.11 A bridge rectifier supplies a resistance load with 2 A dc at 20 V dc. What transformer voltage is needed (rms)?

2.12 A full-wave rectifier is supplied with an rms voltage of 80 V, 60 Hz on each side of the transformer center tap. A 10-μF capacitor is used as a filter and the load takes 50 mA dc. What dc voltage is being obtained at the load?

2.13 A transformer with 250–0–250 V rms, 60-Hz secondary supplies a full-wave rectifier having a load of 1500 Ω. Determine the dc load voltage and current, the PRV rating needed for the diodes, and the average current of each diode. Draw the circuit.

2.14 A transformer with 250 V rms, 60-Hz secondary supplies a bridge rectifier having a dc load of 1500 Ω. Determine the dc load voltage and current, the PRV rating of the diodes, and the average current in each diode.

2.15 A silicon diode is used in a half-wave rectifier circuit operating from a 20 V rms, 60-Hz transformer. The load is a 10-Ω resistance. What is the peak diode current? What is the average load current? What should the PRV rating of the diode be? What is the rms value of the ripple voltage?

2.16 A full-wave rectifier is supplied at 60 Hz by a transformer with a center tap. With a 400-μF filter capacitor, calculate

(a) The ripple

(b) The rms ripple voltage

(c) The rms voltage rating each side of center tap for the transformer, for a load taking 400 mA at 50 V dc.

2.17 Operating from a 60-Hz line, a transformer supplies a peak voltage of 140 V each side of the center tap to a full-wave rectifier. The capacitor filter has a dc output voltage of 120 V and a dc current of 50 mA. Calculate the value of C used in the filter.

2.18 An RC filter supplies 80 V dc with 1.5 per cent ripple. What is the rms ripple voltage?

2.19 A rectifier transformer operates at 60 Hz and supplies a full-wave rectifier with a capacitor filter of 100 μF. The load takes 200 mA at 50 V dc.

(a) What is the no-load (R very large) output voltage?

(b) What should be the rms voltage rating of the transformer, measured to center tap?

(c) What is the per cent of ripple at full load?

2.20 A 60-Hz transformer having a 40–0–40 V rms secondary supplies a full-wave rectifier circuit. The load is 100 Ω with a filter capacitor of 3000 μF. Find

(a) The dc load voltage

(b) The rms ripple voltage across the load

(c) The PRV rating for the diodes.

2.21 A transformer on a 60-Hz line supplies a bridge rectifier. A capacitor filter and a load of 24 Ω gives a ripple of 1.2 per cent at a dc voltage of 40 V.
(a) What rms transformer voltage is needed?
(b) What capacitance is needed for the filter?
(c) What should be the voltage rating for the filter capacitance?

2.22 With a full-wave rectifier at 60 Hz, a π filter supplies 154 V dc to a load. With an ac voltmeter having a large series capacitor, the voltage measured across the load is 127 mV rms.
(a) What is the ripple factor?
(b) If the dc load current is 0.250 A, what value of $C = C_1 = C_2$ is being used in the filter? $L = 10$ H.

2.23 Determine the ripple factor for a bridge rectifier circuit at 60 Hz with a π filter having $C_1 = C_2 = 100$ μF, $L = 7.5$ H, and a load current of 500 mA at 60 V dc.

2.24 Determine the ripple factor for a bridge rectifier circuit at 60 Hz with a capacitor filter of 2000 μF and a load taking 500 mA at 6 V dc.

2.25 The circuit of Fig. 2.19 uses a transformer with 120–0–120 V_{rms} secondary.
(a) What is the magnitude and polarity of voltage at A with no load?
(b) Find the same quantities at B, also with no load.
(c) If the load at B is 1000 Ω, what is the ripple per cent there, 60-Hz supply?

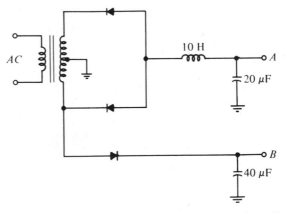

Figure 2.19

2.26 What is the dc voltage available at no load from a voltage-doubler circuit supplied by a source of 120 V, 60 Hz, at no load?

2.27 A bridge rectifier operating at 60 Hz is to supply 1 A at 35 V, with a ripple less than 0.5 V rms. Specify C and the transformer voltage rating, the average diode current rating, and the PRV rating.

2.28 In Problem 2.27, what surge protection resistor should be used if the diode surge current rating is 30 times the average current?

2.29 A bridge rectifier with π filter uses two 40-μF capacitors and one 10-H inductor of zero resistance. The transformer voltage is 300 V rms, 60 Hz and the dc load current is 150 mA. Find the dc output voltage and the ripple.

2.30 The waveform of Fig. 2.20 is applied to the ammeter shown at Fig. 2.20(b), where M is a PMMC average reading instrument with 1-mA full-scale reading. If the diode is ideal, what does the meter read if its resistance is 9000 Ω.

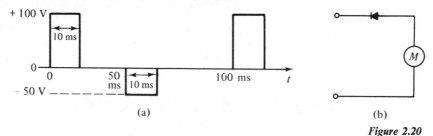

(a)   (b)

Figure 2.20

3

Models for Circuits

Transistors and tubes are called *active devices* because they convert energy from dc to ac at signal frequencies. Simple models are needed for the active devices, that can be used in circuit analysis along with resistors, capacitors, and inductors. With such models we can design our electronic systems on paper and determine how they should be operated for best performance. Breadboarding and checking of the circuits can follow in the laboratory. Without preliminary calculation with models, the laboratory work becomes an exercise in frustration of the cut-and-try variety.

Fortunately, those who work with electrical circuits have shown us how to develop models for circuits that contain energy sources, and are active. While seemingly rather abstract, we can use these methods to replace actual active devices with a simple circuit model that performs in an equivalent manner.

3.1 The Black Box Concept

Consider a device as simple as a relay. It has an input pair of terminals or an *input port* and an output pair of terminals or an *output port*. If we apply a proper voltage (or signal) to the input port, we discover that a connection appears across the output port. If we know what happens at the output port for every input condition, we do not need to know the internal design to utilize the relay in a control system. The relay box might be painted black to hide its internal details, with just the ports exposed. Then the box of Fig. 3.1(a) might represent the relay as a *two-port network*. A simpler *one-port network* is shown at Fig. 3.1(b).

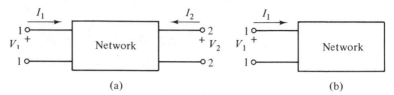

(a)   (b)

Figure 3.1 (a) A two-port network; (b) one-port network.

A great many of our electrical devices can be thought of as "black boxes" with one or two ports. We are able to apply them because of our knowledge of the electrical performance at the ports; we do not have to know what happens internally. For instance, we buy transistors by calling for a type number that assures us that the output will respond to the input in a certain way. We need not ask how the transistor is built internally; thus we buy transistors as black boxes.

We can study the performance of the one-port box of Fig. 3.2(a) by means of the current and voltage at the port. The values obtained indicate the response of the box to a connected source or load. Now, consider the second box in Fig. 3.2(b). If for every value of applied voltage $V_A = V_B$ we

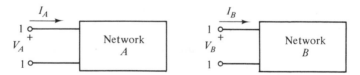

Figure 3.2 Two one-port electrical networks.

find that current $I_A = I_B$, we would conclude that the electrical networks inside the boxes were the same in *external* performance. One box could be replaced with the other without noticing any difference at the ports and we would say that the boxes are *equivalent*. We need not know anything about the internal operation or connections of the boxes. We conclude

> *that a circuit is equivalent to a second circuit if the second circuit can be substituted for the first without change in the currents and voltages appearing at the ports.*

Now, if we can devise a circuit model that operates at input and output ports the same as does a transistor, then we can use that circuit model in place of the actual transistor.

In the networks of Fig. 3.1, the conventional current and voltage polarities are indicated. The currents moving inward are made positive at both ports of the two port so that a network can be reversed or changed end-for-end without altering an analysis.

3.2 Active One-Port Models:
the Voltage-Source Circuit

Suppose that we connect a voltmeter at the port of Fig. 3.3(a), and discover that a voltage V_1 is present. We conclude that the box contains a source of electrical energy and houses an *active network*. It would be useful to determine a circuit equivalent for this box.

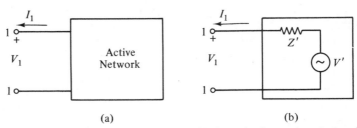

Figure 3.3 (a) An active one-port network; (b) the series form of equivalent circuit.

We know that some internal voltage source is present and we speculate that the source might have some associated impedance. We then have two unknowns, a source $\mathbf{V}'$ and an impedance $\mathbf{Z}'$, that can only be connected in two ways—in series or in parallel. We might start with assumption of the series form at Fig. 3.3(b). We need two algebraic equations to determine the two unknown quantities; these equations might be obtained by connecting two arbitrary loads at the port and measuring the resultant voltage and current.

Probably the simplest loads we can find are an open circuit and a short circuit; the latter will cause no damage as long as we work only on paper! We measure the voltage at the port of Fig. 3.3(a) without load and call that value $\mathbf{V}_1 = \mathbf{V}_{oc}$. Using the circuit of Fig. 3.3(b) and an open-circuit load, we measure $\mathbf{V}'$ at the port. For equivalence of the two boxes we must have

$$\mathbf{V}_{oc} = \mathbf{V}' \tag{3.1}$$

Now apply a short circuit to port 1,1 in Fig. 3.3(a) and we measure a current $\mathbf{I}_1 = \mathbf{I}_{sc}$; similarly with a short circuit on the port of (b) we measure $\mathbf{I}_1 = \mathbf{I}'_1$. For equivalence of the two boxes our second condition is

$$\mathbf{I}_{sc} = \mathbf{I}'_1 \tag{3.2}$$

With the short circuit in place across the 1,1 terminals of Fig. 3.3(b), we can write a circuit equation

$$\mathbf{V}' - \mathbf{Z}'\mathbf{I}' = 0$$

Substitution of the necessary conditions for equivalence as given by Eq. 3.1 and 3.2, we have

$$\mathbf{V}_{oc} - \mathbf{Z}'\mathbf{I}_{sc} = 0$$

and

$$\mathbf{Z}' = \frac{\mathbf{V}_{oc}}{\mathbf{I}_{sc}} \qquad (3.3)$$

Therefore we have in Fig. 3.3(b) an equivalent circuit for the box at (a) if

$$\mathbf{V}' = \mathbf{V}_{oc} \qquad (3.4)$$

$$\mathbf{Z}' = \frac{\mathbf{V}_{oc}}{\mathbf{I}_{sc}} \qquad (3.5)$$

The equivalent circuit includes a voltage generator V' equal to the open-circuit voltage at the port of the original network and a series impedance $\mathbf{Z}'$ determined from the open-circuit voltage and the short circuit current. This circuit in Fig. 3.3(b) can be made equivalent to any one-port network and is known as the *voltage-source equivalent circuit* or the *Thevenin equivalent circuit*.

3.3 Active One-Port Models: the Current-Source Circuit

Let us place a load Z_L at the output port of the voltage-source equivalent circuit in Fig. 3.4(a). A voltage equation written around the loop is

$$\mathbf{V}' - \mathbf{Z}'\mathbf{I}_L - \mathbf{V}_L = 0 \qquad (3.6)$$

Division by $\mathbf{Z}'$ and rearrangement of the terms gives

$$\frac{\mathbf{V}'}{\mathbf{Z}'} = \mathbf{I}_L + \frac{\mathbf{V}_L}{\mathbf{Z}'}$$

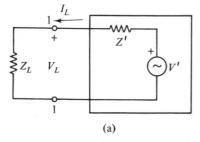 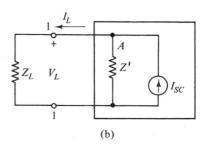

(a)	(b)

Figure 3.4 (a) Thevenin's series equivalent circuit and load; (b) Norton's parallel equivalent circuit and load.

But $\mathbf{V}'/\mathbf{Z}' = \mathbf{I}_{sc}$ by Eq. 3.4 and 3.5 and we can call $\mathbf{V}_L/\mathbf{Z}' = \mathbf{I}_A$ so that the equation above becomes

$$\mathbf{I}_{sc} = \mathbf{I}_L + \mathbf{I}_A \tag{3.7}$$

This equation is that of a current summation at point A for the circuit in Fig. 3.4(b). The circuit in the box consists of a current generator $\mathbf{I}_{sc}$ and a parallel impedance $\mathbf{Z}'$, as derived for the voltage-source circuit.

The circuit in Fig. 3.4(b) was derived from the voltage-source circuit at (a) and so we have another active one-port equivalent circuit, *the current-source equivalent* or the *Norton equivalent circuit.* By mathematics we have found the parallel form of circuit that we originally stated as possible.

Since the voltage-source circuit and the current-source circuit are equivalent, transference from one form to the other is often used in circuit simplification.

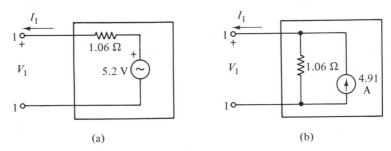

(a)   (b)

Figure 3.5 Circuits for the example.

Example: We find that with open circuit at 1,1 of a box the voltage is 5.2 V. With a load of 2 Ω connected to those terminals the current is 1.7 A. Determine the circuit values for active one-port equivalent circuits.

Since we have two unknowns in the box, we need two equations for solution. From the problem data,

$$V_{oc} = V' = 5.2 \text{ V}$$

is one relation. The second relation comes from the equivalent circuit in Fig. 3.4(a) as

$$V' - IZ' - IZ_L = 0$$

With I given as 1.7 A for $Z_L = 2$ Ω and $V' = 5.2$ V, we have

$$5.2 - 1.7Z' - 1.7 \times 2 = 0$$
$$1.7Z' = 5.2 - 3.4 = 1.8$$
$$Z' = \frac{1.8}{1.7} = 1.06 \text{ Ω}$$

Therefore our voltage source should be that in Fig. 3.5(a), with $V' = 5.2$ V and $Z' = 1.06$ Ω.

We can calculate I_{sc} as

$$I_{sc} = \frac{V'}{Z'} = \frac{5.2}{1.06} = 4.91 \text{ A}$$

and the current-source circuit is that in Fig. 3.5(b) with $I_{sc} = 4.91$ A and $Z' = 1.06 \ \Omega$.

3.4 Maximum Power Output

Many of our electronic power sources, such as microphones, are expensive and develop very little power output. For these generators we try to maximize the power output per unit of investment by operating such sources under conditions that will give the *maximum power output* to a load.

The voltage source in Fig. 3.6(a) represents any active one-port circuit; for simplicity we assume that the load is resistive and $Z' = R'$. If we use an open-circuit load at 1,1, the output current will be zero and there will be zero power output. Likewise, if we use a short-circuit load at 1,1, the output resistance is zero and there will be zero power output. These are the limits

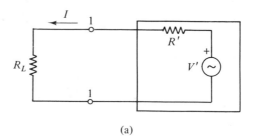

(a)

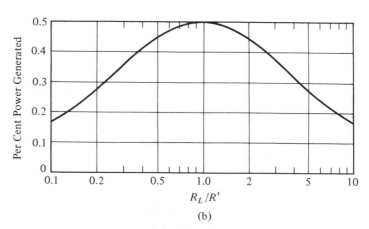

(b)

Figure 3.6 (a) A voltage source and load R_L; (b) output power to the load versus R_L/R'.

for the load and if we experimentally try other load values, the curve in Fig. 3.6(b) develops. This curve shows that the maximum power is delivered from the active generator circuit at 1,1 when

$$R_L = R' \tag{3.8}$$

This is the condition of a *matched resistive load*.

With the same current in R_L and R', the load power is $P_L = I^2 R_L$ and the internal power loss is $P_G = I^2 R'$. With $R_L = R'$, these powers are maximum and equal. The power transferred to the load is only 50 per cent of the total power generated, however, or the output efficiency is 50 per cent. Needless to say, industrial and domestic power systems cannot afford to waste one-half of their output in the generating plant and do not operate with matched loads; such systems use loads much higher than the matching value to achieve high output efficiency.

Example: For the voltage-source equivalent circuit in Fig. 3.5(a) the current I is

$$I = \frac{V'}{R' + R_L}$$

and for the matched load we have $R' = R_L$ so that

$$I = \frac{V'}{2R'} = \frac{5.2}{2 \times 1.06} = \frac{5.2}{2.12} = 2.45 \text{ A}$$

The power delivered to the matched load is

$$P_L = I^2 R_L = 2.45^2 \times 1.06 = 6.37 \text{ W}$$

The total power generated is

$$P_G = I^2 (R' + R_L) = 2.45^2 \times 2.12 = 12.73 \text{ W}$$

and the efficiency of delivery of power to the load is 50 per cent.

3.5 The Two-Port Network

We also make frequent use of a transmission type of network in which there are input and output pairs of terminals, as for the *two-port network* symbolized by the black box in Fig. 3.7(a). There are four undetermined quantities at the ports of this network, namely V_1, I_1, V_2, and I_2. Consequently, if we are to find an equivalent circuit for any network in the box, we shall need four measured parameters.

We can relate four parameters with four terminal voltages and currents by a pair of equations, and electronic usage has shown these desirable:

$$V_1 = h_i I_1 + h_r V_2 \tag{3.9}$$

$$I_2 = h_f I_1 + h_o V_2 \tag{3.10}$$

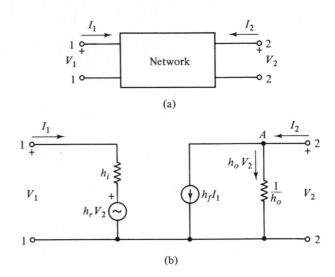

Figure 3.7 (a) A two-port black box; (b) the *h*-parameter equivalent circuit.

where these are alternating voltages and currents. The subscripts of the *h* parameters are a standard usage.

Borrowing from the method already used for the one-port circuit, we can take measurements at the ports with open- and short-circuit loads and choose the internal elements of the equivalent circuit to yield the same terminal measurements.

With a short circuit at 2,2 we have $V_2 = 0$ and, inserting that condition in Eq. 3.9 and 3.10, we find

$$V_1 = h_i I_1$$

$$I_2 = h_f I_1$$

and so we are able to define h_i and h_f by the following measurements made at the 1,1 terminals:

$$h_i = \frac{V_1}{I_1}\bigg]_{2,2 \text{ short}} = \text{short-circuit input resistance} \qquad (3.11)$$

$$h_f = \frac{I_2}{I_1}\bigg]_{2,2 \text{ short}} = \text{short-circuit forward current gain} \qquad (3.12)$$

Taking measurements at the 2,2 port with an open circuit at the 1,1 port means that $I_1 = 0$ and using that condition in Eq. 3.9 and 3.10, we have

$$V_1 = h_r V_2$$

$$I_2 = h_o V_2$$

Then we define h_r and h_o by the following measurements made at the 2,2 terminals:

$$h_r = \left. \frac{V_1}{V_2} \right]_{1,\,1 \text{ open}} = \text{open-circuit reverse voltage gain} \tag{3.13}$$

$$h_o = \left. \frac{I_2}{V_2} \right]_{1,\,1 \text{ open}} = \text{open-circuit reciprocal output resistance} \tag{3.14}$$

The defined h parameters include an input resistance, the reciprocal of the output resistance, and two dimensionless ratios; consequently, the h coefficients are called the *hybrid parameters*.

With alternating voltages and currents the parameters h_i and h_o should be stated in units of impedance and reciprocal impedance but they are usually considered as resistive.

3.6 The h-Parameter Equivalent Circuit

With h parameters corresponding to the measurements, Eq. 3.9 and 3.10 will act as a mathematical equivalent for any two-port network. Our remaining step is to find an actual circuit that can be used to represent that original box.

The hybrid equations were set up as

$$V_1 = h_i I_1 + h_r V_2 \tag{3.15}$$

$$I_2 = h_f I_1 + h_o V_2 \tag{3.16}$$

Equation 3.15 appears as a sum of voltage terms and, with V_1 at the port, the right-hand terms represent the sum of voltages inside the box. The $h_i I_1$ term is a resistance voltage drop and is shown as due to h_i at port 1,1 in Fig. 3.7(b). The second term represents a *controlled voltage generator*, whose output is dependent on the voltage at the output port. The parameter h_r is the constant of proportionality. Equation 3.15 is simulated by the series circuit at port 1.

Equation 3.16 states the sum of currents at A in Fig. 3.7(b). Current I_2 enters at the port and $h_o V_2$ leaves through resistance $1/h_o$. The second term is a leaving current due to a *controlled current generator*, whose output current is proportional to the input current at 1,1; h_f is the constant of proportionality. We have the circuit connected to port 2 representing Eq. 3.16.

Figure 3.7(b) is an equivalent circuit for any two-port network, defined in h parameters. It should be noted that it comprises two one-port equivalents, with a voltage-source equivalent circuit appearing in the input loop and a current-source circuit appearing in the output loop.

Example: Find the h parameters for the circuit in Fig. 3.8(a), and draw the equivalent circuit.

A short circuit at 2,2 connects the 12- and 15-Ω resistors in parallel and

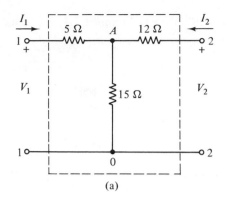

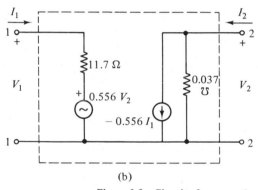

Figure 3.8 Circuits for example.

the input resistance at 1,1 is due to 5 Ω plus the parallel value, that is,

$$\text{(s.c.)} \quad h_i = \frac{V_1}{I_1} = 5 + \frac{12 \times 15}{12 + 15} = 11.7 \, \Omega$$

With the short circuit at 2,2, current I_1 divides at A. Then

$$V_{AO} = I_1 \frac{12 \times 15}{12 + 15}$$

Also

$$-I_2 = \frac{V_{AO}}{12} = I_1 \frac{15}{12 + 15}$$

We can then write

$$\text{(s.c.)} \quad h_f = \frac{I_2}{I_1} = -\frac{15}{27} = -0.556$$

With an open circuit at 1,1 there is no current in the 5-Ω resistor and so the output resistance measured at 2,2 is

$$r_o = 12 + 15 = 27 \, \Omega$$

$$\text{(o.c.)} \quad h_o = \frac{1}{r_o} = \frac{1}{27} = 0.037 \text{ mho}$$

Still with an open circuit at 1,1, the voltage across the 15-Ω resistor is V_1.
Then

$$I_2 = \frac{V_2}{12 + 15}$$

$$V_1 = 15 I_2 = \frac{15 V_2}{12 + 15}$$

and so

$$(\text{o.c.}) \quad h_r = \frac{V_1}{V_2} = \frac{15}{27} = 0.556$$

The circuit parameters are entered on the circuit diagram in Fig. 3.8(b).

3.7 Power in Decibels

In the communications field it is usual to measure power gain in *decibels*.
Devised by the telephone industry and originally named the *bel* for Alexander
Graham Bell, the decibel is one-tenth as large and of a more useful magnitude.
The decibel (dB) is defined as

$$\text{number of dB} = 10 \log \frac{P_2}{P_1} \tag{3.17}$$

where *log* indicates a logarithm to the base 10.

The ear hears sound intensities on a logarithmic intensity scale and since
many amplifiers supply an audio output, the use of a logarithmic power scale
becomes reasonable and convenient. In fact, it was intended that a 1-dB step
be the minimum intensity change that is just noticeable by the human ear
but experiment has found that the minimum noticeable intensity change is
nearer 2.5 dB.

To illustrate the use of the decibel, assume the output of an amplifier
under one condition is 3.5 W with an input of 0.12 W. The power gain,
measured in decibels, is

$$\text{dB} = 10 \log \frac{3.5}{0.12} = 10 \log 29.17$$

$$= 10 \times 1.46 = 14.6\text{-dB power gain}$$

If the amplifier output is increased to 7.0 W, then the decibel gain is

$$\text{dB} = 10 \log \frac{7.0}{0.12} = 10 \log 58.3$$

$$= 10 \times 1.76 = 17.6\text{-dB power gain}$$

Doubling of the power gain is shown by a 3-dB change.

An amplifier with a power gain of 90 has

$$\text{dB} = 10 \log \frac{90}{1} = 10 \times 1.95 = 19.5 \text{ dB}$$

and the amplifier has a gain of 19.5 dB. It is followed by a second amplifier with a power gain of 250 and

$$dB = 10 \log \frac{250}{1} = 10 \times 2.40 = 24 \text{ dB}$$

The overall power gain is

$$90 \times 250 = 22,500$$

and in decibels this is

$$\begin{aligned} dB &= 10 \log 22,500 = 10 \log (2.25 \times 10^4) \\ &= 10 (\log 2.25 + 4) = 10(0.352 + 4) \\ &= 43.5 \end{aligned}$$

But $43.5 = 19.5 + 24$ and we see that the overall gain of *amplifiers in cascade* can be calculated by adding the respective gains in decibels as

$$\text{overall gain, dB} = dB_1 + dB_2 + dB_3 + \ldots \qquad \textbf{(3.18)}$$

While defined as a *power ratio*, the decibel is also used for absolute power measurement when a reference or *zero level* is stated for P_1. A variety of zero levels has been used; the one now commonly employed is 0.001 W (1 mW). Consequently we can state the output level of the amplifier with 7.0-W output as

$$\begin{aligned} 10 \log \frac{7}{0.001} &= 10 \log (7 \times 10^3) \\ &= 10(\log 7 + 3) = 10(0.84 + 3) \\ &= 38.4 \text{ dB above 1 mW} \end{aligned}$$

This unit is sometimes given as dBm (decibel referred to one milliwatt), to indicate the 1-mW zero level.

The amplifier with 3.5-W output has decibel output

$$\begin{aligned} 10 \log \frac{3.5}{0.001} &= 10 \log (3.5 \times 10^3) \\ &= 10(0.54 + 3) = 35.4 \text{ dB above 1 mW} \end{aligned}$$

Taking the difference of the 7.0-W output and the 3.5-W output,

$$38.4 - 35.4 = 3.0 \text{ dB}$$

which was the original difference in the amplifier outputs. If this figure had been negative, a loss would have been indicated.

We may wish to find the dBm of a microphone with 0.0000062-W output. With zero level at 0.001 W, we write

$$dBm = 10 \log \frac{0.0000062}{0.001} = 10 \log 0.0062$$

Obtaining the logarithm of a number less than unity need pose no problems.

Using scientific notation we can employ our regular methods as

$$\text{dBm} = 10 \log (6.2 \times 10^{-3}) = 10 (\log 6.2 - 3)$$
$$= 10 (0.792 - 3) = -22.1$$

and the result is stated as 22.1 dB *below* 1 mW.

Power is given by $P = V^2/R$ or $I^2 R$. The decibel power gain can be found through use of either of these expressions. For instance,

$$\text{gain in dB} = 10 \log \frac{V_2^2/R_2}{V_1^2/R_1}$$

If $R_2 = R_1$, then

$$\text{gain in dB} = 10 \log \left(\frac{V_2}{V_1}\right)^2$$

But we have

$$10 \log \left(\frac{V_2}{V_1}\right)^2 = 10 \left(\log \frac{V_2}{V_1} + \log \frac{V_2}{V_1}\right)$$

and we find that the gain can be written

$$\text{gain in dB} = 20 \log \frac{V_2}{V_1} \qquad\qquad (3.19)$$

We commonly calculate amplifier gains by use of Eq. 3.19 as a voltage ratio or as the current ratio

$$\text{gain in dB} = 20 \log \frac{I_2}{I_1} \qquad\qquad (3.20)$$

3.8 Comments

The concept of an equivalent circuit, equal in terminal voltages and currents to any actual network of unknown internal elements and connections, is a very useful means of simplifying our frequently complex electrical circuits. The voltage source (Thevenin circuit) and the current source (Norton circuit) serve as equivalents for one-port active networks and, as controlled sources, have already appeared as parts of the *h*-parameter two-port equivalent circuit.

The latter circuit will be employed to represent the transistor or tube in circuit analysis. With the proper values for the *h* parameters, we can represent the transistor as a circuit of resistances, capacitances, and generators and can write and solve circuit equations that demonstrate the amplifying effect of the transistor.

REVIEW QUESTIONS

3.1 What is a "black box" intended to represent?

3.2 What do we need to know about the internal connections of a black box?

3.3 Why can we say that a voltage-source circuit is equivalent to a black box of one port?

3.4 What are the two measurements needed to determine the circuit elements of a voltage-source equivalent circuit?

3.5 How are the circuit elements of a voltage-source circuit related to a current-source circuit?

3.6 If the one-port box contains no energy source or is passive, draw the equivalent circuit.

3.7 Why are we interested in obtaining the maximum power output from some generators? Name two such sources.

3.8 What is meant by matching a load?

3.9 What is the output power efficiency of a generator with a matched load?

3.10 What kind of terminal loads are used in determination of the h parameters for a two-port equivalent circuit?

3.11 Why are the assumed currents directed inward at the ports of a two-port network?

3.12 Sometimes the use of a short circuit leads to generator damage. Can you suggest an alternative means of obtaining data to calculate the h parameters without using a short circuit?

3.13 Why are the h parameters called hybrid parameters?

3.14 What is the major property of a controlled source?

3.15 Why are h_r and h_f called control parameters?

3.16 What is one advantage of rating amplifiers in decibels of gain?

3.17 What is the usual reference level when we state that an amplifier output is at $+30$ dB?

3.18 What is the meaning of a circuit input of -30 dBm?

3.19 Determine the logarithm of 0.00672.

3.20 Can you suggest why h_f is negative to h_r in the example in Sec. 3.6?

3.21 Why is the decibel well suited to measurements along telephone lines?

3.22 The noise level of a jet engine is measured at $+137$ dB. What is the decibel noise level of 2 jet engines? Of 10 engines?

PROBLEMS

3.1 In Fig. 3.9(a) we find $V_1 = 107$ V at open circuit; with a load of 200 Ω we find $V_1 = 49$ V. Determine the circuit elements for a voltage-source equivalent circuit.

3.2 Determine a current-source equivalent circuit for the circuit of Problem 3.1.

3.3 Under short circuit, Fig. 3.9(a) has an output current of 8.7 mA. With a load of 10,000 Ω we find the port 1,1 voltage to be 57 V. Draw and label the circuit elements for a voltage-source equivalent circuit.

3.4

    (a) What is the load that should be used for maximum power output with the circuit of Problem 3.3?

    (b) What is the maximum power output available from the 1,1 terminals?

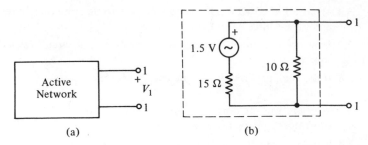

Figure 3.9

3.5 What load should be used at the 1,1 port to obtain maximum power output from the circuit in Fig. 3.9(b)?

3.6 Determine the element values for a current-source circuit equivalent to the circuit in Fig. 3.9(b).

3.7 By use of the basic definitions applied to the circuit in Fig. 3.10(a), calculate the *h* parameters.

3.8 Calculate the *h* parameters for the two-port circuit in Fig. 3.10(b). Draw the equivalent circuit.

3.9 A generator of 1 V is connected to 1,1 in Fig. 3.10(b). Draw an equivalent voltage-source circuit at the 2,2 port and label the circuit elements with their values.

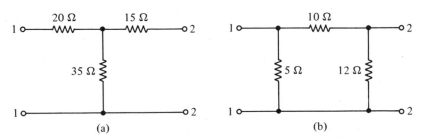

Figure 3.10

3.10 Draw and label the circuit elements for a voltage-source equivalent circuit for the circuit in Fig. 3.11(a).

3.11 Transform the circuit in Fig. 3.11(a) into a current-source equivalent. Draw the circuit.

3.12 What is the matching load for the circuit in Fig. 3.11(a)?

3.13 Determine the voltage-source equivalent circuit for the circuit in Fig. 3.11(b) at the 1,1 port. What is the matching load? What is the maximum available power output to a load?

3.14 Determine the current-source equivalent circuit for Fig. 3.11(c). Draw the equivalent voltage-source circuit.

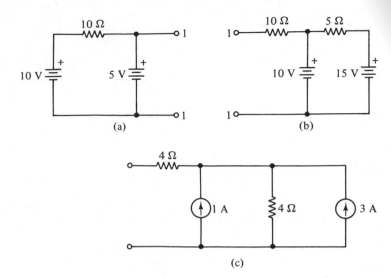

(a)

(b)

(c)

Figure 3.11

3.15 A microphone has an output at -56 dBm. It supplies an amplifier that is to have its output at $+37$-dBm level.
(a) What is the amplifier gain required in decibels?
(b) What is the amplifier power output in watts?
(c) What is the amplifier input in watts?

3.16 An amplifier takes 5 mW input power and has an output of 32 W.
(a) What gain is present in the amplifier in decibels?
(b) What is the dBm level of the output?
(c) What is the dBm level of the input?

3.17 An amplifier of 34-dB gain is connected in cascade with an amplifier of 28-dB gain. If 1 μW is the input to the first amplifier, what is the output power in watts from the second amplifier?

3.18 A radio receiver has an input resistance of 70 Ω. The antenna supplies a current of 0.2 μA to that input. The electrical power to the loudspeaker is 12 W. Find
(a) The decibel gain of the receiver
(b) The decibel level of the input signal, 1-mW zero level

3.19 A transistor is found to have $h_i = 28\ \Omega$, $h_r = 4 \times 10^{-4}$, $h_f = 0.98$, and $h_o = 0.6 \times 10^{-6}$ mho. In a circuit equivalent to that in Fig. 3.7(b), find V_2 if $I_1 = 1.5\ \mu$A. *Hint*: Use the h-parameter equations.

4

Junction Transistors as Amplifiers

In 1948 Bardeen and Brattain found that the current through a forward-biased semiconductor junction could control the current to a reverse-biased contact mounted nearby. The result was the first solid-state control device and was called a *transistor*, as a contraction of the words *transfer resistor*. Such a transistor employs both holes and electrons in the conduction process and is said to be *bipolar*. Shockley later developed the *field-effect transistor* that employs only holes or electrons; it is therefore called a *unipolar* device. This will be discussed in Chapter 6.

4.1 Transistor Voltage and Current Designations

Agreement on abbreviations used for transistor voltages and currents is needed for accurate understanding of transistor circuits. We shall use lower-case letters to designate time-varying quantities and capital letters for dc quantities and for rms values of ac signal voltages and currents. Illustrated in Fig. 4.1 are some of the variations to be expected; these include

Quantity	Example
Total current or voltage	i_C, v_B
AC signal, rms value	I_c, V_b
DC value	I_C, V_C
Supply or bias	V_{CC}, V_{BB}

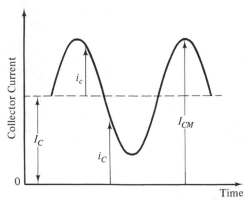

Figure 4.1 Transistor current notation.

The subscripts c or C, e or E, and b or B are used to identify the transistor internal elements; the *collector*, the *emitter*, and the *base*, respectively.

4.2 The Junction Transistor

Junction transistors are three-layer sandwiches of n and p materials, as illustrated in Fig. 4.2. These transistors are made in two types, dependent on the arrangement of the materials as *pnp* or *npn*. Performance of the two types is similar. In one the majority carriers are holes and in the other they are electrons. The bias voltages are also reversed in the two types.

Junctions are present at the interfaces between the n and p materials. The three layers are designated as *emitter*, *base*, and *collector*. The first junction between emitter and base is operated with forward bias. The second junction between base and collector is operated with reverse bias. For the *pnp* unit the base and collector are maintained negative to the emitter; while for the *npn* unit the base and collector are positive to the emitter, as shown in Fig. 4.2.

The *pnp* unit will be used for discussion of transistor operation. The emitter material is heavily doped in comparison to the base and, with forward bias, holes move from the p emitter to the hole-poor n-base region. With a very thin base, in the range of a few thousandths of a millimeter, the holes are attracted by the negative collector voltage. The reverse bias on the second junction makes that junction one of easy flow for holes from the n-base region and the holes are swept across the junction to the collector. Thus we have established a current of holes from the emitter to the collector. Positive holes are shown by the arrows as inward at the emitter and outward at the collector terminal.

The base width is made very narrow compared to the average distance the free holes move before recombination occurs in an n material. Therefore, while a few holes meet electrons and recombine in the base, 90 per cent or

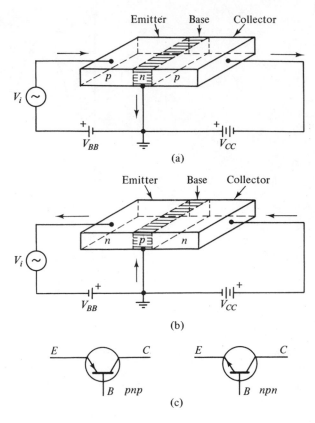

Figure 4.2 (a) Junction transistor, *pnp*; (b) transistor, *npn*; (c) circuit symbols.

more of the injected holes will travel across the base and reach the collector.
A small flow of electrons will occur in the base lead, to resupply the negative
charge lost to recombination in the base. This flow creates an outward base
current, as shown by the arrow in Fig. 4.2(b).

The hole current from the emitter to the base is varied by the forward
voltage across the junction, in accordance with V in the diode equation of
Section 1.11. Control of the emitter-base voltage by a signal voltage causes a
signal variation in the hole flow to the base and a similar variation in the collector current. The transistor is a *control element* in which the emitter-base
voltage controls the collector output current.

The input power is low because of the low resistance of the forward-
biased emitter-base junction, typically a few hundred ohms. The output circuit carries the same current, however, but typically will deliver it to loads of
thousands of ohms. It is thus possible to have power amplification with the
transistor.

The circuit symbols in Fig. 4.2(c) show the forward direction of emitter-base current by the arrow on the emitter element; the direction of the arrow distinguishes between *pnp* and *npn* transistors.

The *npn* transistor could be discussed in like manner by reversing the bias potentials to maintain forward- and reverse-biased junctions and by considering the emitter current as composed of electrons. The arrows on the transistor of Fig. 4.2(b) indicate the directions of forward current.

4.3 The Reverse Saturation Current in Transistors

The reverse-biased base-collector junction of the transistor is to be isolated by opening the emitter connection, as in Fig. 4.3. The main current of holes to the collector is removed but there still exists the usual saturation current of a reverse-biased junction, due to thermal pairs generated in the materials. Holes generated in the base material are swept to the collector by the negative voltage there and electrons generated in the collector are swept to the base.

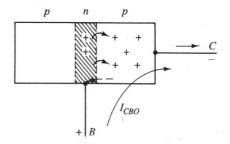

Figure 4.3 Condition for I_{CBO}.

The resultant current can be measured with the emitter circuit open as I_{CBO} (meaning *current, collector-to-base, emitter open*). Current I_{CBO} is always present with or without emitter current and is additive to the major i_C component. Current I_{CBO} is highly temperature sensitive and the operating temperature must be limited to keep I_{CBO} small with respect to the desired control current component coming from the emitter.

4.4 The Volt-Ampere Curves of a Transistor

The output curves, Fig. 4.4(b), are the fundamental means of explaining the control characteristics of a transistor. These are curves of actual currents and related voltages and permit us to predict transistor performance. They will

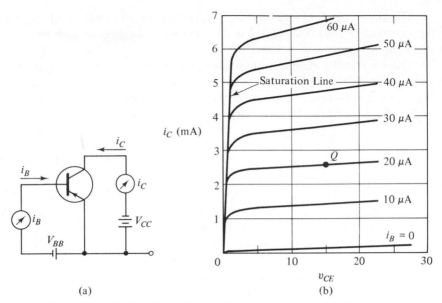

Figure 4.4 (a) Transistor circuit; (b) output curves of a silicon transistor at 25°C (77°F).

provide a basis for the circuit models that will simulate the action of the transistor in amplifier analysis.

The slope of the curves for any i_B value is given by

$$\frac{\Delta i_C}{\Delta v_{CE}}$$

and dimensionally this is the reciprocal of a resistance. Accordingly, we can measure the *collector resistance* of the transistor by the reciprocal of the slope as

$$r_C = \frac{\Delta v_{CE}}{\Delta i_C} \tag{4.1}$$

The low slope of the curves shows that r_c is high; it appears relatively constant for a given base current but becomes lower for increased values of base current.

At values of collector-emitter voltage of 1 V or less the curves merge into the *saturation line*, at the left in Fig. 4.4(b). Control of current by the base is lost. The reciprocal of the slope is again a resistance, that of the collector-base junction as a diode. This minimum resistance of the collector-base junction is known as $R_{CE(\text{sat})}$. It is largely due to the material resistance in the collector region.

The collector current is controlled by variation of the base current, which is, in turn, determined by the base-emitter voltage v_{BE}. In the region bounded by the saturation line, the line for $i_B = 50$ μA, $v_{CE} = 25$ V, and the line for $i_B = 0$, the characteristics are uniformly spaced and regular, the slopes are relatively constant, and this implies constancy of the various transistor parameters. This bounded region will be the region of operation for this transistor in many types of amplifiers.

Cutoff of the transistor occurs for $i_B \cong 0$, which is also the line for $i_C = I_{CEO}$ (current, collector-to-emitter, base open). For silicon transistors this current is very small, typically 2 μA, and the curve for $i_B = 0$ will lie very close to the abscissa. For a germanium unit the typical I_{CEO} value might be 100 to 200 μA.

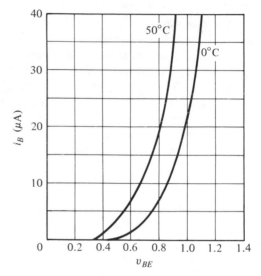

Figure 4.5 Input characteristics for the silicon transistor of Fig. 4.4.

The input curve of the transistor appears in Fig. 4.5. This shows the relation between base current and base-emitter voltage in the forward-biased junction. This curve looks similar to a forward diode curve, which it is. There is considerable variation of v_{BE} with temperature, as indicated.

4.5 The Current Amplification Factors

In Fig. 4.2 we showed the currents that were physically present at the transistor terminals; they reversed when we changed from a *pnp* transistor to an *npn* transistor. To avoid continuing confusion and to allow us to calculate for a circuit and then use either type of transistor, we are going to define all three transistor currents arbitrarily as positive inward, as shown in Fig. 4.6. That is,

$$i_E + i_C + i_B = 0 \tag{4.2}$$

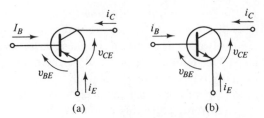

Figure 4.6 Defined currents and voltages: (a) *pnp* transistor; (b) *npn* transistor.

and obviously at least one current must be negative and reversed to satisfy Kirchhoff's current law.

We now can develop several useful transistor parameters. The relation between collector and emitter currents is a measure of the efficiency with which the charges are transported through the base. A change in emitter current will differ from the resultant collector current change by reason of the recombinations occurring in the base. We measure the efficiency of transport at constant collector-base voltage and state it as the ratio of a small change in collector current Δi_C to the small change in emitter current Δi_E. The parameter is useful for small changes or small signals and is called the *small-signal collector-emitter current amplification factor*. This has a magnitude of

$$\alpha = \frac{\Delta i_C}{\Delta i_E}\bigg]_{v_{CB}=\text{constant}} \tag{4.3}$$

Recombination current is kept small through use of a thin base layer and the value of α lies in the range of 0.90 to 0.99.

A good approximation to α can be obtained by using the steady current values at a point on the output curves; that is, we can employ

$$\alpha \cong \frac{i_C}{i_E} = h_{FB} \tag{4.4}$$

as the static or *dc collector-emitter current amplification factor*.

A second amplification factor relates a small change in collector current Δi_C caused by a change in base current Δi_B. This is determined at a constant collector-emitter voltage and is designated as the *small-signal collector-base current amplification factor*, written as

$$h_{fe} = \frac{\Delta i_C}{\Delta i_B}\bigg]_{v_{CE}=\text{constant}} \tag{4.5}$$

Since i_B is small by design, the factor h_{fe}[1] ranges from 20 to 200.

Again, we can approximate h_{fe} by using the steady currents at a point on the output curves; that is,

$$h_{FE} = \frac{i_C}{i_B} \cong h_{fe} \tag{4.6}$$

---

[1] Also called β in some literature but that symbol will not be used here because of conflict with the notation used in feedback amplifiers.

Equation 3.12, with i_C as an output and i_B as an input value, will explain the choice of h_f as a symbol for this amplification factor or current ratio.

Example: At $v_{CE} = 15$ V and $i_B = 20$ mA, the curve in Fig. 4.4 shows $i_C = 2.6$ mA. Using Eq. 4.2,

$$i_E + i_C + i_B = 0$$
$$i_E + 2.6 + 0.02 = 0$$
$$i_E = -2.62 \text{ mA}$$

Consequently, Eq. 4.4 shows for the approximate magnitude of α that

$$\alpha = \left| \frac{2.60}{2.62} \right| = 0.992$$

at point Q on the characteristics.

Example: The constant value of v_{CE} requires that changes be made along a vertical line in Fig. 4.4. Using $v_{CE} = 15$ V, and with changes from the point at Q, with $i_B = 20$ μA down to $i_B = 10$ μA, the corresponding change in i_C is from 2.6 to 1.4 mA. Then,

$$h_{fe} = \frac{(2.6 - 1.4) \text{ mA}}{(0.02 - 0.01) \text{ mA}} = \frac{1.2}{0.01} = 120$$

If we use the dc value as an approximation, we have at the same point Q

$$h_{FE} = \frac{2.6 \text{ mA}}{0.02 \text{ mA}} = 130 \cong h_{fe}$$

4.6 Relations between the Amplification Factors

The algebraic sum of the currents is equal to zero by Eq. 4.2 and so

$$\Delta i_E + \Delta i_C + \Delta i_B = 0 \tag{4.7}$$

Equation 4.3 was written for the magnitude of α but in Fig. 4.6 we can see that i_E and i_C are oppositely directed so that we write

$$\Delta i_E = -\frac{\Delta i_C}{\alpha} \tag{4.8}$$

Inserting this definition in Eq. 4.7,

$$-\frac{\Delta i_C}{\alpha} + \Delta i_C = -\Delta i_B$$

$$\Delta i_C \left(1 - \frac{1}{\alpha} \right) = -\Delta i_B$$

$$\frac{\Delta i_C}{\Delta i_B} = \frac{\alpha}{1 - \alpha} \tag{4.9}$$

We recognize the left side of the equation as the definition of h_{fe} and so we may write

$$h_{fe} = \frac{\alpha}{1 - \alpha} \tag{4.10}$$

Equation 4.10 may be rearranged to give

$$\alpha = \frac{h_{fe}}{1 + h_{fe}} \tag{4.11}$$

Both Eq. 4.10 and 4.11 will be very useful in later discussions. Another useful equation can be obtained from Eq. 4.10 by writing

$$(1 - \alpha)h_{fe} = \alpha$$

Inserting Eq. 4.11 and canceling h_{fe} leads to

$$1 - \alpha = \frac{1}{1 + h_{fe}} \tag{4.12}$$

4.7 The Load Line and Q Point

A simple transistor amplifier appears in Fig. 4.7. Writing a voltage equation around the output loop gives

$$v_{CE} = V_{CC} - Ri_C \tag{4.13}$$

This is the equation of a straight line. The transistor output curve family also involves the variables v_{CE} and i_C; by plotting Eq. 4.13 on the family of transistor curves, the *dc load line* obtained will locate all possible values of v_{CE} and i_C for the series circuit of transistor and load R at a given V_{CC}.

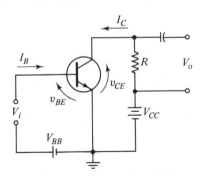

Figure 4.7   A transistor amplifier.

To draw the load line for Eq. 4.13 we use the axis intercepts. Insertion of $i_C = 0$ in Eq. 4.13 gives an x-axis intercept at $x = V_{CC}$; use of $v_{CE} = 0$ gives y-axis intercept at $y = V_{CC}/R$. In Fig. 4.8 a load line is drawn for $V_{CC} = 20$ V, $V_{CC}/R = 40$ mA, and $R = 20/0.04 = 500$ Ω.

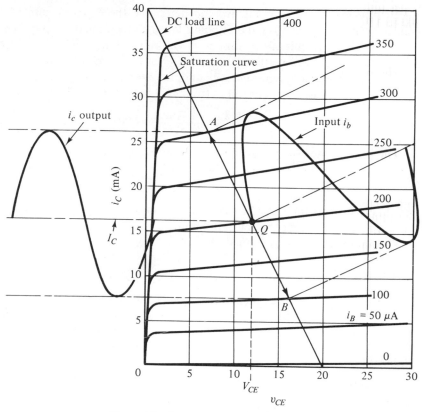

Figure 4.8 Output curves of a silicon transistor at 25°C (77°F).

The load line is the locus for all voltages and currents that can exist with the load and transistor in series. If alternating input signals are used as shown, a *zero axis point* or no-signal point Q must be chosen on the load line. The Q point is also called the *quiescent point* since the circuit is quiet or $V_i = 0$. We rather arbitrarily choose a Q point and decide on the *bias currents and voltages*, the steady V_{CE}, the steady collector current I_C, and the base current I_B. We use these biases to place the signal variations in a region of the curves where the transistor will operate in most linear fashion. The bias sources also supply the dc energy that the transistor converts to the ac signal. Circuits for establishing the bias currents and voltages will be discussed in Chapter 5.

For true reproduction of the input sine current signal, the output ac wave must be derived from equal i_B excursions along the load line, each side of the Q point. As the signal voltage moves positive, the operating point moves along the load line up to A at the positive peak, back to Q, and down to B at the negative peak of the i_B signal input. The output current i_C varies correspondingly, as shown along the i_C axis in Fig. 4.8.

Having selected a Q point, we can determine the h parameters of the transistor in the region around that point. We shall then use those parameters in the equivalent circuit of the transistor. The value of h_{f_e} can be found from a measurement along a vertical line in Fig. 4.8 at constant v_{CE}. The change in i_C per unit change in i_B represents h_{f_e} and, for example, a change of i_B from 200 to 150 μA causes a change in i_C from 16.2 to 11.7 mA. Then

$$h_{f_e} = \frac{(16.2 - 11.7) \text{ mA}}{(0.20 - 0.15) \text{ mA}} = 90$$

The output short-circuit reciprocal resistance is defined as

$$h_{oe} = \frac{I_2}{V_2} = \frac{\Delta i_C}{\Delta v_{CE}}$$

and this represents the slope of a constant i_B curve in Fig. 4.8; it is apparent that h_{oe} is constant over the central region.

The input resistance is defined as

$$h_{ie} = \frac{V_1}{I_1} = \frac{\Delta v_{BE}}{\Delta i_B}$$

This relation represents the reciprocal slope of the input curve in Fig. 4.5, as an example. The input signal variation about the Q point must be kept small for h_{ie} to be considered constant. The requirement for constant values of the h parameters is the reason that h_{f_e} and the other h parameters are defined as *small-signal parameters*.

The parameter h_{re} can be found as the slope of a curve relating v_{CE} and v_{BE} for constant i_C.

4.8 The Basic Transistor Amplifiers

The transistor is an active amplifier element with three external leads, one connecting to the emitter, one to the base, and one to the collector. We shall use it in circuits as a two-port device with four terminals. As such, one of the transistor leads must be made common to the input and output ports. The choice is arbitrary and leads to the three connections in Fig. 4.9. These are the *common-emitter* (C-E) circuit, the *common-base* (C-B) circuit, and the *common-collector* (C-C) circuit.

These three circuits differ in performance and application. For example, we shall find that the C-E circuit is best for power gain. The C-B circuit is a load-matching device that permits a low-resistance source to be efficiently connected to a high-resistance output circuit and results in a voltage gain near unity. The C-C circuit reverses the resistance transformation property and gives a current gain near unity. Other differences in the characteristics of these three circuits will become apparent with further study.

To indicate the common input-output electrode and the use of the C-E,

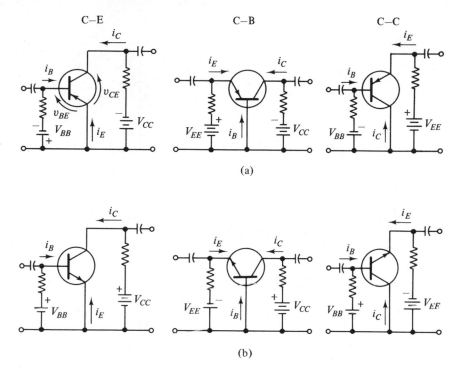

C–E   C–B   C–C

(a)

(b)

Figure 4.9 Defined currents in C-E, C-B, and C-C connections with (a) *pnp* transistors; (b) *npn* transistors.

C-B, or C-C amplifier circuit, a second subscript is employed with the *h* parameters, as h_{fe}, h_{ib}, h_{oc}. In this text we intend to use only the *h* parameters for the common-emitter configuration; however, conversion equations for all three configurations are presented in Sec. 4.14.

4.9 Simplification of the Equivalent C-E Circuit

Any linear two-port circuit can be replaced with the *h*-parameter circuit of Chapter 3. We have just shown how the transistor can be connected as a two-port circuit element. Therefore, we can replace the C-E connected transistor in Fig. 4.10(a) with the *h*-parameter circuit in Fig. 4.10(b). The values of the *h* parameters can be taken from the manufacturer's data or from the characteristic curves but greater accuracy will result if they are measured at the selected *Q* point.

A simple common-emitter amplifier is drawn in Fig. 4.11(a), including the bias voltages V_{BB} and V_{CC} that establish the *Q* point. In Fig. 4.11(b) we have

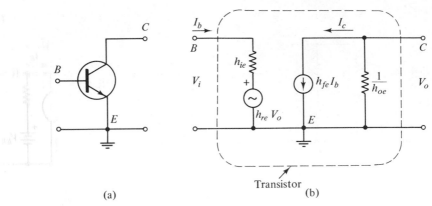

Figure 4.10 (a) The C-E transistor as a two-port element; (b) the *h*-parameter equivalent circuit as a two-port element.

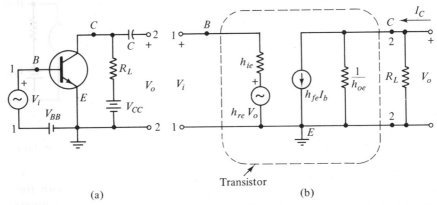

Figure 4.11 (a) C-E transistor amplifier; (b) *h*-parameter equivalent for (a).

replaced the transistor with its equivalent circuit between *B*, *C*, and *E*. Since no ac signal voltage can appear across the dc bias sources, they are eliminated from the circuit. The output blocking capacitor is chosen so large that no ac signal voltage appears there and it too may be removed from the equivalent circuit. Later we shall consider its effects. With voltages V_i and V_o as rms values, we now have the complete *ac equivalent circuit* shown in Fig. 4.11(b).

The *h*-parameter circuit equations of Sec. 3.6 apply to this transistor amplifier between ports 1,1 and 2,2. Using voltage and current designations from Fig. 4.11 we have

$$V_i = h_{ie}I_b + h_{re}V_o \qquad\qquad (4.14)$$
$$I_c = h_{fe}I_b + h_{oe}V_o \qquad\qquad (4.15)$$

Because of the direction of I_c and the assigned polarity of the voltage across the load, we must write

$$I_c = -\frac{V_o}{R} \qquad\qquad (4.16)$$

Using this relation in Eq. 4.15, we have

$$-\frac{V_o}{R} = h_{fe}I_b + h_{oe}V_o$$

Solving for the input current I_b and inserting the result in Eq. 4.14, we have an equation involving V_i and V_o as variables. We can write the equation for the *voltage gain* as

$$A_{ve} = \frac{V_o}{V_i} = \frac{-1}{\dfrac{h_{ie}(1 + h_{oe}R)}{h_{fe}R} - h_{re}}$$

(4.17)

The performance of the amplifier can be better understood if we can find a less complicated expression for the voltage gain. Fortunately it is possible to simplify the circuit and the resultant equations with small error.

Generator $h_{re}V_o$ is a control source that feeds back some of the output voltage to the input circuit. The effect is usually small and we are justified in assuming $h_{re} = 0$ and in dropping the generator from the equivalent circuit.

Values of $1/h_{oe}$ approximate $100,000\ \Omega$ and as such are negligible in effect when in parallel with load R values of 1000 to 10,000 Ω. The product $h_{oe}R$ then ranges from 0.01 to 0.1 and $h_{oe}R$ can be dropped in comparison to unity in the denominator of Eq. 4.17.

By making $h_{re} = 0$ and $h_{oe} = 0$, we obtain the simplified equivalent circuit for the C-E amplifier in Fig. 4.12(a). The equation for the voltage gain is simplified from Eq. 4.17 to

$$A_{ve} = \frac{V_o}{V_i} = -\frac{h_{fe}R}{h_{ie}}$$

(4.18)

The two assumptions made above may seem arbitrary; however, we should not expect close agreement between computed values and laboratory practice because of the considerable variation in the transistor parameters between units of the same type. Values given by the manufacturer are average parameter values. For example, while h_{fe} may have an average value of 100, the actual values of this parameter may range from 50 to 200 in transistors of the same type.

The simplified equivalent circuit is amply justified by the saving in work and in increased understanding of transistor circuit operation.

4.10 The Transconductance g_m

Equation 4.18 for the C-E amplifier voltage gain employs the ratio h_{fe}/h_{ie}. Reference to the *h*-parameter definitions shows us that

$$\frac{h_{fe}}{h_{ie}} = \frac{I_c/I_b}{V_{be}/I_b} = \frac{I_c}{V_{be}}$$

(4.19)

The quantity I_c/V_{be} is the reciprocal of resistance or a *conductance* with units in *mhos*. It represents the output current obtained per unit of input voltage. Because it relates a current in the output circuit to a voltage in the input circuit, the quantity is a *transfer* conductance, called a *transconductance* and given the symbol g_m, where

$$g_m = \frac{I_c}{V_{be}} = \frac{h_{fe}}{h_{ie}} \tag{4.20}$$

With $g_m = 0.01$ mho $= 10{,}000$ μmhos, Eq. 4.20 indicates that we should have an output current of 10 mA ac for each ac volt input.

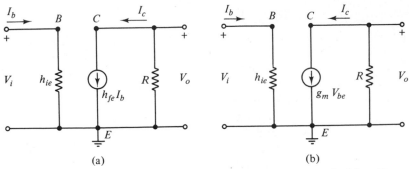

(a)   (b)

Figure 4.12   (a) Simplified C-E equivalent circuit (b) same as (a) with $g_m V_{be}$ as the source.

In the equivalent circuit in Fig. 4.12(a), we see that the generator current and the collector current are the same and

$$I_c = h_{fe} I_b$$

Using Eq. 4.20, we have

$$h_{fe} I_b = g_m V_{be} \tag{4.21}$$

We use $g_m V_{be}$ as the value of the current source in a revised C-E equivalent circuit in Fig. 4.12(b). The two circuits of the figure are completely interchangeable, with one having its current source controlled by input current and the other by the input voltage.

4.11 The Common-Emitter Amplifier

We can now use the simplified g_m model of the C-E amplifier in Fig. 4.12(b), to evaluate the performance of that amplifier circuit.

Voltage Gain

We first express the output voltage as

$$V_o = -I_c R = -g_m R V_{be}$$

and from this can obtain the voltage gain as

$$A_{ve} = \frac{V_o}{V_i} = -g_m R \tag{4.22}$$

Current Gain

The current in the load R is

$$I_c = g_m V_{be} = h_{fe} I_b$$

by Eq. 4.21 and so

$$A_{ie} = \frac{I_c}{I_b} = h_{fe} \tag{4.23}$$

This is the current gain of the transistor itself.

Input Resistance

By observation of the circuit, we have

$$R_{ie} = \frac{V_i}{I_b} = \frac{V_{be}}{I_b} = h_{ie} \tag{4.24}$$

for the resistance facing the input signal source.

Output Resistance

The resistance measured at the output 2,2 port is found by reducing the signal source to zero. With $V_i = V_{be} = 0$, the collector current source is also zero, or open. From the 2,2 port we see an open circuit but in practice we know that $1/h_{oe}$ is present there as a large resistance. Therefore

$$R_{oe} = \frac{1}{h_{oe}} \tag{4.25}$$

Power Gain

The power gain of the circuit is defined as

$$\text{power gain} = \text{P.G.} = |A_{ve} A_{ie}| \tag{4.26}$$

$$= g_m h_{fe} R = \frac{h_{fe}^2 R}{h_{ie}} \tag{4.27}$$

This result is usually converted to decibels. The transconductance g_m may be considered as a transistor figure of merit and its effect can be seen in the several gain expressions.

4.12 Performance of a C-E Amplifier

In a C-E amplifier we shall employ a *pnp* transistor with the following parameters:

$$h_{ie} = 900 \ \Omega \qquad\qquad h_{fe} = 25$$
$$h_{re} = 2.1 \times 10^{-4} \cong 0 \qquad h_{oe} = 16 \times 10^{-6} \ \text{mho} \cong 0$$

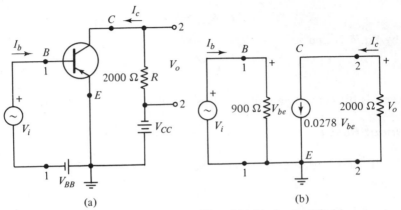

Figure 4.13 (a) Common-emitter amplifier; (b) with the simplified *h*-parameter equivalent circuit.

The load R will be chosen as 2000 Ω.

With the circuit in Fig. 4.13, we have

$$g_m = \frac{h_{fe}}{h_{ie}} = \frac{25}{900} = 0.0278 \ \text{mho}$$

The terminal resistances are

$$R_{ie} = h_{ie} = 900 \ \Omega$$
$$R_{oe} = \frac{1}{h_{oe}} = 62,500 \ \Omega$$

The gains can be calculated as

$$A_{ve} = -g_m R = -0.0278 \times 2000 = -55.5$$
$$A_{ie} = h_{fe} = 25$$
$$\text{P.G.} = 25 \times 55.5 = 1387$$

In decibels this becomes

$$\text{P.G.}_{\text{dB}} = 10 \log 1387 = 10 \times 3.14 = 31.4 \ \text{dB}$$

Had we used the exact expression for voltage gain, Eq. 4.17, we would have found the gain to be -53.8 compared with -55.5 above. The use of the approximate circuit and gain expression results in an error of only 3 per cent in the voltage gain.

We can conclude that the C-E amplifier has moderate input and output resistances and good voltage and current gains. It is found to have the highest power gain of the three basic transistor amplifier circuits and is widely applied for this reason.

4.13 Relation between A_i and A_v

We can rearrange the C-E gain expression as

$$A_{ve} = -g_m R = -\frac{h_{fe}}{h_{ie}} R = -A_i \frac{R}{h_{ie}} \tag{4.28}$$

by reference to Eq. 4.20 and 4.23. Equation 4.28 is actually a general relation between voltage gain and current gain in amplifiers, which can be stated as

$$A_v = -A_i \frac{\text{load resistance}}{\text{input resistance}} = -A_i \frac{R}{R_i} \tag{4.29}$$

This result may often simplify the calculation of the several circuit gain figures.

4.14 Conversion of the h Parameters

The form of the h-parameter equivalent circuit is the same for any two-port circuit, and therefore is the same for the C-E, C-B and C-C amplifiers. The parameter values, however, will vary with the choice of common element and Table 4.1 shows these relationships.

Since the common-emitter parameters are most readily available, our circuit analyses will be carried out with those parameters.

TABLE 4.1   C-B and C-C Parameters as Functions of C-E Parameters

$h_{ib} = \dfrac{h_{ie}}{1 + h_{fe}} \cong \dfrac{h_{ie}}{h_{fe}}$	$h_{ic} = h_{ie}$
$h_{ob} = \dfrac{h_{oe}}{1 + h_{fe}} = \dfrac{h_{oe}}{h_{fe}}$	$h_{oc} = h_{oe}$
$h_{fb} = \dfrac{-h_{fe}}{1 + h_{fe}} = -\alpha$	$h_{fc} = -(1 + h_{fe}) \cong -h_{fe}$
$h_{rb} = \dfrac{h_{ie}h_{oe} - h_{re}h_{fe}}{1 + h_{fe}}$	$h_{rc} \cong 1$

C-E and C-C Parameters as Functions of C-B Parameters

$h_{ie} = \dfrac{h_{ib}}{1 - \alpha}$	$h_{ic} = \dfrac{h_{ib}}{1 - \alpha}$
$h_{oe} = \dfrac{h_{ob}}{1 - \alpha}$	$h_{oc} = \dfrac{h_{ob}}{1 - \alpha}$
$h_{fe} = \dfrac{\alpha}{1 - \alpha}$	$h_{fc} = \dfrac{-1}{1 - \alpha}$
$h_{re} = \dfrac{h_{ib}h_{ob} - h_{rb}\alpha}{1 - \alpha}$	$h_{rc} \cong 1$

4.15 The Common-Base Transistor Amplifier

For the common-base (C-B) amplifier in Fig. 4.14(a), the input signal V_i is supplied between emitter and base. The input current is I_e and is much larger than the input current for the C-E circuit. This implies that we have a lower resistance transistor input circuit. We use the dashed box shown in Fig. 4.14(b) as the *h*-parameter equivalent circuit for the transistor, with common-base *h* parameters.

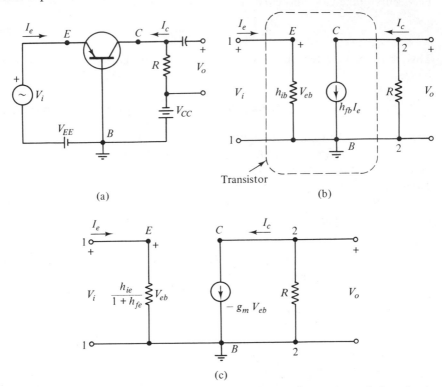

(a)

(b)

Transistor

(c)

Figure 4.14 (a) Common-base amplifier; (b) the *h*-parameter equivalent circuit; (c) equivalent g_m circuit.

Use of a conversion factor from Table 4.1 allows us to write for the collector current

$$I_c = -\alpha I_e = \frac{-h_{fe}}{1 + h_{fe}} I_e \qquad (4.30)$$

From the C-B circuit in Fig. 4.14(b), we have

$$V_{eb} = h_{ib} I_e$$

and using the conversion factor for h_{ib}

$$V_{eb} = \frac{h_{ie}}{1 + h_{fe}} I_e \qquad (4.31)$$

We can determine the transconductance by taking the ratio of Eq. 4.30 to 4.31, to give us

$$g_m = \frac{I_c}{V_{eb}} = -\frac{\dfrac{h_{fe}}{1 + h_{fe}} I_e}{\dfrac{h_{ie}}{1 + h_{fe}} I_e} = -\frac{h_{fe}}{h_{ie}} \qquad (4.32)$$

The transconductance for the C-B circuit has the same magnitude as for the C-E circuit; the negative sign arises because we are using V_{eb} for the input voltage instead of V_{be}.

Using $-g_m$ and $h_{ib} = h_{ie}/(1 + h_{fe})$ we draw the g_m model for the C-B amplifier shown in Fig. 4.14(c). The performance of the C-B circuit can then be calculated, using the common-emitter parameters.

Voltage Gain

We have from the circuit

$$I_c = -g_m V_{eb} \qquad (4.33)$$

Because of the polarity assigned to the output voltage, we write it as

$$V_o = -I_c R = g_m R V_{eb}$$

Since the input to the transistor is also the signal input, or $V_{eb} = V_i$, we have for the voltage gain

$$A_{vb} = \frac{V_o}{V_i} = g_m R \qquad (4.34)$$

Current Gain

The load current is given by Eq. 4.33 and if we use Eq. 4.31 for V_{eb}, we can write I_c as

$$I_c = -\frac{g_m h_{ie}}{1 + h_{fe}} I_e$$

The current gain can be obtained from this relation, as

$$A_{ib} = \frac{I_c}{I_e} = -\frac{g_m h_{ie}}{1 + h_{fe}} \qquad (4.35)$$

We often find that $h_{fe} \gg 1$ and the unity term in the denominator can be neglected, giving

$$A_{ib} \cong -\frac{g_m}{h_{fe}/h_{ie}} = -1 \qquad (4.36)$$

as a limiting value.

Input Resistance

The input resistance at the 1,1 port is

$$R_{ib} = \frac{h_{ie}}{1 + h_{fe}} \simeq \frac{h_{ie}}{h_{fe}} = \frac{1}{g_m} \quad (\Omega) \tag{4.37}$$

This is a low resistance, as predicted.

Output Resistance

We make $V_i = 0$ and effectively this reduces the current source to zero or an open circuit. The output resistance measured at 2,2 appears infinite but in practice it represents $1/h_{ob} = (1 + h_{fe})/h_{oe}$, which is usually over 1 MΩ.

Power Gain

This is obtained from A_{vb} and A_{ib} as

$$\text{P.G.} = |A_{vb}A_{ib}| = g_m R \tag{4.38}$$

The current gain approximates unity, with 180° of phase shift. The voltage gain is equal to that of the C-E amplifier in magnitude. The major reason for using this circuit is to obtain maximum power transfer through impedance matching. This is accomplished by matching the low-resistance input circuit to a low-resistance signal source and the high-resistance output circuit to a high-resistance load. Since the current from input to output is essentially equal, the circuit behaves in a manner similar to a water pump, forcing water from a low-pressure system to a high-pressure system.

4.16 Performance of the C-B Amplifier

Using the same transistor as in the C-E amplifier example, we find the performance of the C-B circuit with a load of 2000 Ω.

$$h_{ie} = 900 \ \Omega \qquad h_{fe} = 25$$
$$h_{re} \cong \text{negligible} \qquad h_{oe} = 16 \times 10^{-6} \text{ mho}$$
$$\cong \text{negligible}$$

We calculate $g_m = h_{fe}/h_{ie} = \frac{25}{900} = 0.0278$ mho.
The voltage gain is

$$A_{vb} = g_m R = 0.0278 \times 2000 = 55.5$$

The current gain is

$$A_{ib} = -\frac{g_m h_{ie}}{1 + h_{fe}} = -\frac{0.0278 \times 900}{26} = -0.96$$

The power gain is

$$\text{P.G.} = |A_{vb}A_{ib}| = 55.5 \times 0.96 = 53.4$$
$$\text{P.G.}_{dB} = 10 \log 53.4 = 10 \times 1.73 = 17.3 \text{ dB}$$

The input resistance is

$$R_{ib} = h_{ib} = \frac{h_{ie}}{1 + h_{fe}} = \frac{900}{26} = 35 \, \Omega$$

and the output resistance is

$$R_{ob} = \frac{1}{h_{ob}} = \frac{1 + h_{fe}}{h_{oe}} = \frac{26}{16 \times 10^{-6}} = 1.62 \text{ M}\Omega$$

This is very large, as assumed.

4.17 The Common-Collector Amplifier

The common-collector (C-C) amplifier in Fig. 4.15(a) is also called an *emitter follower*. This name is justified when we write the voltage equation around the input loop in Fig. 4.15(a) as

$$V_i - V_{be} - V_o = 0 \qquad (4.39)$$

Making the input signal V_{be} small, we have

$$V_i \cong V_o \qquad (4.40)$$

The output voltage is approximately equal to, or follows, the input voltage. The voltage gain is, of course, near unity.

The transistor is in the normal B, C, E connection and between those terminals we simply substitute the g_m model of the transistor, as in Fig. 4.15(b). The input resistance to the emitter is h_{ie} and the current source is $g_m V_{be}$.

From the load circuit

$$V_o = (I_b + I_c)R \qquad (4.41)$$

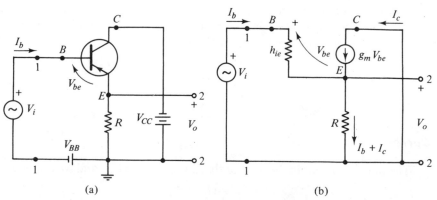

(a) (b)

Figure 4.15 (a) Common-collector amplifier; (b) g_m equivalent circuit.

by definition in Sec. 4.10, we have

$$I_c = g_m V_{be} = h_{fe} I_b \tag{4.42}$$

Using Eq. 4.42 in 4.41, we have

$$V_o = (I_b + h_{fe} I_b)R = (1 + h_{fe})RI_b \tag{4.43}$$

Using Eq. 4.43 and $V_{be} = h_{ie} I_b$ in the input loop equation, Eq. 4.39, we have

$$V_i = [h_{ie} + (1 + h_{fe})R]I_b$$

Input Resistance

The input resistance at the 1,1 port of the amplifier is $R_{ic} = V_i/I_b$, which can be evaluated from the result above as

$$R_{ic} = \frac{V_i}{I_b} = h_{ie} + (1 + h_{fe})R \tag{4.44}$$

Since $h_{fe} \gg 1$ is a usual condition, we then have

$$R_{ic} \cong h_{ie} + h_{fe}R \tag{4.45}$$

which is a large resistance, much greater than h_{ie}.

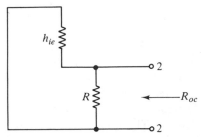

Figure 4.16 For the output resistance of the C-C amplifier.

Output Resistance

To obtain the output resistance at 2,2 we short-circuit the signal source V_i, which reduces the control generator to zero or an open circuit. The circuit at the output port then becomes that shown in Fig. 4.16, with h_{ie} in parallel with R, so

$$R_{oc} = \frac{h_{ie}R}{h_{ie} + R} \tag{4.46}$$

which is a low resistance, less than h_{ie}.

Current Gain

The output current in the load of the C-C amplifier is I_e and we know that

$$I_e = -(I_b + I_c)$$

from the Kirchhoff current law. But $I_c = h_{fe}I_b$ from Eq. 4.42 and the current gain can be obtained as

$$A_{ic} = \frac{I_e}{I_b} = -\frac{I_b + h_{fe}I_b}{I_b} = -(1 + h_{fe})$$

$$\cong -h_{fe} \qquad (4.47)$$

Voltage Gain

The voltage gain can be derived by use of the principle of Sec. 4.13, written as

$$A_{vc} = -A_{ic}\frac{R}{R_{ic}}$$

$$= \frac{h_{fe}R}{h_{ie}\left(1 + \dfrac{h_{fe}}{h_{ie}}R\right)}$$

using A_{ic} from Eq. 4.47 and R_{ic} from Eq. 4.45. The expression above then becomes

$$A_{vc} = \frac{g_mR}{1 + g_mR} \qquad (4.48)$$

which has a value less than but near unity.

One use of the emitter follower is as a unity voltage gain amplifier to isolate one circuit from another; this might be the case if we supply a telephone line with the output of the amplifier. The C-C circuit is also widely used because of its high value of input resistance, which is much larger than can be obtained from a C-E amplifier. The C-C amplifier output also provides a match to a low-resistance load since R need be only 1000 Ω or so. This impedance transformation property, from a high input resistance to a low output resistance, is the inverse of the action of the C-B circuit.

4.18 *Performance of the C-C Amplifier*

Using the same transistor employed for the C-E amplifier, and with $R = 2000\ \Omega$, typical C-C performance can be calculated. The value of $g_m = 0.0278$ mho, as before.

$$h_{ie} = 900\ \Omega \qquad h_{fe} = 25$$
$$h_{re} \cong \text{negligible} \qquad h_{oe} = 16 \times 10^{-6}\ \text{mho}$$
$$\cong \text{negligible}$$

The voltage gain is

$$A_{vc} = \frac{g_mR}{1 + g_mR} = \frac{0.0278 \times 2000}{1 + 55.5}$$
$$= 0.984$$

and this approaches unity and has no phase shift.

The current gain is $-h_{fe} = -25$, showing a $180°$ phase shift.
The power gain is

$$\text{P.G.} = |A_{vc}A_{ic}| = 0.984 \times 25 = 24.6$$
$$\text{P.G.}_{\text{dB}} = 10 \log 24.6 = 10 \times 1.39$$
$$= 13.9 \text{ dB}$$

This is a low value of power gain.

The input resistance is given by use of Eq. 4.44 as

$$R_{ic} = h_{ie} + (1 + h_{fe})R$$
$$= 900 + 26 \times 2000 = 52,900 \; \Omega$$

Using the approximate relation in Eq. 4.45, we have

$$R_{ic} \cong h_{ie} + h_{fe}R = 50,900 \; \Omega$$

and the difference is not significant.

The output resistance is

$$R_{oc} = \frac{h_{ie}R}{h_{ie} + R} = \frac{900 \times 2000}{2900} = 621 \; \Omega$$

which is quite low, as predicted. This design would be well suited to operate from a high-resistance microphone into a telephone line.

4.19 Comparison of Amplifier Performance

Table 4.2 is introduced to summarize the predicted performances of the three basic amplifiers and to allow comparison of their respective capabilities.

TABLE 4.2 **Comparison of Basic Transistor Amplifiers**

	C-E	C-B	C-C
Input resistance, R_i	h_{ie}	$\dfrac{h_{ie}}{1 + h_{fe}}$	$h_{ie} + h_{fe}R$
Output resistance, R_0	$\dfrac{1}{h_{oe}}$	$h_{fe}\dfrac{1}{h_{oe}}$	$\dfrac{h_{ie}R}{h_{ie} + R}$
Current gain, A_i	h_{fe}	$\cong -1$	$-h_{fe}$
Voltage gain, A_v	$-g_mR$	g_mR	$\cong 1$
Power gain	$h_{fe}g_mR$	g_mR	$\dfrac{h_{fe}g_mR}{1 + g_mR}$
Power gain as a ratio to that of the C-E amplifier	1	$\dfrac{1}{h_{fe}}$	$\dfrac{1}{1 + g_mR}$

Table 4.3 adds the data from the several examples, again for comparison purposes. The data show that the C-E circuit has nominal input and output

resistances and the highest power gain. For those reasons it is the most generally used amplifier circuit. The C-B circuit finds application as an impedance matcher from low to high values, while the C-C circuit isolates the input from the output and matches high to low impedances.

TABLE 4.3 **Typical Transistor Amplifier Performance**

Transistor:	$h_{ie} = 900\ \Omega$	$h_{fe} = 25$	
	h_{re} = negligible	$h_{oe} = 16 \times 10^{-6}$ mho	
	$g_m = 0.0278$ mho	$\cong$ negligible	

	C-E	*C-B*	*C-C*
R_i	900 Ω	35 Ω	52,900 Ω
R_o	62,500 Ω	1.62 MΩ	621 Ω
A_i	25	−0.96	−25
A_v	−55.5	55.5	0.984
P.G.	1387	55.5	24.6
P.G. $_{dB}$	31.4 dB	17.3 dB	13.9 dB

4.20 Transistor Manufacturing Techniques

The manufacture of transistors employs the same basic purification and crystal growth processes already described for diodes. Subsequent processing of the semiconductor material leads to *grown junction* transistors, *alloyed junction* transistors, and *mesa* and *planar* transistors.

Grown junction transistors are produced by crystal growth from a doped bath of molten semiconductor, as diagrammed in Fig. 4.17(a). Change of the predominant impurity from *p* to *n* and back to *p* is made at precise moments to yield layers of the desired conduction properties.

Wafers are then cut from the crystal to include sections of the junction regions. Critical resistance requirements for the several regions are hard to meet and the method is not widely used.

Alloyed junction transistors are made by fusing a pellet of the desired impurity element onto both sides of a base chip, as in Fig. 4.17 (b). With an *n*-silicon base, *p* impurity pellets of indium will alloy with the silicon to form *pn* junctions at the interfaces, as for the alloyed diode. Precise control of time and temperature of the fusion process yields transistors in which the critical base width can be very thin. The remainder of the impurity pellets serves as the emitter and collector contacts.

The process of diffusion is widely used in the manufacture of mesa and planar transistors. A substrate wafer of desired resistivity is coated with a

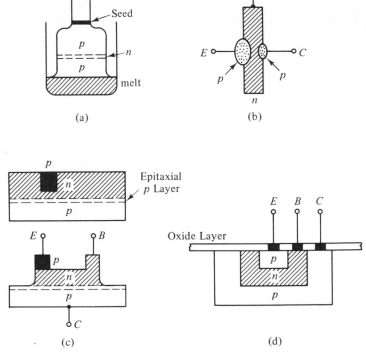

Figure 4.17 (a) Grown *pnp* transistor material; (b) alloyed *pnp* junction transistor; (c) diffused mesa transistor; (d) planar transistor.

layer of a doping element, by deposition from a gaseous atmosphere containing silicon and the desired doping material. Exposure to high temperature for a limited time follows while the doping atoms migrate to the desired depth in the wafer. If an *n*-silicon wafer is used with boron as the diffusing impurity, the surface of the wafer will become *p* material to the desired depth. A second diffusion with an *n* impurity changes some of the *p* material back to *n* for an emitter layer. The result is illustrated in Fig. 4.18. The depths of diffusion are a few thousandths of a millimeter.

The diffusion process can be confined to very small areas by use of photographically applied chemical surface masks. The surface of the silicon is coated with a very thin layer of silicon dioxide, followed by a layer of photoresist lacquer. The mask is placed over the wafer in accurate register and identical areas for all the transistors are exposed to ultraviolet light. Where exposed to light the photoresist is chemically changed and made inactive. The unexposed areas are chemically etched away along with the underlying silicon dioxide, leaving openings to areas of the silicon wafer. The silicon of these

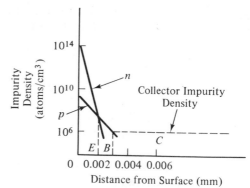

Figure 4.18 Diffusion process to form p base and n emitter.

electrode areas is available for further impurity diffusion. The process is repeated with different masks, resulting in superimposed electrode areas. The masks are similar to negatives, produced as much-reduced photographs of large-scale drawings. The masked areas are very small and hundreds of transistors can be produced on one silicon wafer cut from a 2- or 3-cm diameter crystal.

In the process of *epitaxial* growth (*epi*, upon; *taxi*, arrange), a doped crystal region is grown upon the wafer with the same atom arrangement and crystal structure. Growth usually occurs from a vapor containing silicon and a doping material at an elevated temperature. The impurity concentration can differ from that of the substrate, giving lower or higher resistivity in a thin layer.

Use of a high-resistivity collector wafer raises the breakdown voltage but also increases the saturation resistance and the collector voltage drop at high current. By epitaxial growth, it is possible to place a high-resistivity but thin collector layer on a low-resistivity wafer, giving both high breakdown voltage due to the high-resistivity material at the junction and a low saturation resistance due to the bulk of the collector material. The use of an epitaxial layer is shown in the mesa transistor in Fig. 4.17(c).

The *mesa transistor* may use an epitaxial layer on a p wafer, followed by an n diffusion to form the base region. A second diffusion is applied to a small area through an unmasked opening in the silicon dioxide and a small p region formed for the emitter. Excess n material is then etched away to reduce the size and electrical capacitance of the collector-base junction. The name *mesa* is derived from the similarity of the completed transistor to the desert mesas of the western United States.

The *planar transistor* in Fig. 4.17(d) is produced by diffusion of an n impurity to the desired depth in the p chip through openings in a silicon dioxide layer, masked as described above. The process is repeated and a p impurity is diffused to form the emitter region in the previously diffused base.

Silicon dioxide is added to reduce surface leakage and aluminum is evaporated through openings in the dioxide layer to form the necessary element connections.

4.21 Comments

The junction transistor represents two back-to-back diodes, with the signal voltage controlling the current through the forward-biased input junction. This current is extracted through the reverse-biased output junction as a current source.

In Chapter 3 we developed algebraic methods of active circuit analysis and a general h-parameter equivalent circuit model. This circuit, being equivalent to any active linear circuit, is used as an ac equivalent for the transistor by assignment of h parameters measured at the ports of the actual transistor. The transistor becomes a set of circuit elements and a source, and for ac signal analysis we have no concern for holes, electrons, junctions, and biases. Since the output of a transistor is a varying current and we require a voltage for input to a cascaded transistor, we use a load resistor to provide conversion between output current and output voltage.

There are three basic amplifier forms, the C-E, C-B, and C-C, for which we employ the same equivalent circuit model. This is the g_m model for the active source, in which the internal current is made dependent on the input signal voltage. This g_m model will be applied for the field-effect transistor and vacuum tube as well.

REVIEW QUESTIONS

4.1 Draw the circuit symbol for a *pnp* transistor; show and label the three assumed currents and the three voltages.

4.2 Repeat Question 4.1 for an *npn* transistor.

4.3 What is meant by I_C, I_B, V_{be}, I_c, I_b, V_{BB}, I_E, V_{CC}?

4.4 Explain the meaning of *npn* and *pnp* when applied to transistors.

4.5 What is the collector polarity with respect to the emitter in a *pnp* transistor?

4.6 What bias polarities should be applied to the base and collector of an *npn* transistor, with the emitter as reference?

4.7 What kind of charges transit the base in an *npn* transistor?

4.8 Why are the charges injected into the base able to reach the collector terminal in a properly biased transistor?

4.9 Why is the impurity concentration made much greater in the emitter than in the base?

4.10 Why does I_{CBO} vary with temperature?

4.11 What should be the allowable maximum temperature of a silicon transistor to keep I_{CBO} small?

4.12 The transistor is sometimes said to be a current-operated device; use the diode law and explain that it is also a voltage-operated device.

4.13 What currents and voltages are related by the output characteristics of a transistor with common emitter?

4.14 On the input curve of Fig. 4.5, choose a Q point and identify it. What should V_{BB} be to obtain that Q point? What would I_B then be?

4.15 What is the cause of I_{CBO}?

4.16 For a C-E transistor, define h_{FE} in words.

4.17 Define the current gain α in words.

4.18 What is the approximate relationship between h_{FE} and h_{fe} at usual operating temperatures?

4.19 Define the saturation resistance for a C-E transistor.

4.20 Explain how the h parameters are determined from the characteristic curves.

4.21 Why can h_{FB} never be greater than unity?

4.22 If α is given, how do you determine h_{fe}?

4.23 If $h_{FE} = 60$ and $I_C = 15$ mA, what is the base current? What is the emitter current?

4.24 What physical measurement of the transistor should be reduced to increase h_{FE}?

4.25 State one reason for choice of a C-B amplifier.

4.26 State one reason for choice of a C-C amplifier.

4.27 What is the major advantage of the C-E amplifier over the other forms?

4.28 Define A_v, A_i, and P.G. What does the negative sign on a voltage gain figure mean?

4.29 Why is the input resistance of a C-B amplifier so much smaller than that of a C-E amplifier?

4.30 Compare the input resistance of a C-C amplifier to that of a C-B amplifier.

4.31 Why is the C-C circuit called an emitter follower?

4.32 A C-B circuit is sometimes called a collector follower; explain.

4.33 What circuit element value should be changed to make A_{vc} approach unity in a C-C circuit?

4.34 How should a transistor be chosen to accomplish the purpose of Question 4.33?

4.35 What assumption has been made to reduce the C-E amplifier voltage gain to $-g_m R$?

4.36 Describe the fabrication of an alloy junction transistor.

4.37 What is meant by the epitaxy process?

4.38 What is a mesa transistor?

PROBLEMS

4.1 For the transistor in Fig. 4.8, find I_B for a Q point at $I_C = 15$ mA, with $V_{CC} = 12$ V and a load of 500 Ω.

4.2 Find I_B for the transistor in Fig. 4.8, with $V_{CC} = 30$ V and $R = 1200\ \Omega$, $V_{CE} = 15$ V.

4.3 For the transistor in Fig. 4.8, the dc load line is drawn between $V_{CC} = 20$ V, $I_C = 0$ and $I_C = 30$ mA, $V_{CE} = 0$.
(a) What value of R is being used in the circuit?
(b) What is I_C if the Q point is selected at $I_B = 100\ \mu$A?

4.4 Determine h_{FE} for $V_{CE} = 20$ V, $I_B = 250\ \mu$A for the transistor in Fig. 4.8. At the same Q point, what is the value of α?

4.5 A silicon transistor has $\Delta i_C = 1.80$ mA for $\Delta i_E = 1.89$ mA. What change in i_B will produce an equivalent change in i_C?

4.6 The transistor described by Fig. 4.4 and 4.5 is biased by $V_{BB} = 0.9$ V at 0°C. Determine I_C and V_{CE} for $R = 6000\ \Omega$, $V_{CC} = 30$ V. *Hint*: Find the Q point from Fig. 4.5.

4.7 A transistor in a C-B circuit has $I_B = 105\ \mu$A, $I_C = 2.05$ mA. Determine h_{FE}; also find I_E.

4.8 For the transistor of Problem 4.7, we note that when i_B changes $+27\ \mu$A, i_C changes $+0.65$ mA. Find α and h_{fe}.

4.9 The h parameters for a transistor are measured as

$$h_{ie} = 800\ \Omega \qquad g_m = 0.035\ \text{mho}$$
$$h_{re} = \text{negligible} \qquad h_{oe} = 9 \times 10^{-6}\ \text{mho}$$

With $R = 5000\ \Omega$ and neglecting h_{oe}, find A_v, A_i, R_i, and the power gain in decibels in a C-E circuit.

4.10 Determine R_{ie}, A_{ve}, and A_{ie} for a load of 1000 Ω in a C-E circuit with a transistor having

$$h_{ie} = 1500\ \Omega \qquad h_{fe} = 40$$
$$h_{re} = \text{negligible} \qquad h_{oe} = 5 \times 10^{-6}\ \text{mho}$$

What error is created in A_{ve} by neglect of h_{oe}?

4.11 Find the power gain in decibels when the transistor of Problem 4.10 is used in the C-B circuit with $R = 5000\ \Omega$.

4.12 Repeat Problem 4.10 for the C-C circuit. Also find R_{oc}.

4.13 With the transistor of Problem 4.9, determine A_{vb}, A_{ib}, R_{ib}, and P.G.$_{\text{dB}}$ in the C-B circuit with $R = 1500\ \Omega$.

4.14 In a C-C amplifier we use $R = 1500\ \Omega$ with a transistor having

$$h_{ie} = 1800\ \Omega \qquad g_m = 0.018\ \text{mho}$$
$$h_{re} = \text{negligible} \qquad h_{oe} = \text{negligible}$$

Find A_{vc}, A_{ic}, R_{ic}, R_{oc}, and P.G.$_{\text{dB}}$. If $h_{oe} = 9 \times 10^{-6}$ mho, find the per cent error in neglecting it in computing A_{ic}.

4.15 A transistor having $h_{fe} = 65$, $h_{ie} = 850\ \Omega$, $h_{re} =$ negligible, and $h_{oe} =$ negligible delivers an output of 6 V across a 1500-Ω load resistance in a C-E amplifier.
(a) What is the ac input current?
(b) What input voltage is needed?

4.16 A transistor having $h_{ie} = 1200\ \Omega$, $h_{fe} = 32$, $h_{oe} =$ negligible, and $h_{re} =$ negligible is used for Q_1 and Q_2 in the circuit in Fig. 4.19(a). Determine the current gain for these transistors in cascade with $R = 2000\ \Omega$. Find the overall A_v.
Hint: The load of Q_1 is the input resistance of Q_2.

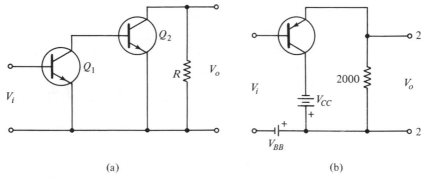

(a)   (b)

Figure 4.19

4.17 We have a signal source with an internal resistance of 150 Ω connected to a C-B amplifier. What value of input resistance should the transistor have for power matching? With an input signal of $V_{eb} = 0.1$ V, what value of g_m should the transistor have to give an output of 12 V across a load of 2500 Ω?

4.18 A signal source of 100,000 Ω internal resistance is to be power matched by the input circuit of a C-C amplifier. If $h_{ie} = 800\ \Omega$ and $g_m = 0.025$ mho, what value of R should be used? What is the resulting A_v? P.G.$_{\text{dB}}$?

4.19 Draw the g_m equivalent circuit for Fig. 4.19(b) when using a transistor having $h_{ie} = 1250\ \Omega$, $h_{fe} = 35$, $h_{re} =$ negligible, and $h_{oe} =$ negligible. What output resistance will be measured at the 2,2 port? What is the voltage gain?

5

DC Bias for the Transistor

Transistor amplifiers must be operated with steady bias voltages and currents to provide the desired operating conditions for the emitter-base and collector-base junctions. The bias sources also supply the circuit energy.

The transistor converts only a part of the dc input energy to ac signal output; the remainder is dissipated as heat and must be removed. This thermal loss and the resultant temperature rise place a limit on transistor operation.

The maintenance of the steady currents and voltages through circuit design will be discussed in this chapter. Biasing circuits are added to the amplifiers of Chapter 4, hopefully without reducing amplifier performance. We find, however, that obtaining stability for the bias collector current often causes a loss in performance.

5.1 Choice of the Quiescent Point

On the output characteristics, the usable region for transistor operation is defined by the maximum ratings of the transistor. In Fig. 5.1 this usable region is bounded by the following:

1. The horizontal line at maximum allowable collector current.

2. The saturation line for the transistor near $v_{CE} = 0$, where the base current loses control.

3. The cutoff line at $i_B = 0$ near the abscissa.

4. The maximum allowable value of collector-emitter voltage to prevent avalanching in the reverse-biased junction.

5. The hyperbolic curve set by the maximum safe transistor power dissipation.

The maximum dissipation curve results from the transistor power rating for maximum allowable temperature, and the curve is drawn for

$$P_d = V_{CC(\max)} I_{C(\max)} \tag{5.1}$$

In Fig. 5.1 this maximum dissipation curve is drawn for $P_d = 200$ mW. For large power output the Q point will be located close to but below this maximum rating curve.

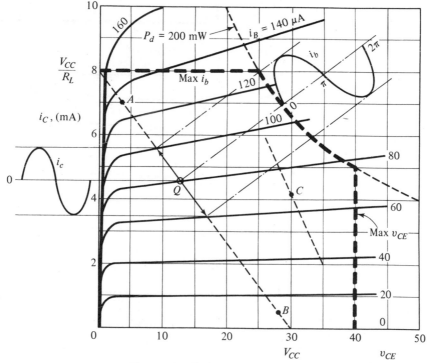

Figure 5.1 Choice of a Q point for a transistor; $R = 3750 \, \Omega$.

For linear operation, the Q point should be placed near the central portion of the region where the curves are most linear and uniformly spaced. For small-signal linear operation, there is no unique location for the Q point but signal swings along the load line should remain within the linear region. Any Q point in a region having constant h_i, h_f, and h_o values will give equal performance for a given load.

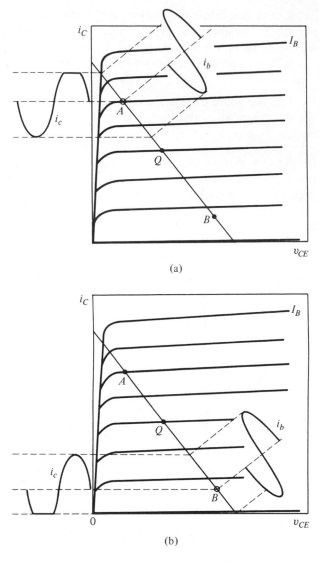

Figure 5.2 (a) Positive peak clipping; (b) negative peak clipping.

A sine input signal applied to a Q point at A, Fig. 5.2(a), can cause output distortion at the positive peak, due to transistor saturation. A sine input signal applied at the Q point at B, Fig. 5.2(b), can drive the transistor into cutoff on the negative swing and distortion of output waveform results. At a central location at Q on the output characteristics, saturation or cutoff distortion is not likely to result for small-signal amplitudes. For this purpose we define a *small signal* as one whose amplitude is small when compared to the base-emitter bias voltage or current at the Q point. For $V_{BE} = 2$ V, we might consider an input sine signal of ± 0.2-V peak, or less, to be a small signal.

5.2 Variation of the Q Point

The current ratio h_{FE} varies with temperature over a considerable range for a given transistor, as shown in Fig. 5.3(a). In addition, h_{FE} is not closely controlled in manufacture and a typical rating might be $h_{FE} = 60$ but with a range from 30 to 110. Since

$$I_C = h_{FE}I_B$$

it can be seen that the Q point at $I_C = 4.6$ mA, for $I_B = 80$ μA, could not be maintained for different transistors. For a fixed-bias current of 80 μA, the Q-point current would range from 2.4 to 8.8 mA for various transistors of the same type. The latter value would lead to a saturation situation.

In small-signal amplifiers there will usually be sufficient load resistance in the collector circuit to hold I_C within safe limits. A change in h_{FE} will cause a change in I_C and I_B, however. This may result in distortion through movement of the Q point toward cutoff on the input curve, Fig. 5.3(b). Cutoff distortion may also be created by further movement of the base-emitter bias point due to temperature changes. Fixed-base bias voltage, $V_{BB} = 0.84$ V, will allow the base bias current to change from 40 to 240 μA as the temperature changes from 0 to 50°C.

The design of the base circuits should provide some stabilization of the Q-point currents against changes in h_{FE} and v_{BE} due to manufacturing variations in transistors or to changes in temperature.

The transistor reverse current I_{CBO} is a sensitive temperature function, nearly doubling for each 10°C (18°F) rise in transistor temperature. In power amplifiers this can cause damaging increases in I_C because of the small amount of resistance in the collector circuit. Methods of protection of the transistor against such damaging currents will be discussed in Chapter 12.

To convert readily from one transistor current to another, we show their relationships in Table 5.1.

TABLE 5.1 **Transistor Current Relations**

	To	I_B	I_C	I_E
From	Multiply by			
I_B		1	h_{FE}	$1 + h_{FE} \cong h_{FE}$
I_C		$\dfrac{1}{h_{FE}}$	1	$\dfrac{1 + h_{FE}}{h_{FE}} \cong 1$
I_E		$\dfrac{1}{1 + h_{FE}}$	$\dfrac{h_{FE}}{1 + h_{FE}} \cong 1$	1

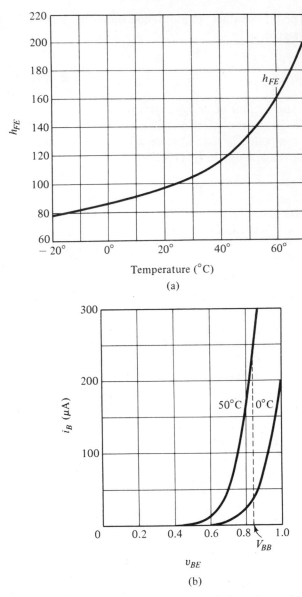

Figure 5.3 (a) Variation of h_{FE} for a silicon transistor; (b) variation of v_{BE} with temperature.

5.3 Fixed Transistor Bias

The circuit in Fig. 5.4(a) is perhaps the simplest form for base-current bias but it is the least satisfactory in compensation for changes in h_{FE}. It can, however, effectively reduce bias current changes due to temperature variation of V_{BE}. Writing a voltage equation around the input loop of Fig. 5.4(b), we have

$$V_{CC} - R_B I_B - V_{BE} = 0$$

and we solve for the base current as

$$I_B = \frac{V_{CC} - V_{BE}}{R_B} \qquad (5.2)$$

But because $V_{CC} \gg V_{BE}$, we have

$$I_B \cong \frac{V_{CC}}{R_B} \qquad (5.3)$$

This shows that changes in V_{BE} due to temperature have little effect on the bias current I_B if V_{CC} is large. The choice of R_B then determines the bias current for the base.

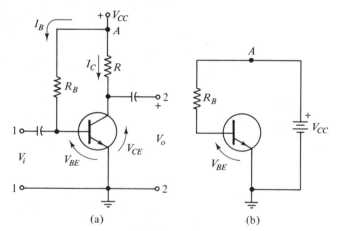

Figure 5.4 Fixed-bias circuit analysis.

The following equation may be written for the output loop:

$$V_{CC} - R I_C - V_{CE} = 0 \qquad (5.4)$$

But by definition of h_{FE}

$$I_C = h_{FE} I_B$$

and so transposing and substituting in Eq. 5.4 we have

$$\begin{aligned} V_{CE} &= V_{CC} - R I_C \\ &= V_{CC} - h_{FE} R I_B \end{aligned} \qquad (5.5)$$

With V_{CC} and R fixed and I_B a constant bias current, then Eq. 5.5 shows that a change in h_{FE} must change V_{CE} and this causes a shift in the Q point.

It would be necessary to adjust the bias current by changing the value of R_B in each amplifier completed in a production run. Readjustment of R_B

would be necessary to compensate for variations of h_{FE} every time a new transistor were placed in the circuit.

It should be obvious that a better bias circuit is needed, one that will prevent shifts of the original Q point for changes in h_{FE} and V_{BE}.

5.4 The Four-Resistor Bias Circuit

The bias circuit in Fig. 5.5(a) uses four resistors and provides improved stability of the Q point. The input circuit resistors R_1 and R_2 are used to give fixed bias, while the emitter resistor R_E provides a bias voltage that varies with I_C. This compensates somewhat for changes of h_{FE}.

Consider first the input circuit, which is redrawn in Fig. 5.5(b). We assume $I_B \ll I_1$ or that $I_1 \cong I_2$. If I_B is in the microampere range and if I_1 and I_2 are adjusted to the fractional milliampere range, then our assumption is valid. A suitable selection is to make $R_1 + R_2 = 60{,}000$ to $100{,}000 \, \Omega$.

At the B, E terminals we can replace R_1 and R_2 and the voltage V_B with a voltage-source equivalent circuit. With R_1 and R_2 in series and I_B negligible, the open-circuit voltage at B, E is given by the voltage-divider ratio

$$V_{OC} = V_B = \frac{R_2}{R_1 + R_2} V_{CC} \tag{5.6}$$

If we short-circuit the B, E terminals, the short-circuit current becomes

$$I_{sc} = \frac{V_{CC}}{R_1} \tag{5.7}$$

Then, by our rules for deriving the voltage-source equivalent circuit,

$$R' = R_B = \frac{V_{oc}}{I_{sc}} = \frac{V_B}{V_{CC}/R_1} = \frac{R_1 R_2}{R_1 + R_2} \tag{5.8}$$

This is the value for R_1 and R_2 in parallel. The resultant voltage-source circuit is drawn at the input in Fig. 5.5(c).

Writing a voltage equation around the input as redrawn in Fig. 5.5(d), we have

$$-V_B + R_B I_B + V_{BE} - R_E I_E = 0 \tag{5.9}$$

We can replace I_E with

$$I_E = -(I_B + I_C) = -(1 + h_{FE})I_B$$

and have

$$-V_B + R_B I_B + V_{BE} + (1 + h_{FE})I_B = 0$$

Solving for the base current, we find

$$I_B = \frac{V_B - V_{BE}}{R_B + (1 + h_{FE})R_E} = \frac{V_B - V_{BE}}{R_B} \left[\frac{1}{1 + \dfrac{(1 + h_{FE})R_E}{R_B}} \right] \tag{5.10}$$

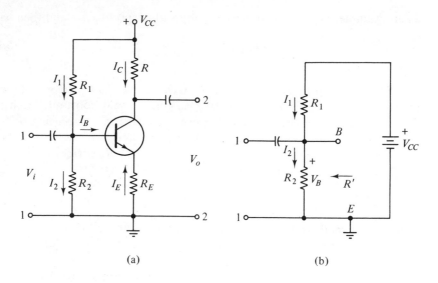

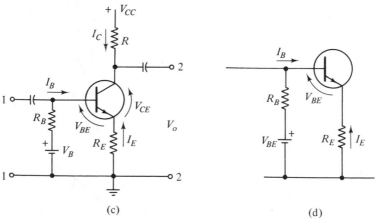

(c)   (d)

Figure 5.5 (a) The four-resistor bias circuit; (b), (c), and (d) reduced circuits.

For the fixed-bias circuit

$$I_C = h_{FE}I_B \tag{5.11}$$

and for the four-resistor bias circuit, I_C is h_{FE} times the I_B value in Eq. 5.10, leading to

$$I_C = \frac{V_B - V_{BE}}{R_B}\left[\frac{h_{FE}}{1 + \frac{(1 + h_{FE})R_E}{R_B}}\right]$$

We can make I_C less affected by temperature-induced changes of V_{BE} by making V_B larger. Since V_{BE} approximates 0.5 V, we may make $V_B = 3$ V, as an example. We may also neglect unity with respect to h_{FE} and then have

$$I_C \cong \frac{V_B}{R_B}\left[\frac{h_{FE}}{1 + h_{FE}\dfrac{R_E}{R_B}}\right] \qquad (5.12)$$

With the fixed-bias circuit, I_C was given by Eq. 5.11, and was directly affected by changes in h_{FE}, with the bias current held constant. With Eq. 5.12 for the four-resistor circuit, however, we have h_{FE} appearing both in the numerator and the denominator. It may be seen that the sensitivity of I_C to changes in h_{FE} can be reduced if we choose R_E/R_B correctly.

If we make $R_E = 0$, however, then Eq. 5.12 reduces to

$$I_C = \frac{V_B}{R_B}h_{FE} \qquad (5.13)$$

and I_C will vary directly as h_{FE}, which is the undesirable situation in the fixed-bias circuit. In fact, making $R_E = 0$ in the circuit in Fig. 5.5(c) reduces it to the fixed-bias circuit.

If we make $R_E/R_B \cong 1$, then the $h_{FE}(R_E/R_B)$ term in the denominator would be the influencing factor. The h_{FE} terms in numerator and denominator would approximately cancel and I_C would be independent of h_{FE}, which is

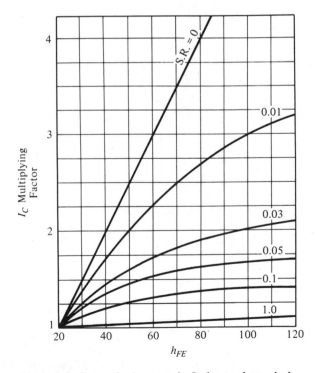

Figure 5.6 Curves for increases in I_C due to change in h_{FE}.

the result desired. R_E must be small in magnitude or we lose signal gain and dc power, however, and R_B must be large with respect to h_{i_e} of the transistor with which it is in parallel or we lose gain again. Thus some compromise between ideal bias stability and gain performance must be made.

The sensitivity of I_C to variations in h_{FE}, for various values of R_E/R_B, is plotted in Fig. 5.6. Values of R_E/R_B in the range of 0.05 to 0.1 usually lead to satisfactory gain and reasonable stability against changes in h_{FE}. With the R_E/R_B ratio equal to 0.1, the value of I_C will increase 70 per cent when h_{FE} changes from 20 to 100. Without the compensation provided by R_E, the value of I_C would increase as shown by the curve for $R_E/R_B = 0$, or 500 per cent for the same h_{FE} variation.

Thus we have the design conditions for the four-resistor bias circuit:

$$V_B \cong 3 \text{ V}$$

$$\frac{R_E}{R_B} \cong 0.05 \text{ to } 0.1$$

We call R_E/R_B, or its equivalent in other circuits, the *stabilizing ratio*. This ratio serves as an index of stability, which increases as I_C is made more stable by the circuit design. Figure 5.6 functions as a universal curve for predicting such stability as a function of the stabilizing ratio.

5.5 Design of a Fixed-Bias Circuit

The fixed-bias circuit in Fig. 5.4(a) will be designed, using the transistor with characteristics of Fig. 5.7, with $h_{FE} = 100$. A Q point at $V_{CE} = 10$ V, $I_C = 2$ mA is chosen as a desirable location in the middle of the linear region of the curves. We usually assume that $V_{CC} = 2V_{CE}$ and so we have $V_{CC} = 20$ V.

The load resistor can be found from the voltage equation around the output loop, Eq. 5.4;

$$V_{CC} - RI_C - V_{CE} = 0$$

$$R = \frac{V_{CC} - V_{CE}}{I_C} \qquad (5.14)$$

$$= \frac{20 - 10}{0.002} = 5000 \ \Omega$$

The remaining resistor is R_B and the base current is

$$I_B = \frac{I_C}{h_{FE}}$$

$$= \frac{0.002}{100} = 20 \times 10^{-6} \text{ A} = 20 \ \mu\text{A}$$

This value could also be read from the Q-point location at $V_{CE} = 10$ V,

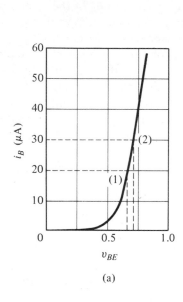
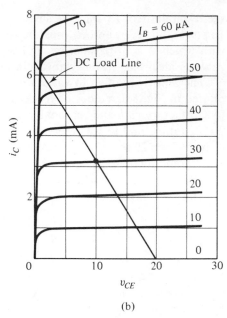

Figure 5.7 Curves for the design example.

$I_C = 2$ mA in Fig. 5.7(b). Rewriting Eq. 5.2 we have

$$R_B = \frac{V_{CC} - V_{BE}}{I_B} \qquad (5.15)$$

For a base current of $I_B = 20$ μA at the Q point, we can go to the input curves of Fig. 5.7(a) and find that V_{BE} at (1) must be 0.65 V. Then

$$R_B = \frac{20 - 0.65}{20 \times 10^{-6}} = \frac{19.35}{20 \times 10^{-6}}$$
$$= 967{,}000 \ \Omega \cong 1 \ \text{M}\Omega$$

and our circuit design is completed.

In the analysis of the circuit, an assumption was made that a large V_{CC} would stabilize the circuit against changes of V_{BE} with temperature. This can be shown by allowing V_{BE} to be zero volts, in which case

$$R_B = \frac{20}{20 \times 10^{-6}} = 1 \ \text{M}\Omega$$

Then we can make $V_{BE} = 1$ V and find that

$$R_B = \frac{20 - 1}{20 \times 10^{-6}} = 0.95 \ \text{M}\Omega$$

The required values of R_B at 1 MΩ, 0.967 MΩ, or 0.95 MΩ all lie within the tolerance zone of a 1-MΩ, ± 10 per cent resistor so that changes due to extreme values of V_{BE} cause less circuit change than might be expected to occur in normal production runs from the tolerance on R_B.

With

$$I_C = h_{FE}I_B \tag{5.16}$$

and I_B fixed by our circuit design at 20 μA, however, the collector current I_C will vary directly with changes in h_{FE}. Since the latter parameter is sensitive to temperature and to variation between transistors of the same type, the fixed-bias circuit gives no stabilization of the Q point with respect to h_{FE} changes.

5.6 Design of a Bias-Stabilized C-E Amplifier

A bias-stabilized network for a *C-E* amplifier, using the transistor in Fig. 5.7, will be designed. This transistor has $h_{ie} = 650\ \Omega$ and $h_{FE} = 100$. Previously it was pointed out that there is no unique location for the Q point for a linear amplifier since with a small signal and constant h_{fe}, h_{ie}, and h_{oe} values over a region of the transistor curves we would have equal amplifier performance at many Q points in the central linear region. Arbitrarily we select a Q point at $I_C \cong 3.2$ mA along the $I_B = 30\ \mu$A line. A supply voltage less than the rated $V_{CE(\max)}$ should be chosen. A value of 20 V was selected.

The values for the four resistors, one being the load resistor, are to be determined for the circuit in Fig. 5.8(a). There are more circuit unknowns than there are relations to solve and so we must call on engineering experience to determine some of the design values.

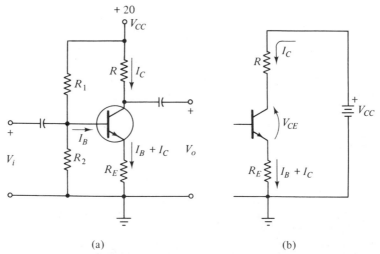

(a)   (b)

Figure 5.8 (a) Four-resistor C-E amplifier; (b) output circuit and currents.

Through the collector circuit, isolated at Fig. 5.8(b), we write a voltage equation

$$V_{CC} - RI_C - V_{CE} - R_E(I_B + I_C) = 0$$
$$V_{CC} - V_{CE} = (R + R_E)I_C + R_E I_B$$

But since $I_B \ll I_C$, we can then simplify the above to

$$V_{CC} - V_{CE} \cong (R + R_E)I_C \qquad (5.17)$$

We now have one equation and three unknowns, V_{CE}, R, and R_E. Experience indicates that the voltage across the transistor at the Q point ought to be about one-half of the supply voltage, expressed mathematically as

$$V_{CE} \cong \frac{V_{CC}}{2} \qquad (5.18)$$

Then Eq. 5.17 becomes

$$\frac{V_{CC}}{2} = I_C(R + R_E)$$

We have now reduced the unknowns to two.

For our particular amplifier, with $I_C = 3.2$ mA, we can write

$$10 = 0.0032(R + R_E)$$
$$R + R_E = \frac{10}{0.0032} \cong 3100 \ \Omega$$

Since usual resistors are accurate to ± 10 per cent, we round off the numerical result.

We can now draw the dc load line for $R + R_E$ on the transistor characteristics, from $V_{CC} = 20$ V to $i_C = V_{CC}/(R + R_E) = 6.5$ mA, as intercepts. The Q point is placed at $I_B = 30 \ \mu$A, $V_{CE} = 10$ V as selected.

The division of $R + R_E = 3100 \ \Omega$ between collector and emitter circuits must now be determined. The emitter resistor was added to stabilize the steady collector current against variations in h_{FE}. This purpose would dictate a large value for R_E. The dc voltage across R_E reduces the voltage across R, however, and the latter voltage limits the peak swing of the signal voltage across the load. Thus R_E should not be large. As a rule, we make the emitter voltage V_E about 10 to 20 per cent of the supply voltage and for this design we shall choose

$$V_E = 0.1 V_{CC}$$
$$= 2 \text{ V} \qquad (5.19)$$

We then have

$$R_E = \frac{V_E}{I_E} \cong \frac{V_E}{I_C}$$
$$= \frac{2}{0.0032} \cong 600 \ \Omega \qquad (5.20)$$

Consequently,

$$R = 3100 - 600 = 2500 \ \Omega$$

and the output circuit is complete.

The base voltage V_B is

$$V_B = V_E + V_{BE} \tag{5.21}$$

We go to the input curve, Fig. 5.7(a), and for a base current $I_B = 30 \ \mu A$, we find at (2) that V_{BE} is 0.70 V. With

$$V_E \cong R_E I_C$$
$$= 600 \times 0.0032 = 1.92 \ V \cong 2.0 \ V \tag{5.22}$$

Then, using Eq. 5.21,

$$V_B = 2.0 + 0.7 = 2.7 \ V$$

We are able to write two design equations involving resistors R_1 and R_2:

$$I_1 \cong I_2 = \frac{V_{CC}}{R_1 + R_2} \tag{5.23}$$

$$V_B = \frac{R_2}{R_1 + R_2} V_{CC} \tag{5.24}$$

We hope to make $I_1 \geq 10 I_B$ so that the current I_B can be neglected in passing through R_1, as we assumed in the circuit analysis. Therefore, we choose $I_1 \cong I_2$ as $10 \times 30 \ \mu A = 0.3 \ \text{mA}$ and from Eq. 5.23 we have

$$R_1 + R_2 = \frac{V_{CC}}{I_1}$$
$$= \frac{20}{3 \times 10^{-4}} \cong 67{,}000 \ \Omega \tag{5.25}$$

From Eq. 5.24 we can find R_2 as

$$R_2 = (R_1 + R_2) \frac{V_B}{V_{CC}}$$
$$= 67{,}000 \frac{2.70}{20} = 9000 \ \Omega \tag{5.26}$$

Then

$$R_1 = 67{,}000 - 9000 = 58{,}000 \ \Omega$$

The design of the circuit is now complete and it is shown in Fig. 5.9.

The bias resistor R_2 is in parallel with $h_{ie} = 650 \ \Omega$ of the transistor, and the shunting effect of R_2 is negligible.

That I_C is stabilized can be determined. The value of R_B can be found from Eq. 5.8 as

$$R_B = \frac{R_1 R_2}{R_1 + R_2}$$
$$= \frac{58 \times 10^3 \times 9 \times 10^3}{67 \times 10^3} = 7800 \ \Omega$$

The stabilizing ratio R_E/R_B is

$$\text{S. R.} = \frac{R_E}{R_B} = \frac{600}{7800} = 0.077$$

Reference to Fig. 5.6 for $h_{FE} = 100$ shows that the collector current can change only by a factor of $1/1.6$ if h_{FE} falls from 100 to 20. Thus our basic design choices are shown to be satisfactory, using the stabilization criterion of R_E/R_B.

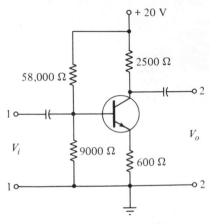

Figure 5.9 Completed design for the four-resistor bias circuit.

5.7 Voltage Feedback Bias

The circuit in Fig. 5.10 uses an emitter resistor for stabilizing I_C and adds further control through variation of I_B. If I_C increases, the RI_C' voltage drop increases, lowering the voltage at the collector. Resistor R_f supplies base current from the collector and as I_C increases, the base current falls. This change in I_B tends to oppose the original change in I_C.

Around the input loop, shown in heavy lines in Fig. 5.10, we have the voltage equation

$$V_{CC} - RI_C' - R_f I_B - V_{BE} - R_E I_C' = 0 \qquad (5.27)$$

Since I_B is small compared to I_C, we can neglect the branching of I_B at point A at the collector and say

$$I_C' \cong I_C \qquad (5.28)$$

Rearranging Eq. 5.27,

$$V_{CC} - V_{BE} = (R + R_E)I_C + R_f I_B$$

We can drop V_{BE} as small compared to V_{CC}. Then, using $I_B = I_C/h_{FE}$, we can write the equation in terms of I_C as

$$V_{CC} = \left(R + R_E + \frac{R_f}{h_{FE}}\right)I_C$$

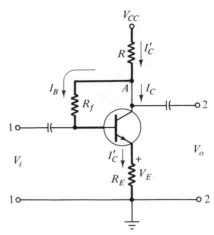

Figure 5.10 The voltage-feedback bias circuit.

We can solve for I_C and obtain

$$I_C = V_{CC} \frac{1}{R + R_E + \dfrac{R_f}{h_{FE}}}$$

Dividing out the R_f term, we now have

$$I_C = \frac{V_{CC}}{R_f} \left[\frac{h_{FE}}{1 + \dfrac{h_{FE}(R + R_E)}{R_f}} \right] \tag{5.29}$$

This expression for the collector current can be compared to Eq. 5.12 for the four-resistor bias circuit. We see that they are the same in form but differ in the stabilizing ratio, which now is

$$\text{S. R.} = \frac{R + R_E}{R_f} \tag{5.30}$$

The effect of this value of stabilizing ratio can be assessed by reference to Fig. 5.6. Since both R and R_E are included in the numerator, the division of resistance between emitter and collector circuits has no effect on stability. Resistor R_f can be determined from

$$V_{CE} = R_f I_B + V_{BE}$$
$$R_f = \frac{V_{CE} - V_{BE}}{I_B} \simeq \frac{V_{CE}}{I_B} \tag{5.31}$$

It is usually possible to make the stabilizing ratio and the stability for the four-resistor bias network larger than for the voltage-feedback circuit since R_f is relatively large. The voltage-feedback circuit with R_f saves one resistor, however, and avoids the power dissipation due to the current I_1 in the R_1, R_2 network. This point may be significant in battery-operated equipment.

5.8 Design of a Voltage-Feedback Bias Circuit

We shall use a transistor with the following values: $I_C = 2.3$ mA, $I_B = 30$ μA, $V_{BE} = 0.5$ V, and $h_{FE} = 75$ in a C-E circuit with $V_{CC} = 15$ V.

Making $V_{CE} = V_{CC}/2 = 7.5$ V, we find

$$R + R_E = \frac{V_{CC}}{2I_C}$$

$$= \frac{15}{2 \times 0.0023} \cong 3200 \ \Omega$$

With $R_E I_C = 0.15 V_{CC} = 2.25$ V, we can calculate that

$$R_E \cong \frac{2.25}{I_C} = \frac{2.25}{0.0023} \cong 1000 \ \Omega$$

and $R = 3200 - 1000 = 2200 \ \Omega$.

Using Eq. 5.31,

$$R_f = \frac{V_{CE} - V_{BE}}{I_B}$$

$$= \frac{7.5 - 0.5}{30 \times 10^{-6}} = 230,000 \ \Omega$$

This completes the circuit design.

The stabilizing ratio is

$$\text{S. R.} = \frac{R + R_E}{R_f}$$

$$= \frac{3200}{230,000} = 0.014$$

From Fig. 5.6 we see that the I_C value will change about 275 per cent for a 500 per cent change in h_{FE}.

This figure for stability ratio is not so good as can be obtained for the four-resistor network, where we might have $R_B = 10,000 \ \Omega$. Then

$$\text{S. R.} = \frac{R_E}{R_B} = \frac{1000}{10,000} = 0.1$$

In the voltage-feedback circuit the value of R_f is determined by base current and R_f is large. The large R_f value reduces S.R. as shown.

5.9 Bias for the Emitter Follower

The bias circuit for the emitter follower, Fig. 5.11, will now be considered. We start with the four-resistor network but make $R = 0$. Resistor R_E now becomes the total load resistance.

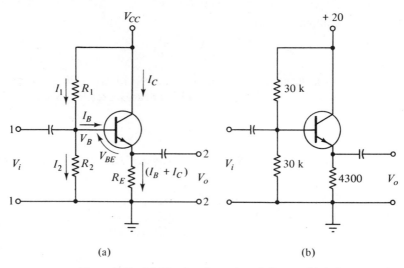

Figure 5.11 (a) Bias for the emitter follower; (b) design example.

Using $V_{CC}/2$ as the voltage across the load R_E, we have

$$R_E = \frac{V_{CC}}{2(I_B + I_C)} \cong \frac{V_{CC}}{2I_C} \qquad (5.32)$$

and

$$V_E \cong R_E I_C = \frac{V_{CC}}{2} \qquad (5.33)$$

The voltage V_B at the base of the transistor must be

$$V_B = V_{BE} + V_E \qquad (5.34)$$

The input resistance of the emitter follower is very large and I_B will be very small. Then $I_1 \cong I_2$ and the voltage-divider ratio for R_1 and R_2 will accurately determine V_B. Then

$$V_B = \frac{R_2}{R_1 + R_2} V_{CC} \qquad (5.35)$$

We have only this single relationship with two unknown resistors. Again we must rely on experience. A value of 60,000 Ω may be arbitrarily chosen for $R_1 + R_2$, after which we can solve for R_1 and R_2 separately.

We can use Eq. 5.12 for determining I_C since that equation was obtained for the four-resistor bias network that we are designing but with $R = 0$. That is,

$$I_C = \frac{V_B}{R_B}\left[\frac{h_{FE}}{1 + h_{FE}\dfrac{R_E}{R_B}}\right] \qquad (5.36)$$

and the stabilizing ratio is

$$\text{S. R.} = \frac{R_E}{R_B}$$

Due to the large value employed for R_E in the emitter follower, the stability ratio usually exceeds 0.1. This ensures excellent stability of I_C in the emitter follower circuit.

5.10 Design of the Emitter Follower Circuit

Let us use a transistor with $h_{FE} = 100$, $V_{BE} = 0.5$ V, $V_{CC} = 20$ V, and $I_C = 2.3$ mA in an emitter follower (C-C) amplifier.

We find R_E from the Q-point data,

$$R_E = \frac{V_{CC}}{2I_C} = \frac{20}{2 \times 0.0023} \cong 4300 \ \Omega$$

It follows that $V_E = V_{CC}/2 = 20/2 = 10$ V and by Eq. 5.34 we find V_B as

$$V_B = 0.5 + 10 = 10.5 \text{ V}$$

We choose $R_1 + R_2 = 60,000 \ \Omega$ and can find R_2 from Eq. 5.35 as

$$R_2 = (R_1 + R_2)\frac{V_B}{V_{CC}} \tag{5.37}$$

$$= 60,000\frac{10.5}{20} = 31,500 \cong 30,000 \ \Omega$$

Then

$$R_1 = 60,000 - 30,000 = 30,000 \ \Omega$$

The circuit design is complete.

With $R_1 = R_2 = 30,000 \ \Omega$, the value of R_B is

$$R_B = \frac{R_1 R_2}{R_1 + R_2} = \frac{3 \times 10^4 \times 3 \times 10^4}{6 \times 10^4}$$

$$= 15,000 \ \Omega$$

The stabilizing ratio is

$$\text{S. R.} = \frac{R_E}{R_B} = \frac{4300}{15,000} = 0.287$$

Figure 5.6 shows that, with this large ratio, excellent stability of I_C is provided against changes in h_{FE}.

The complete circuit appears in Fig. 5.11(b).

5.11 Comments

We can now design complete amplifier circuits, including the necessary bias networks. These use only the source V_{CC}, avoiding the cost of separate base and collector sources. Hopefully, these bias networks do not affect the gain or other performance figures of our amplifiers. The use of the emitter resistance R_E often reduces the gain, however, and this must be allowed for in overall amplifier design. The use of the current-wasting resistors R_1 and R_2 introduces a cost in additional power. The effective value for R_1 and R_2, known as R_B, shunts the input circuit of the transistor and wastes some of the input current, again reducing the gain. But the offsetting improvement in stability of the Q point and the elimination of individual amplifier adjustment in production are very significant.

We have discovered that payment in some form is extracted elsewhere for every advantage gained in a circuit.

Three basic bias circuits have been studied and the stability ratios are compared as

	S.R.	
Fixed bias	h_{FE}	(Eq. 5.16)
Four-resistor bias	R_E/R_B	(Eq. 5.12)
Voltage feedback	$(R + R_E)/R_f$	(Eq. 5.30)

Because of the flexibility in choice of R_E and R_B, the four-resistor bias network usually is the most stable form.

REVIEW QUESTIONS

5.1 What is meant by the Q point? What factors do we consider in locating the Q point?

5.2 Give two reasons for h_{FE} not being at the rated value.

5.3 Why does the collector power dissipation bound the operating region of a transistor?

5.4 Why does the maximum voltage bound the operating region of a transistor?

5.5 What is meant by the saturation line of a transistor?

5.6 What are the end points or intercepts of the dc load line?

5.7 How is the slope of the load line related to the resistance in the output circuit?

5.8 Why must the Q point be located on the dc load line?

5.9 Explain how an ac signal or I_b value shifts the operating point along the load line.

5.10 How does I_{CBO} vary with temperature?

5.11 Name one advantage and one disadvantage of the fixed-bias circuit.

5.12 What is meant by fixed bias?

5.13 What is the stabilizing ratio of the four-resistor bias circuit?

5.14 What range of I_C variation could occur in a four-resistor bias circuit with S.R. $= 0.05$?

5.15 How does the fixed-bias circuit stabilize against temperature-caused changes in V_{BE}?

5.16 Name an advantage of the voltage-feedback bias circuit; name a disadvantage.

5.17 How is the emitter voltage usually related to V_{CC} in an emitter follower?

5.18 What is the usual relation between V_E and V_{CC} in a C-E amplifier?

5.19 From the load line, explain why we choose V_{CE} as one-half of V_{CC} to start our circuit design.

5.20 What is the stabilizing ratio for the voltage-feedback circuit?

5.21 Name three forms of payment we make to obtain stability of the Q point.

PROBLEMS

5.1 For the transistor of Fig. 5.7, choose $V_{CC} = 20$ V, $I_C = 4$ mA, and $V_{CE} = 9$ V; draw the load line. What is the value of the load resistance? What is the base current I_B at the Q point?

5.2 For the transistor with characteristics in Fig. 5.1, choose a Q point for $I_B = 80$ μA and $V_{CE} = 12$ V.
(a) What I_C is obtained?
(b) With $V_{CC} = 25$ V, what is the voltage across the load?
(c) What is the load resistance?

5.3 The Q point at C in Fig. 5.1 is used. With $V_{CC} = 40$ V, what load resistance is needed?

5.4 With the fixed-bias circuit in Fig. 5.4(a), find R and R_B to place the Q point at C, Fig. 5.1, with $V_{CC} = 40$ V.

5.5 A fixed-bias circuit is designed as in Fig. 5.12(a).
(a) With $V_{BE} = 0.5$ V, $V_{CE} = V_{CC}/2$, determine the Q-point values of I_B and I_C.
(b) What is h_{FE} of the transistor at the Q point?

5.6 For the circuit in Fig. 5.12(b),
(a) Determine R_B to bring I_C to 1.8 mA, if $h_{FE} = 50$.
(b) What is the stability ratio for the circuit?

5.7 The transistor of Fig. 5.1 is used in the circuit in Fig. 5.13(a), with $V_{CC} = 25$ V, $V_{BE} = 0.5$ V, $I_C = 4.5$ mA. Determine R_1 and R_2 values needed. What is the stabilizing ratio? What percentage will I_C change if h_{FE} changes from 40 to 100?

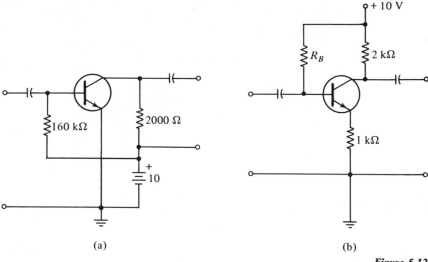

Figure 5.12

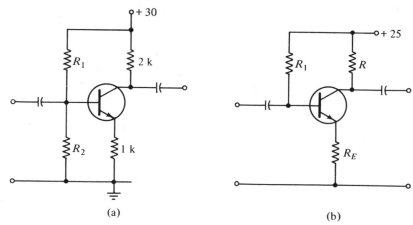

Figure 5.13

5.8 With a Q point at $I_C = 3.5$ mA, $h_{FE} = 60$, $V_{CC} = 25$ V, and $V_{BE} = 0.5$ V, determine the values of the other resistors needed for the circuit in Fig. 5.13(b). What is the stabilizing ratio for the circuit?

5.9 Determine the Q-point currents and voltages for the circuit in Fig. 5.14(a), with $V_{CC} = 12$ V, $V_{BE} = 0.6$ V, and $h_{FE} = 55$.

5.10 Determine the bias currents and transistor voltages for the circuit in Fig. 5.14(b), with $V_{BE} = +0.3$ V and $h_{FE} = 75$. *Note*: The circuit is an emitter follower.

5.11 Determine the design values of the resistances for the circuit in Fig. 5.15(a), with $V_{CC} = 20$ V, $I_C = 6.7$ mA, and $h_{FE} = 60$, using the usual circuit assumptions. Determine V_{CE} and V_E and the stabilizing ratio.

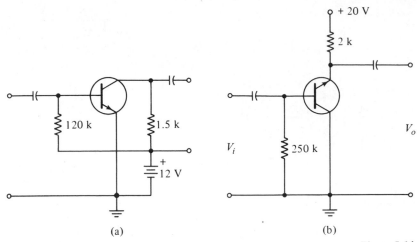

Figure 5.14

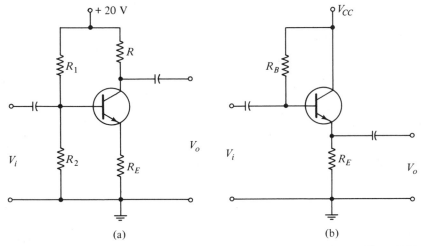

Figure 5.15

5.12 Let $R_E = 0$ in Fig. 5.13(b) and calculate the bias currents and transistor voltages for $R_1 = 150\text{ k}\Omega$, $R = 2.1\text{ k}\Omega$, $V_{CC} = 9\text{ V}$, $V_{BE} = 0.6\text{ V}$, and $h_{FE} = 40$.

5.13 Using the fixed-bias circuit in Fig. 5.4(a),
   (a) Determine the collector-emitter bias voltage (V_{CE}) for $R_B = 250\text{ k}\Omega$, $R = 2\text{ k}\Omega$, $V_{CC} = 12\text{ V}$, $V_{BE} = 0.3\text{ V}$, and $h_{FE} = 64$.
   (b) Find I_C and I_B.

5.14 For the circuit in Fig. 5.13(b), with $R_1 = 250\text{ k}\Omega$, $R = 2\text{ k}\Omega$, $R_E = 500\text{ k}\Omega$, $V_{CC} = 9\text{ V}$, $V_{BE} = 0.5\text{ V}$, and $h_{FE} = 55$,
   (a) Find the dc currents and V_{CE}.
   (b) What is the stabilization ratio for the circuit?

5.15 For the circuit in Fig. 5.13(b), find V_{CE} for $R_1 = 47$ kΩ, $R_E = 750$ Ω, $R = 500$ Ω, $V_{BE} = 0.5$ V, $h_{FE} = 55$, and $V_{CC} = 18$ V.

5.16 What value of emitter resistor would be needed to make the stabilization ratio equal 0.02 for the circuit in Fig. 5.13(b), with $R = 750$ Ω, $R_1 = 50,000$ Ω, and $h_{FE} = 40$?

5.17 Calculate the bias currents and transistor voltages for a four-resistor bias C-E amplifier with $R_1 = 56,000$ Ω, $R_2 = 5000$ Ω, $R_E = 750$ Ω, $R = 6800$ Ω, $V_{CC} = 24$ V, $V_{BE} = 0.6$ V, and $h_{FE} = 50$.

5.18 Design an amplifier circuit as in Fig. 5.13(b) for a transistor having $h_{FE} = 45$, $I_C = 5$ mA, $V_{CC} = 20$ V, and $V_{BE} = 0.5$ V. Make the usual circuit assumptions.

5.19 For the emitter follower circuit in Fig. 5.15(b), find the value of V_E for $R_B = 90$ kΩ, $R_E = 1500$ Ω, $V_{CC} = 25$ V, $V_{BE} = 0.35$ V, and $h_{FE} = 60$. Also find the stabilizing factor and predict the I_C change if h_{FE} changes to 100.

5.20 Design an emitter follower, using the circuit in Fig. 5.15(a) with $R = 0$. The transistor has $h_{FE} = 100$, $V_{BE} = 0.5$ V and $V_{CC} = 15$ V, $I_C = 2.5$ mA. Make the usual assumptions.

6

The Field-Effect Transistor

The *field-effect transistor* (FET) was an early proposal by Shockley that had to wait for the development of new production methods to become a practical device. It is a *unipolar device* because the current is either of holes or of electrons, in contrast to the *bipolar transistor* of *pnp* or *npn* form. The cross section and consequently the resistance of the conducting path in this device may be controlled by a signal voltage applied to a gate electrode.

With a high input resistance of 100 MΩ or more and generating less noise than a bipolar type, the FET is well suited as an input amplifier with low-level signals.

6.1 The Junction Field-Effect Transistor

Figure 6.1(a) illustrates the operating principle of a *junction field-effect transistor* (JFET). The thin slab of semiconductor has contacts at each end, a *source S* for the mobile charges, and a *drain D* for extraction of the charges. A *p* electrode forms a junction on the *n* wafer and is known as the *gate G*; the thin region under the gate is called the *channel*. If *n* material is chosen for the channel, the conduction is by electrons. If *p* material is used, the conduction will be by holes.

For discussion, we shall use *n* material for the channel. With a voltage

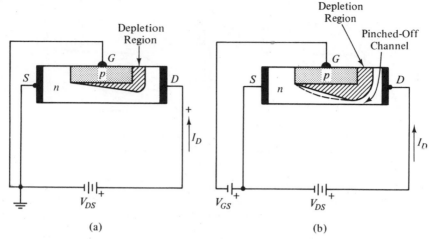

Figure 6.1 (a) Low voltage applied to the FET; (b) in the pinch-off condition.

V_{DS} applied, electrons move in the channel from the source to the drain. There is a progressive voltage drop along the bar and, with the source and gate being connected, points in the channel will be positive to the source and gate. With a positive n channel and a negative p gate, the junction at the gate-channel interface is reverse-biased. As a result, a depletion region forms in the channel as shown. This depletion region appears largely in the n channel because the p-gate material is more heavily doped.

The voltage drop is distributed along the gate and the reverse voltage at the right end of the gate is larger than at the left end. Because of the greater reverse voltage, the depletion region at the right end of the gate is thickened, as shown. The depletion region has no mobile charges and the channel current is confined to the wedge-shaped portion of the channel.

The drain current increases as V_{DS} is raised, resulting in the curve of (1) in Fig. 6.2(a). As there is greater reverse voltage between gate and channel, the depletion region begins to pinch off the channel as in Fig. 6.1(b) and the rate of current increase falls off to (2) on the curve. With further increase in V_{DS} the pinched-off region lengthens and the current curve flattens to (3). This region of essentially constant current, relatively independent of V_{DS}, is called the *pinch-off region*.

Placing the gate at -1 V to the source S further reverse-biases the junction and drives the depletion region into the channel. Saturation of the current occurs at lower values of V_{DS} and at lower currents. In the pinch-off region the current is sensitive to the gate-source voltage V_{GS}; as such, we have characteristics suited to the use of the FET as a control device, shown in Fig. 6.2(b).

The operation is said to be in the *depletion mode*, with the increased

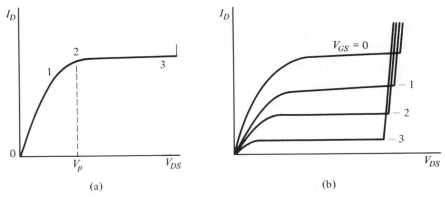

Figure 6.2 (a) Current variation with V_{DS}, with gates connected to source; (b) an output curve family, showing I_D controlled by V_{GS}.

negative gate voltage depleting the channel of charges and lowering the channel current.

The abrupt rise in current above (3) in Fig. 6.2(a) is due to an avalanche-type voltage breakdown in the depletion region between gate and drain and this places an upper limit on the gate-drain voltage. The breakdown voltage is designated BV_{DGO}, the source being open-circuited.

The input resistance is that of the gate-channel reverse-biased junction. The gate area is small so that the leakage resistance is of the order of 100 MΩ. A depletion capacitance C_{GS} appears in parallel. This capacitance is that of the reverse-biased junction, due to the dielectric effect of the depletion region between the gate electrode and the channel. This capacitance may be in the range of 2 to 10 pF.

Currents of milliampere size in the channel can be controlled by the gate voltage. With the input current from the gate extremely small, we obtain significant amplification in a circuit such as that in Fig. 6.3.

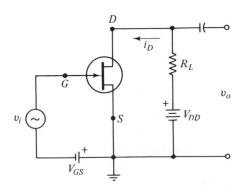

Figure 6.3 The JFET in the common-source amplifier.

6.2 The MOS Field-Effect Transistor

The *metal-oxide-semiconductor* FET (MOSFET) has its gate electrode insu-
lated from the channel by a very thin layer of silicon dioxide. A wafer of high-
resistivity *p* silicon is used and *n* impurity is diffused into a region near the
top surface to form an *n* channel of moderate resistivity. Low-resistance *n*
contacts are diffused through a mask at the channel ends as source and drain.
The surface is covered with a thin layer of silicon dioxide and a small metal
gate electrode deposited over the channel. The thickness of the insulating
layer is usually less than 10^{-3} mm.

Application of a negative bias to the gate drives electrons from the *n*
channel immediately under the gate, depleting the region of free charges.
The thickness of the depletion region varies with gate bias and constricts the
conduction area in the channel. In Fig. 6.4(a) the condition at pinch-off is
shown.

The drain current is large with zero gate voltage and is reduced as the gate
is made more negative. This is the *depletion mode* of operation; a family of
output characteristics is shown in Fig. 6.5(a).

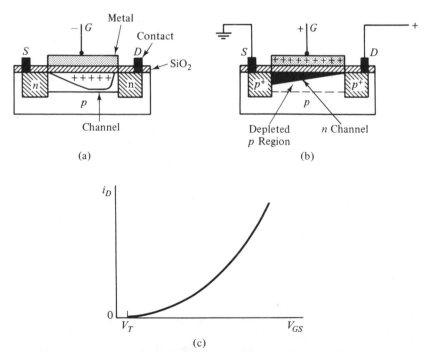

Figure 6.4 (a) Depletion-mode MOSFET at pinch-off; (b) channel configuration
of enhancement mode; (c) transfer curve for an *n*-channel enhancement-mode
transistor.

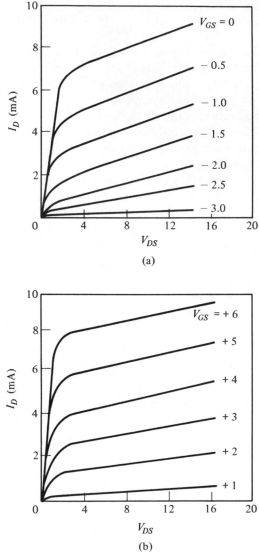

Figure 6.5 (a) MOSFET output characteristics: depletion mode, *n* channel; (b) enhancement mode, *p* channel.

When constructed with a *p* channel, a positive voltage on the gate repels holes from the wafer and a depletion region forms under the gate, as indicated in Fig. 6.4(b). There is no current in the channel. If the gate is made more positive than some threshold V_T, however, the positive gate charge attracts electrons from the negative source and builds an *n* channel between source and drain. This is shown in black in the figure. Current now passes in this channel, with a transfer curve between voltage and current as shown in Fig. 6.4(c).

The positive drain potential readily sweeps out electrons, the channel is narrowed at the drain end, and pinch-off occurs. With zero current at zero V_{GS} and an increasing drain current with increasing gate voltage, we have the *enhancement mode* of FET operation.

The input resistance of the MOSFET is due to the silicon dioxide layer and can be as high as 10^9 to 10^{15} Ω. The gate has a length of about 15×10^{-3} mm and the capacitance between gate and channel is about 1 to 4 pF.

MOSFET construction is particularly adaptable to the processes of impurity diffusion and metal deposition that are carried out in the production of integrated circuits.

6.3 Symbols for the FET

In Fig. 6.6 we show the several circuit symbols used to identify field-effect transistors. In Fig. 6.6(a) and (b) the arrows show the materials used in a manner similar to the diode symbol; at (a) we have a *p* gate to an *n* channel and at (b) there is an *n* gate on a *p* channel. The arrows tell us the polarities of the needed bias voltages. The gate electrode identifies the source by being placed above *S*.

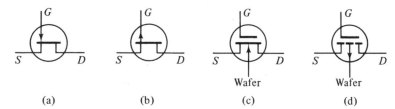

| (a) | (b) | (c) | (d) |

Figure 6.6 JFET: (a) *n* channel; (b) *p* channel. MOSFET: (c) *n*-channel depletion; (d) *p*-channel enhancement.

The isolated gate symbol in Fig. 6.6(c) and (d) identifies the transistor as a MOSFET. The arrows in the wafer leads indicate the materials, again following the diode symbolism; at (c) we have an *n*-channel depletion mode and at (d) there is a *p*-channel enhancement-mode transistor.

6.4 The Load Line for the FET

We may draw a dc load line on the characteristics of an *n*-channel JFET in Fig. 6.7, as done for the bipolar transistor in Sec. 4.7. The circuit in Fig. 6.7(b) is a *common-source* circuit. We write from the output circuit

$$v_{DS} = V_{DD} - Ri_D \qquad (6.1)$$

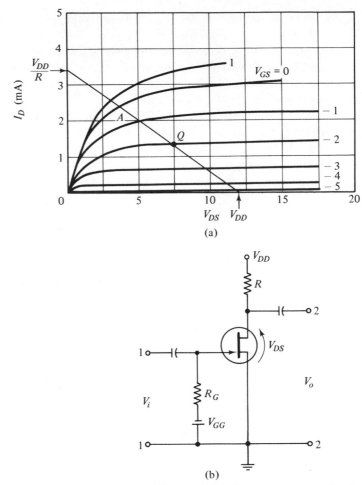

Figure 6.7 (a) Depletion-mode, JFET characteristics; (b) common-source circuit.

which represents a straight line. This *dc load line* can be drawn for an x-axis intercept at V_{DD} and a y-axis intercept at $i_D = V_{DD}/R$. These quantities should be so chosen that the load line traverses a region of uniform transistor characteristics.

The Q point or zero V_i point must lie on the load line for the circuit. A signal applied to the gate will vary v_{GS} and cause the operating point to move from Q up and down along the load line. If the output waveform is to be an undistorted image of the input wave, then the Q point must lie in a region of uniformly spaced output curves. The curve for $v_{GS} = 0$ serves as one limit for the operating region. Point A at that limit should be in the saturation portion of the current curve. Voltage V_{DD} should be less than the maximum rated drain-source voltage for the transistor.

With these limitations, point Q of the figure seems a reasonable choice and a dc load line can be drawn through Q and V_{DD} as an intercept.

We can determine the value of load used as the reciprocal of the slope of the load line. The slope can be measured on the triangle formed by the load line and the axes as

$$\text{slope} = \frac{\Delta i_D}{\Delta v_{DS}} = \frac{V_{DD}/R}{V_{DD}} = \frac{1}{R} \tag{6.2}$$

We then have

$$R = \frac{12 - 0}{0.0034 - 0} \cong 3500 \ \Omega$$

We use the output characteristics to choose a Q point for bias determination for the transistor. The Q-point location gives us the quantities I_D, V_{GS}, and V_{DS}.

6.5 Obtaining Bias for the FET

The channel-resistance and the gate-control properties of FETs are subject to considerable variation among units of the same type. In addition there are changes due to temperature. Bias circuits must be designed so that some self-regulation of the Q-point drain current is obtained. The circuit should be able to compensate when one transistor is substituted for another of the same type but with differing parameters.

The problem is illustrated in Fig. 6.8 where the drain-current versus gate-voltage transfer curves are plotted for three transistors of the same type. The zero bias-drain current I_{DSS} varies from 25 mA for the high unit to 6 mA for the low unit. With *fixed-voltage bias*, $V_{GG} = -1.2$ V, the drain current of the high unit would be 12 mA and that of the low unit would be zero, or the transistor would be in a cutoff condition. Fixed-gate bias is not a satisfactory solution to the problem.

Self-bias, as shown in Fig. 6.9(a), is preferred over fixed bias. This eliminates one voltage source at the cost of two resistors. Resistor R_1 is large, usually 1 MΩ or more, and is present only to fix the average gate voltage at ground. This is accomplished since $i_G \cong 0$ and there is no voltage drop in R_1. The indicated polarities in Fig. 6.9(a) lead to an input circuit equation of

$$V_{GS} + V_S = 0$$
$$-V_{GS} = V_S = R_S I_D \tag{6.3}$$

The negative sign shows that the gate is negative in relation to the source. If I_D increases, then V_S rises and the gate is made more negative. But a more negative gate tends to reduce I_D toward its initial value. Self-regulation of I_D is thus accomplished by this circuit. The amount of stabilization of I_D

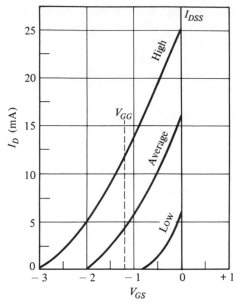

Figure 6.8 Drain current versus gate voltage for several 3N128 FETs.

is proportional to R_S, however; large values of R_S require increased V_{DD} and also result in a reduction of gain.

We may decide that about 15 per cent of V_{DD} should appear across R_S at Q-point current. Then the resistor voltage will be

$$V_S \cong 0.15 V_{DD} \tag{6.4}$$

Due to the flatness of the V_{GS} curve, we can read I_D and can find R_S as

$$R_S = \frac{V_S}{I_D} \quad (\Omega) \tag{6.5}$$

The voltage drop across the transistor may be made equal to $V_{DD}/2$ at the Q point and so we know that

$$R + R_S \cong \frac{V_{DD}}{2I_D} \quad (\Omega) \tag{6.6}$$

Then we can determine the value of the load R. The load line for $R + R_S$ can be drawn from the x intercept at V_{DD} to the y intercept at $V_{DD}/(R + R_S)$. The Q point and V_{DS} are fixed by the intersection of the V_{GS} line and the dc load line.

This completes the design of the self-bias circuit.

The input and output capacitors are blocking capacitors, as before. Since they are expected to represent negligibly small reactances, they will not be considered at this time.

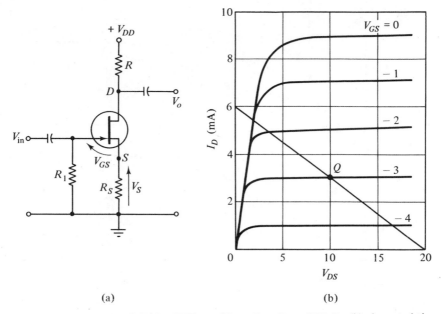

(a)   (b)

Figure 6.9 (a) Self-bias FET amplifier; $R + R_s \cong 4300 \ \Omega$; (b) characteristics and load line.

6.6 The Four-Resistor Bias Network

The circuit in Fig. 6.10(a) requires one more resistor than the self-bias circuit but gives greater design flexibility in that the gate bias can be negative, zero, or positive as required. We choose a Q point in Fig. 6.10(b) and find I_D. Using the arbitrary rule of Eq. 6.6,

$$R + R_s \cong \frac{V_{DD}/2}{I_D} \quad (\Omega) \tag{6.7}$$

The voltage across R_s may be arbitrarily chosen as 20 per cent of the total drop across R and R_s so that

$$R_s I_D \cong \frac{0.20 V_{DD}}{2} = 0.10 V_{DD} \tag{6.8}$$

With R_s determined, we can find the load R.

In the circuit we show V_{GS} with gate positive to the source. If the gate is to be negative, then V_{GS} will carry a negative sign. Around the input loop

$$V_G - V_{GS} - V_S = 0$$
$$V_G = V_{GS} + V_S \tag{6.9}$$

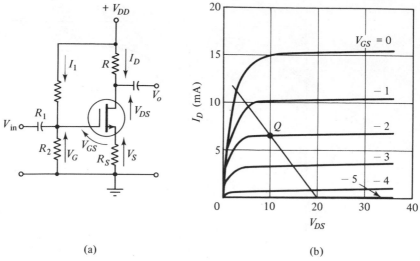

(a)   (b)

Figure 6.10 (a) Four-resistor bias circuit; (b) *n*-channel depletion FET, 3N139.

Since $I_G = 0$, we can use the voltage-divider relation to determine R_1 and R_2. That is,

$$V_G = \frac{R_2}{R_1 + R_2} V_{DD}$$

$$R_2 = (R_1 + R_2) \frac{V_G}{V_{DD}} \qquad (6.10)$$

Again we have two unknowns with only one equation. We arbitrarily select $R_1 + R_2$ to draw a small current from V_{DD}, compared to I_D. Having $R_1 + R_2$ assumed, we can solve Eq. 6.10 for R_2 and then obtain R_1.

This completes the amplifier design. It may seem that much of this procedure is arbitrary but we remember that transistor curve families are drawn for average units and differences between transistors will create circuit variations that make more precise circuit design unneccessary. The self-regulating action of the circuit will usually bring the Q point to within about 10 per cent of the expected I_D value for normal transistor and resistor variations.

6.7 Design Examples

Example 1: Using the FET represented by the curves in Fig. 6.9(b) with a Q-point current $I_D = 3$ mA, with $V_{DD} = 20$ V, find the exact Q-point location and the design values for R, R_1, and R_S for the self-bias circuit of Fig. 6.9(a).

Selection of $I_D = 3$ mA places our Q point somewhere along the $V_{GS} =$

-3 V characteristic. But $-V_{GS} = V_S = 3$ V and, using Eq. 6.5, we find

$$R_S = \frac{V_S}{I_D} = \frac{3}{0.003} = 1000 \ \Omega$$

Then using $(R + R_S)I_D = V_{DD}/2$, we have from Eq. 6.6

$$R + R_S = \frac{V_{DD}}{2I_D}$$

$$= \frac{20}{2(0.003)} = 3300 \ \Omega$$

The load line of the figure is drawn between $V_{DD} = 20$ V and a y intercept at $i_D = V_{DD}/(R + R_S) = 6$ mA. The Q point is at $I_D = 3$ mA, $V_{GS} = -3$ V, and $V_{DS} = 10$ V.

Arbitrarily we select $R_1 = 1$ MΩ so the circuit resistances are

$$R_S = 1000 \ \Omega$$
$$R = 2300 \ \Omega$$
$$R_1 = 1 \ \text{M}\Omega$$

Example 2: The circuit in Fig. 6.10(a) is supplied with a transistor having the characteristics in Fig. 6.10(b). Supply voltage is $V_{DD} = 20$ V and the $V_{GS} = -2$ V line is considered a suitable location for the Q point of this n-channel depletion FET.

We shall choose $(R + R_S)I_D = V_{DD}/2$ and this assumption fixes $V_{DS} = V_{DD}/2$ so that our Q point is located at $V_{GS} = -2$ V, $V_{DS} = \frac{20}{2} = 10$ V. A dc load line for $R + R_S$ is drawn from the x intercept at V_{DD} through the Q point.

The Q point fixes $I_D = 6.5$ mA. Then we have

$$R + R_S = \frac{V_{DD}}{2I_D} = \frac{20}{0.013} \cong 1500 \ \Omega$$

We choose the voltage V_S as 15 per cent of V_{DD} or

$$V_S = R_S I_D = 0.15 \ V_{DD}$$
$$0.0065 R_S = 0.15 \times 20 = 3 \ \text{V}$$
$$R_S = \frac{3}{0.0065} \cong 470 \ \Omega$$

after rounding the resistance to the nearest standard value. Then we have

$$R = 1500 - 470 \cong 1000 \ \Omega$$

We can design the voltage divider to give us the needed value of V_G from

$$V_G = V_S + V_{GS}$$
$$= 3.0 + (-2) = 1.0 \ \text{V}$$

That is, the source is 3.0 V positive to ground and the gate is 1.0 V positive to ground. Thus the gate is -2 V to the source as we determined from the Q point.

Since current through $R_1 + R_2$ results in a power loss, we choose $R_1 + R_2$ to take a current that is small with respect to I_D. Selecting $I_1 = 300\ \mu\text{A}$, we have

$$R_1 + R_2 = \frac{V_{DD}}{0.0003} = \frac{20}{0.0003} \cong 67,000\ \Omega$$

Because $I_G = 0$, the voltage-divider ratio will be accurate in setting V_G and we have

$$V_G = \frac{R_2}{R_1 + R_2} V_{DD}$$

$$R_2 = (R_1 + R_2)\frac{V_G}{V_{DD}}$$

$$= 67,000\frac{1.0}{20} = 3300\ \Omega$$

Then $R_1 = 63,700\ \Omega$.

As the nearest standard resistance values, our circuit will employ

$$R = 1000\ \Omega \qquad R_1 = 3300\ \Omega$$
$$R_S = 470\ \Omega \qquad R_2 = 62,000\ \Omega$$

6.8 The FET as an Amplifier

As with the *pnp* or *npn* bipolar transistors, the FET has three internal electrodes. We want to connect the FET, however, to four circuit terminals or two ports. Some FETs have four leads but one is a second gate that is largely used for bias control. Here our gate symbol indicates the ac signal or control element of the FET.

The three choices of common lead provide three amplifying circuits, the *common-source* circuit, the *common-gate* circuit, and the *common-drain* circuit. The common-drain circuit is also known as a *source follower* and serves the same function as the emitter follower. In the common-gate circuit the high input resistance of the FET is lost.

Depletion-mode operation is more commonly used for amplifiers, while the enhancement mode is used mainly for switching purposes in digital circuits.

6.9 Circuit Characteristics of the FET

The output curves for the FET predict the operation of the devices when used in an amplifier circuit. The JFET and MOSFET curves have the same general shape and as a result the devices may be discussed together. We choose the source as common and the voltages are measured to that electrode, as reference.

As an amplifier, the FET is operated in the pinch-off condition where variations in drain current are dependent almost entirely on gate voltage. The slope of the output curves is

$$\frac{\Delta i_{DS}}{\Delta v_{DS}} = \frac{1}{r_d} \tag{6.11}$$

where r_d is called the *drain resistance* of the FET. The low and nearly constant slope of the curves in the pinch-off region shows that r_d is high and constant, almost independent of V_{DS}. This is a characteristic of a constant-current generator.

The input resistance r_{GS} is considered very large and usually neglected. The effect of the shunting capacitance will be considered in a later chapter.

A typical FET *transfer curve*, relating the drain current i_D to the input signal voltage v_{GS}, is drawn in Fig. 6.11(a). This curve is found to be predicted by

$$i_D = I_{DSS} \left(1 - \frac{v_{GS}}{V_p} \right)^2 \tag{6.12}$$

where I_{DSS} is the value of i_D with the gate shorted to the source S; this current is indicated by the small circle on the ordinate in Fig. 6.11(a). The constant V_p is the pinch-off value of v_{DS}, as at (2) of Fig. 6.2. It is found from a projection of the slope of the transfer curve from I_{DSS}, which intersects the abscissa at a voltage of $V_p/2$.

A useful transistor performance figure is the *transconductance* g_m. This is the change in output current per volt change in input voltage. This is found from the slope of the transfer curve as

$$g_m = \frac{\Delta i_D}{\Delta v_{GS}} \quad \text{(mhos)} \tag{6.13}$$

The slope is a variable and g_m varies with i_D. The extent of this change for a typical FET is plotted in Fig. 6.11(b). The selection of the Q-point current then determines the value of g_m being used for a given FET and circuit.

Equal changes in v_{GS} do not produce equal changes in i_D on the transfer curve because of the curve variation. If we reason that a curve is made up of short straight lines and if we choose a small segment of the curve, however, we can assume that segment is straight. Translated to the FET, this means that we can use the FET for relatively distortionless amplification if we apply only small-signal voltages, causing i_D to vary over only a small segment of the transfer curve; that is, we might restrict the input signal to ± 0.2 V peak-to-peak. Thus the FET is most often employed as a linear amplifier in the input, or small-signal, stages of a system. It is also useful in digital-switching circuits, where amplitude distortion is not important.

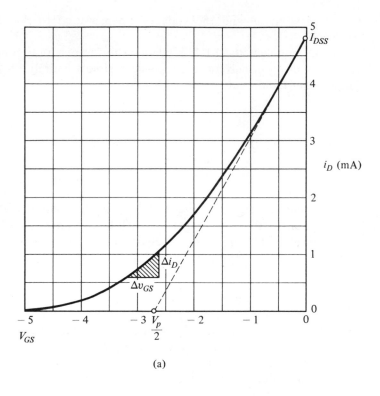

(a)

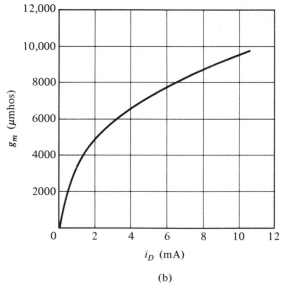

(b)

Figure 6.11 (a) FET transfer curve, depletion-mode FET; (b) variation of g_m with i_D for a 3N128.

Typically, we would expect r_d to be in the range of 5000 to 50,000 Ω for MOSFETs and above 100,000 Ω for JFETs. The value of g_m may be expected in the range of 1000 to 10,000 μmhos. This latter figure means

$$0.010 = \frac{I_d}{V_{gs}}$$

and indicates that we shall have an output current of 10 mA ac for each ac volt input.

6.10 The Equivalent Circuit of the FET

Figure 6.12(a) shows the internal elements of an FET arranged as a two-port network in common-source connection. Too complicated to be readily analyzed, we reduce it to the simple equivalent circuit at Fig. 6.12(b) by consideration of the relative magnitudes of the resistances and reactances. The equivalent circuit that is developed is satisfactory for applications not exceeding frequencies of a few megahertz.

The element r_{gs} is the reverse-biased junction resistance of the JFET or the silicon dioxide layer resistance of the MOSFET. These resistances are so high that we may consider r_{gs} as representing an open circuit.

The series circuit of C_c and r_c represents the capacitance between gate and channel and the series resistance of the channel. Being only a few picofarads and a few hundred ohms, their effect is negligible at low radio frequencies and they are eliminated from the equivalent circuit.

Capacitance C_{gd} is the gate-to-drain capacitance and includes the capacitance of the transistor mounting; being only 1 to 3 pF and of very high reactance we drop the element from present consideration.

It was pointed out that the output current of an FET behaves as that of a current generator and so the internal generator is shown as the current source $g_m V_{gs}$.

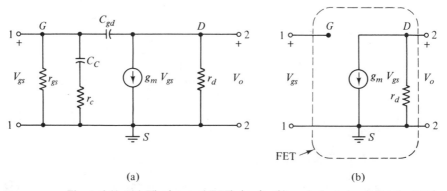

(a)   (b)

Figure 6.12   (a) The internal FET circuit; (b) equivalent circuit for the FET.

Thus in Fig. 6.12(b) we have the same form of equivalent circuit as we previously used for the bipolar transistor; however, the input circuit of the FET takes zero current. The circuit elements are assumed constant around a given Q point and this restricts our operation with ac signals to a linear region of the output characteristics and to small input signals. Since the bias sources are not included, the circuit is useful for analysis with ac signals only.

The effects of C_c and r_c and C_{gd} at higher frequencies will be discussed later.

6.11 The Common-Source Amplifier

In the *common-source amplifier* in Fig. 6.13(a), the input signal $V_i = V_{gs}$ is applied between gate and source and the output V_o is obtained between drain and source. In Fig. 6.13(b) we have replaced the FET between $G,S,$

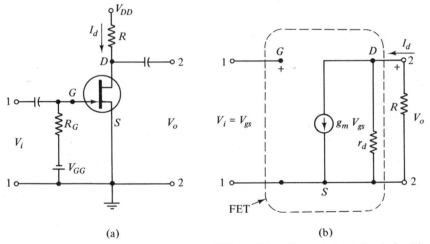

(a)   (b)

Figure 6.13   (a) Common-source FET amplifier; (b) equivalent circuit for (a).

and D with its equivalent circuit consisting of a current source and the internal drain resistance r_d. Resistor R_G is usually so large that it appears as an open circuit. Except for the zero input current, the FET common-source amplifier is similar to the C-E circuit for the bipolar transistor.

Voltage Gain

The parallel effect of r_d and the load R is represented by

$$R_p = \frac{r_d R}{r_d + R} \qquad (6.14)$$

The current $g_m V_{gs}$ produces a voltage V_o across the output as

$$V_o = -g_m V_{gs} R_p = -g_m \frac{r_d R}{r_d + R} V_{gs}$$

The negative sign results from defining the output voltage and current such that $V_o = -RI_d$. The negative sign indicates that a phase reversal has occurred between V_{gs} and V_o.

The voltage gain in the circuit is obtained by dividing by V_{gs}, leading to

$$A_v = \frac{V_o}{V_{gs}} = -g_m \frac{r_d R}{r_d + R} \tag{6.15}$$

Another useful form of this expression can be obtained by dividing numerator and denominator by r_d, giving

$$A_v = -\frac{g_m R}{1 + \dfrac{R}{r_d}} \tag{6.16}$$

Usually $R \ll r_d$ and R/r_d becomes small with respect to one and we then have

$$A_v \cong -g_m R \tag{6.17}$$

which is the same result as obtained for the C-E transistor amplifier.

The voltage gain depends on the selection of a transistor with a high g_m and on the resistance chosen for R.

Output Resistance

The output resistance at the 2,2 port is defined as the resistance at that port with the input signal set to zero.

With $V_i = V_{gs} = 0$ the current of the current source is zero, a condition reached by opening the circuit of that source. The resistance remaining at the 2,2 terminals is r_d and that becomes the output resistance of the amplifier.

The value of r_d is nominally high, perhaps 50,000 Ω, and the gain approximates $A_v = -g_m R$. The circuit performance is similar to that of the C-E amplifier, except for the negligibly small input current taken by the FET.

6.12 The Common-Drain Circuit

For the FET *common-drain circuit* in Fig. 6.14(a), the input signal is applied between gate and the drain and the output is taken between source and drain. As noted previously, the circuit is known as a *source follower*. This is because the output voltage V_o is approximately equal to and varies with the input V_i. In this, the circuit performance parallels that of the emitter follower.

We replace the FET with its equivalent circuit in Fig. 6.14(b). For better understanding, we fold down the upper half of the output circuit about the

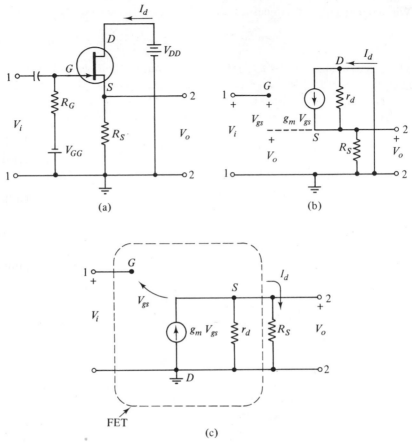

Figure 6.14   (a) Common-drain circuit; (b) and (c) equivalent circuits.

line of the source and in (c) have resistors r_d and R_S in parallel; inspection of (b) shows this to be correct. The current source appears to have turned upside down but its outgoing terminal remains connected to S.

Voltage Gain

Using Eq. 6.14 for the parallel resistance of r_d and R_S, the output voltage is

$$V_o = g_m R_o V_{gs} = g_m \frac{r_d R_S}{r_d + R_S} V_{gs} \qquad (6.18)$$

In this circuit the input signal is connected between gate and drain and from the circuit

$$V_i = V_{gs} + V_o$$
$$V_{gs} = V_i - V_o \qquad (6.19)$$

We make this substitution for V_{gs} in Eq. 6.18 and rearrange the terms so

$$V_o\left(1 + \frac{g_m r_d R_S}{r_d + R_S}\right) = g_m \frac{r_d R_S}{r_d + R_S} V_i \tag{6.20}$$

Multiplying by $r_d + R_S$ and dividing by V_i, the voltage gain from the input at 1,1 to the output voltage across R_S is

$$A_v = \frac{V_o}{V_i} = \frac{g_m r_d R_S}{r_d + R_S + g_m r_d R_s} = \frac{g_m R_S}{1 + \frac{R_S}{r_d} + g_m R_S} \tag{6.21}$$

The drain resistance is usually large with respect to R_S and R_S/r_d will be small with respect to the unity term. The voltage gain is then approximated by

$$A_v \simeq \frac{g_m R_S}{1 + g_m R_S} \tag{6.22}$$

This result would be less than but near unity since $g_m R_S$ is usually larger than one. There is no phase shift between the input and output voltages in this amplifier.

Equation 6.22 has the same form as obtained for the emitter follower. Since the equivalent circuit for the amplifier is the same, we should expect this result.

Output Resistance

As a demonstration of the method, we shall find the output resistance of the transistor at 2,2, with the load R_S open, by use of the impedance relation for a voltage-source equivalent circuit

$$R' = R_o = \frac{V_{oc}}{I_{sc}} \tag{6.23}$$

Written without R_S, Eq. 6.20 gives for V_{oc}

$$V_{oc} = \frac{g_m r_d V_i}{1 + g_m r_d} \tag{6.24}$$

With a short circuit across 2,2, we have $V_o = 0$ and Eq. 6.19 gives $V_{gs} = V_i$. The current in the short circuit is

$$I_{sc} = g_m V_{gs} = g_m V_i \tag{6.25}$$

Dividing Eq. 6.24 by 6.25, we have R_o as

$$R_o = \frac{\frac{g_m r_d V_i}{1 + g_m r_d} V_i}{g_m V_i} = \frac{r_d}{1 + g_m r_d} \tag{6.26}$$

Usually $g_m r_d \gg 1$ and this equation reduces to a simple form,

$$R_o \simeq \frac{1}{g_m} \tag{6.27}$$

Since g_m is in the range of 0.001 to 0.01 mho, the output resistance of the circuit is small.

Therefore the source follower, with only unity voltage gain, serves mainly as an impedance transformer, useful in amplification of a signal from a high-resistance input to a low-resistance output, such as a relay or a telephone line.

The design of the bias circuit for a common-drain amplifier follows the procedure for the four-resistor circuit of Sec. 6.6 with $R = 0$.

Example: Consider a source follower using a JFET with $r_d = 50,000\ \Omega$ and $g_m = 0.0025$ mho, as data given by the manufacturer of the JFET. Find the gain and output resistance with $R_S = 5000\ \Omega$.

Using Eq. 6.21, we have

$$A_v = \frac{g_m R_S}{1 + \frac{R_S}{r_d} + g_m R_S}$$

$$= \frac{0.0025 \times 5000}{1 + \frac{5000}{50,000} + 0.0025 \times 5000}$$

$$= \frac{12.5}{1 + 0.1 + 12.5} = 0.920$$

If we had neglected the R_S/r_d term in the denominator, as in Eq. 6.22, we would have

$$A_v \cong \frac{12.5}{1 + 12.5} = 0.926$$

and the difference is negligible.

The output resistance is

$$R_o \cong \frac{1}{g_m} = \frac{1}{0.0025} = 400\ \Omega$$

6.13 Design Example

We shall perform the design calculation for a common-drain amplifier using the circuit in Fig. 6.10, with $R = 0$. The Q point is set at $I_D = 6.5$ mA with $V_{DD} = 20$ V, as in that figure.

Using the arbitrary rule of Eq. 6.7,

$$R_S I_D = \frac{V_{DD}}{2}$$

and so

$$R_S = \frac{V_{DD}}{2I_D} = \frac{20}{0.013} \cong 1500\ \Omega$$

The voltage V_S is

$$V_S = R_S I_D = 1500 \times 0.0065 = 9.75 \text{ V}$$

This value is not quite the specified value of 10 V because of the rounding of the R_S value.

At the Q point,

$$V_{GS} = -2 \text{ V}$$

and so

$$V_G = V_{GS} + V_S = (-2) + 9.75 = 7.75 \text{ V}$$

positive to ground.

Since the gate current is zero, then resistors R_1 and R_2 serve only as a voltage divider for the gate voltage and we can arbitrarily choose $R_1 + R_2 = 1 \text{ M}\Omega$. Then Eq. 6.10 gives

$$R_2 = (R_1 + R_2)\frac{V_G}{V_{DD}} = 10^6 \times \frac{7.75}{20} = 387{,}500 \ \Omega$$

$$\cong 0.39 \text{ M}\Omega$$

Then

$$R_1 = 10^6 - 0.39 \times 10^6 = 0.62 \text{ M}\Omega$$

again choosing a standard resistor value.

The circuit is complete with

$$R_S = 1500 \ \Omega$$
$$R_1 = 0.62 \text{ M}\Omega$$
$$R_2 = 0.39 \text{ M}\Omega$$

The value of g_m at the Q point can be determined graphically by measuring Δv_{GS} as 1 V, from the -1-V curve to the -2-V curve, and reading the resultant Δi_D along the vertical line for $v_{DS} = 10$ V, through the Q point. The value of Δi_D is read as 3.5 mA and we have

$$g_m = \frac{\Delta i_D}{\Delta v_{GS}} = \frac{0.0035}{1} = 0.0035 \text{ mho}$$

The value of r_d is large, as can be seen from the flatness of the $v_{GS} = -2$ V curve at the Q point and we can calculate the voltage gain from Eq. 6.22:

$$A_v \cong \frac{g_m R_S}{1 + g_m R_S} = \frac{0.0035 \times 1500}{1 + (0.0035 \times 1500)}$$

$$= \frac{5.25}{1 + 5.25} = 0.84$$

The output resistance is

$$R_o \cong \frac{1}{g_m} = \frac{1}{0.0035} = 286 \ \Omega$$

6.14 Using the FET as a Variable Resistance

When operated near point (1) of the curve in Fig. 6.2, the drain-source channel of an FET can be used as a *voltage-variable resistance* (VVR). The device
represents a resistance of several thousand ohms, as shown in Fig. 6.15.
Currents will be of microampere order and voltages will be a few hundred
millivolts.

The device has some application in signal level control and expansion.
The basic elements of a volume control application appear in Fig. 6.15(a),

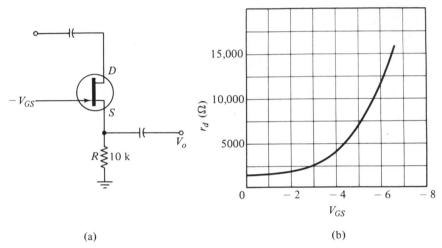

(a)   (b)

Figure 6.15   (a) The FET as a voltage-variable resistance; (b) resistance variation
obtained.

where the voltage across R is the output, as a varying fraction of V_i. The
gate voltage is varied by a rectified voltage derived from the input signal;
as this increases, the FET resistance increases, passing on less of the input
signal through the voltage-divider action of the circuit. The output signal is
then controlled to a more constant level.

6.15 Summary

The unipolar FET, with either *n*- or *p*-conduction properties, controls the
output current magnitude by means of a voltage applied to a gate electrode.
Its most important feature is the very high value of input resistance. It is well
adapted to production by methods of diffusion and metal deposition.

Because of the zero input current, the design of its bias circuits is simpler than for the bipolar transistor but the methods of circuit design are the same. We also find the FET useful in the common-source and source follower circuits, which parallel the C-E and emitter follower circuits in performance and applicability.

Because of its curved transfer curve, the FET can give linear amplification only with small signals. This makes it most adaptable to input circuits.

REVIEW QUESTIONS

6.1 Why is the FET called a unipolar transistor?

6.2 Describe how pinch-off is obtained in an n-channel JFET.

6.3 What is the construction difference between a JFET and a MOSFET?

6.4 Explain how a depletion-mode MOSFET operates.

6.5 What is meant by "enhancement mode"?

6.6 What is meant by "depletion mode"?

6.7 What is the significance of the arrow on the gate of a JFET symbol? Of the arrow on the MOSFET symbol?

6.8 What range of resistances can we expect in the gate-to-channel path of a MOSFET?

6.9 What is the reason for limiting V_{DG} for an FET?

6.10 Why do we limit FET operation to small signals?

6.11 The transistor in Fig. 6.7(a) has $V_{DD(\text{max})} = 20$ V. Bound the region for acceptable Q-point location for a linear amplifier.

6.12 Relate r_d and g_m to the characteristic curves of an FET.

6.13 Why is self-regulation of the drain current important in the common-source circuit?

6.14 Compare the voltage gain and output and input resistances of the common-source and common-drain circuits.

6.15 The common-source circuit is the counterpart of what bipolar transistor circuit? Why?

6.16 The common-drain circuit is the counterpart of what bipolar transistor circuit? Why?

6.17 An FET has $g_m = 6000$ μmhos at the Q point. In the common-drain circuit, what output resistance will be obtained?

PROBLEMS

6.1 For the curves of Fig. 6.5(a), with $V_{DD} = 20$ V, $R = 4000$ Ω, and $V_{GS} = -1.5$ V, find the Q-point current I_D.

6.2 Determine R_S if the circuit of Problem 6.1 is to be self-biased.

6.3 Design a self-bias circuit for a common-source amplifier using the transistor in Fig. 6.9(b), with $V_{DS} = 12$ V, $I_D = 3$ mA at the Q point, and $V_{DD} = 20$ V. Determine R and R_S.

6.4 An FET, Fig. 6.9(b), is in a common-source circuit with Q point at $V_{DS} = 7.5$ V, $I_D = 4.0$ mA, and $V_{DD} = 15$ V. Find R_S, R, V_{GS}, R_1, and R_2 for $R_1 + R_2 = 1.25$ MΩ.

6.5 The 3N158 FET, with characteristics given in Fig. 6.7(a), is used in a common-source circuit with $V_{DS} = 10$ V, $I_D = 1.4$ mA at the Q point. What should be the values of R, V_{DD}, and V_{GG} for a good circuit design?

6.6 Design a four-resistor bias circuit as in Fig. 6.10, using $V_{DD} = 30$ V, $V_{GS} = -2$ V. Select I_D and suitable values for the four circuit resistors.

6.7 A depletion-mode FET has $I_{DSS} = 3$ mA, $V_p = -4.25$ V; find I_D for $V_{GS} = -2.5$ V by the square-law relation of Eq. 6.12.

6.8 A depletion-mode FET obeys the square-law relation of Eq. 6.12 with $I_{DSS} = 8.4$ mA, $V_p = -3.0$ V. Find I_D at $V_{GS} = -1.0, -2.0, -3.0$, and 0 V and plot the transfer curve.

6.9 Find g_m at each V_{GS} value listed in Problem 6.8 by assuming ΔV_{GS} changes of 0.2 V and computing the resultant ΔI_D. Plot a curve of g_m versus I_D.

6.10 What R should be specified for the circuit in Fig. 6.13(a) to obtain $A_v = -40$ with $g_m = 6000$ μmhos and $r_d = 40{,}000$ Ω? What is the output resistance?

6.11 Given $V_{DD} = 30$ V, $R_2 = 0.1$ MΩ, and $R_1 = 2$ MΩ for the transistor in Fig. 6.10, find R_S to place Q at $I_D = 5$ mA, $V_{DS} = 15$ V, and $V_{GS} = -2$ V.

6.12 Determine R for a JFET self-biased amplifier circuit as in Fig. 6.7(b), for a voltage gain of -25, with $r_d = 40$ kΩ, $g_m = 0.005$ mho. Also, select a suitable value for R_G.

6.13 Find R, R_S, and R_1 for the circuit in Fig. 6.9(a), with a Q point for the transistor at $I_D = 4.0$ mA, $V_{DS} = 7.5$ V.

6.14 An FET with $r_d = 50$ kΩ, $R_S = 2$ kΩ, and $g_m = 2500$ μmhos is used in a common-drain circuit. What is the voltage gain? What is the output resistance?

6.15 Calculate the ac voltage gain of a common-source FET amplifier in the circuit in Fig. 6.7(b), with $g_m = 0.0045$ mho, $r_d = 50$ kΩ, and $R = 20$ kΩ.

6.16 What value of transistor g_m is needed in the circuit in Fig. 6.7(b) to provide a signal gain of -40 if $r_d = 60{,}000$ Ω and R is chosen as $20{,}000$ Ω and R_G is 1 MΩ?

6.17 A signal $V_i = 2$ mV is applied to the FET amplifier in Fig. 6.7(b) with $g_m = 2500$ μmhos, $r_d = 40$ kΩ, $R = 10$ kΩ, and $R_G = 1$ MΩ. Find the ac output voltage V_o.

6.18 What value of load resistor R_S would match the output of a common-drain amplifier with $g_m = 0.0025$ mho, $r_d = 40{,}000$ Ω? Find the voltage gain with that resistor.

7

The Vacuum Tube

The *triode*, or three-element form of vacuum tube, was the first electronic control device. It employs a heated electron-emitting *cathode*, a control *grid* of wire mesh, and an electron-collecting plate or *anode*, assembled in an evacuated glass, metal, or ceramic envelope. Additional grids are added to improve the electrostatic shielding between grid and anode, leading to the five-element *pentode*.

Because of the need for heating power to induce emission of electrons, the large size and fragility, and the limited life, the vacuum tube has been superseded by the transistor at frequencies below 500 MHz and power levels below 200-W output. Many millions of tubes remain in service, however, so we give a brief discussion of tube and circuit theory here.

7.1 Circuit Notation for the Vacuum Tube

As for the transistor, there is an established system of nomenclature for circuit variables employed with vacuum tubes. The symbols lack the formality of those for the transistor, however.

Again, we use lowercase letters to designate instantaneous values of varying currents or voltages and capital letters to denote rms or dc values.

Subscripts b and p indicate anode circuit quantities and c and g indicate grid circuit variables. For example,

$i_b =$ instantaneous anode current
$I_b =$ quiescent value of anode current
$I_p =$ rms value of the signal component of anode current
$v_c =$ instantaneous grid-cathode voltage, also v_{gk}
$V_{CC} =$ grid circuit bias voltage
$V_{BB} =$ anode circuit supply voltage

Some tubes have more than one grid and numerical subscripts are used, with the grid nearest the cathode as number 1.

7.2 The Triode

The drawing in Fig. 7.1(a) shows the construction of the internal elements of triodes used at audio and low radio frequencies. There is a cylindrical nickel anode A, a helical grid G, and a central electron-emitting cathode K. The tube in Fig. 7.1(b) is designed for high-power transmitting service with air cooling and that in Fig. 7.1(c) is for high-frequency radio reception and illustrates a *planar* form of cathode and grid.

If the free electrons within a metal are given sufficient energy, they are able to overcome the surface binding forces and can be emitted into space. The releasing energy can be supplied by heat in *thermionic emission*, by

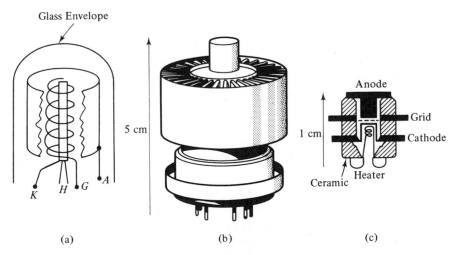

Figure 7.1 (a) Low-frequency triode structure; (b) Eimac 8874 triode, 1000-W peak output up to 500 MHz; (c) small ceramic triode, usable to 450 MHz.

radiant energy in *photoelectric emission,* or by bombardment by atomic particles in *secondary emission,* or the electrons can be pulled out of the surface by the attractive force of a strong electric field, as in *cold-cathode emission.*

For thermionic emission, the emitting cathode is most often a small nickel cylinder coated with a layer of barium or strontium oxides and heated to about 600 to 800°C (1000 to 1400°F). A low voltage is applied to a heater wire of tungsten to raise the nickel cylinder to emitting temperature. Such a cathode is said to be *indirectly heated.* Some tungsten wire filaments, impregnated with thorium, are used and operated at temperatures of 1600 to 1700°C (3000°F). Wire filaments are heated by a current through the wire and are said to be *directly heated.*

The electrons from the heated cathode reach the anode only when the

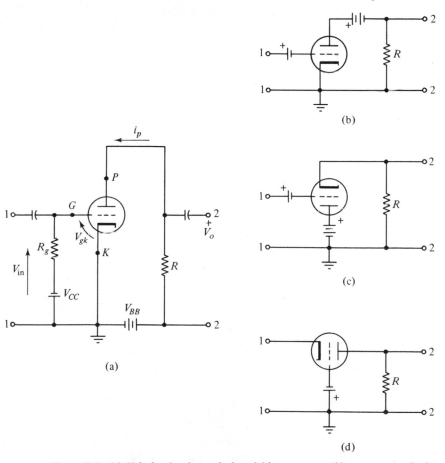

Figure 7.2 (a) Triode circuit symbol and bias sources; (b) common-cathode circuit; (c) cathode follower circuit; (d) grounded-grid circuit.

anode is positive to the cathode and the triode conducts in only one direction. In transit the electrons pass through the openings in the wire grid. If the grid is negative to the cathode, electrons are repelled and the anode current is reduced. If the grid is made less negative, the anode current increases. It is seen that the negative potential between grid and cathode controls the electron flow and consequently the current through the tube.

Practically no electrons can reach the negative grid and hence there is no grid current. As a result the triode has an infinite input resistance at frequencies through the audio range. Effects of the capacitances between the tube elements become important at frequencies of a megahertz or more and will be considered in Chapter 8.

The circuit symbol for a triode and the applied bias potentials are shown in Fig. 7.2. Again we have three internal electrodes and wish to connect them to four terminals or two output ports. As a result of the choice of common electrode, we have three basic circuits. These are usually designated as the *common-cathode circuit*, the common-anode circuit or *cathode follower*, and the *grounded-grid circuit* with the grid common. All are diagrammed in Fig. 7.2.

Our conventional current is again defined as positive inward and shown as i_p from anode to cathode in the figure.

7.3 Triode Characteristics

The output curve family for a typical triode is drawn in Fig. 7.3(a). The region of linear, equally spaced curves is bounded by $V_c = 0$ and by the maximum rated anode loss or dissipation to maintain a safe operating temperature. The loss limit is determined by

$$P_d = V_b I_b \quad \text{(W)} \tag{7.1}$$

and a limiting hyperbola is drawn for $P_d = 0.6$ W in the figure.

Opposite changes of anode voltage and grid voltage are offsetting in effect on the anode current. The anode current is more sensitive to the grid voltage since the grid is closer to the cathode than is the anode. We measure this sensitivity by the *amplification factor* μ, defined as the ratio of anode voltage change to grid voltage change needed to keep i_b constant. That is,

$$\mu = -\frac{\Delta v_b}{\Delta v_c}\Bigg]_{i_b = \text{constant}} \tag{7.2}$$

The negative sign arises because the anode voltage and the grid voltage must change in opposite directions to maintain constant current.

Values of μ for triodes range from 3 to over 100.

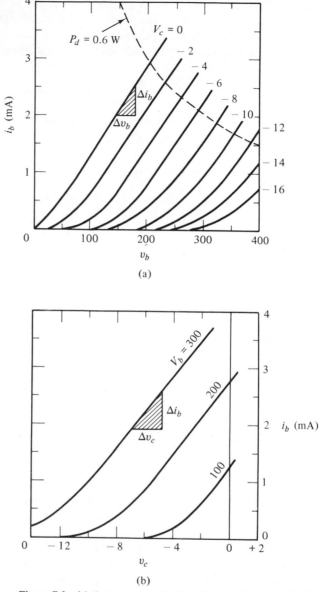

Figure 7.3 (a) Output curve family; (b) transfer curve family.

The reciprocal of the slope triangle of Fig. 7.3(a) defines the *plate resistance* r_p as

$$r_p = \frac{\Delta v_b}{\Delta i_b}\bigg]_{v_c=\text{constant}} \tag{7.3}$$

This is the resistance presented by the triode to ac signal currents, and typical values range from a few hundred to over 100,000 Ω.

The transfer curves, relating input voltage and output current, are shown in Fig. 7.3(b). From the slope triangle shown we can define the *transconductance* g_m as

$$g_m = \frac{\Delta i_b}{\Delta v_c}\bigg]_{v_b=\text{constant}} \quad \text{(mhos)} \tag{7.4}$$

Transconductance is the control factor for the output generator, as it was for the transistor. The magnitude of g_m for vacuum tubes will range from 1000 to over 40,000 μmhos.

The product of the magnitudes is

$$g_m r_p = \frac{\Delta i_b}{\Delta v_c}\frac{\Delta v_b}{\Delta i_b} = \frac{\Delta v_b}{\Delta v_c} = \mu \tag{7.5}$$

or

$$\mu = g_m r_p \tag{7.6}$$

This shows that the three coefficients are related. A low-μ tube will have a low plate resistance and a high-μ tube will have a high plate resistance.

7.4 The Pentode

In the triode we find a small capacitance C_{gp} of 2 to 5 pF between grid and anode and a similar small capacitance between grid and cathode. At frequencies of 1 MHz or more in the common-cathode circuit, the reactance of C_{gp}

$$X_{C_{gp}} = \frac{1}{2\pi f C_{gp}}$$

becomes small enough so that it feeds back an appreciable current from the high signal voltage at the anode to the lower signal voltage at the grid. This feedback current causes instability of the circuit gain and possible oscillation. The use of the triode in the common-cathode circuit is therefore limited to the audio frequencies.

This defect of the triode was overcome by adding two shielding grids between the control grid and the anode. Such a tube contains a cathode, a control grid G_1, a screen grid G_2, a suppressor grid G_3, and the anode; the resultant five-element tube is called a *pentode*. The pentode is useful to frequencies of several hundred megahertz because G_2 and G_3, as electrostatic shields, eliminate the capacitance between control grid and anode. A circuit symbol and a common-cathode circuit for a pentode are shown in Fig. 7.4(a).

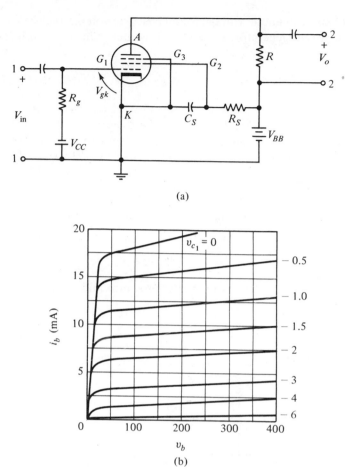

Figure 7.4 (a) Pentode in common-cathode circuit; (b) output characteristics, 6EA8 pentode, $V_{G_2} = 125$ V.

Grid G_2 is maintained at ground potential for signal frequencies by the large *bypass capacitor* C_s. A positive voltage is placed on G_2, the screen grid, and this accelerates the electrons to a high velocity. Thus G_2 is able to accelerate the electrons with its positive dc voltage but still serves as a grounded screen for signal frequencies. The suppressor grid G_3 is connected to the cathode so as to serve as another electrostatic screen between grid and anode. The high velocity electrons that pass through the openings in G_2 coast through G_3 and reach the positive anode.

The parameters for the pentode, μ, g_m, and r_p, remain as defined for the triode. Because of the low slope of the output curves, the plate resistance is found to approximate 1 MΩ. The anode current is almost constant with changes in anode voltage for a given grid voltage and the pentode behaves as a current generator. In general, it is similar in characteristics to the FET.

7.5 The Equivalent Circuits

In Chapter 3 we developed the idea of an h equivalent for any two-port circuit, and in Fig. 7.2 we drew the basic connections for a triode as a two-port element. Therefore, the h equivalent can represent the triode as in Fig. 7.5. For the h-parameter circuit we had Eq. 3.15 and 3.16:

$$V_1 = h_i I_1 + h_r V_2 \tag{7.7}$$
$$I_2 = h_f I_1 + h_o V_2 \tag{7.8}$$

and we can rewrite these equations with vacuum-tube quantities as

$$V_i = h_i I_c + h_r V_o \tag{7.9}$$
$$I_p = g_m V_{gk} + \frac{V_o}{r_p} \tag{7.10}$$

That is, $r_p = 1/h_o$ as the plate resistance and we previously showed $h_f I_1 = g_m V_1$.

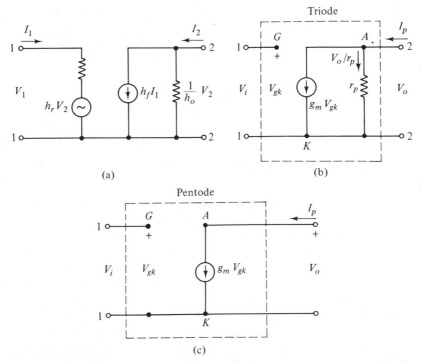

Figure 7.5 (a) h-Parameter model from Chapter 4; (b) equivalent circuit for the ideal triode; (c) pentode equivalent circuit.

We reasoned that a negative grid attracts no electrons and so $I_c = 0$ for the triode. At low frequencies there is no feedback of voltage from output to input and so $h_r = 0$. The input circuit is open and Eq. 7.9 is meaningless.

Equation 7.10 represents the output circuit of the triode, as in Fig. 7.5(b). We have a load current I_p, dividing into current V_o/r_p through the plate resistance, and a controlled current source $g_m V_{gk}$, with its current dependent on the *grid-to-cathode* ac voltage V_{gk}. The circuit in Fig. 7.5(b) represents the triode for ac and is the *current-source equivalent circuit* for the triode, as a small-signal linear amplifier.

A pentode equivalent circuit is derived by the same procedure used for the triode. The current taken when values of r_p exceed 500,000 Ω is negligible, compared to the current taken in usual loads, however, and r_p may be dropped from the circuit with negligible effect. We have the pentode equivalent circuit in Fig. 7.5(c), for small signals at audio and low radio frequencies.

The form of the equivalent circuits for triode and pentode is identical to the models previously developed for the bipolar transistor and the FET.

7.6 The Triode Common-Cathode Amplifier

With the cathode as the common element we have the common-cathode triode amplifier in Fig. 7.6. We replace the triode with its equivalent circuit between G, A, and K in Fig. 7.6(b). Except for the notation, the circuit is

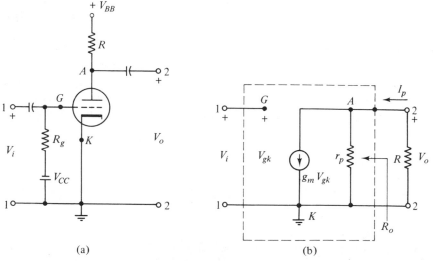

(a)   (b)

Figure 7.6 (a) Common-cathode triode amplifier; (b) equivalent circuit for (a).

identical to those already analyzed for the bipolar transistor and the FET so we shall merely state the results of circuit analysis here.

Voltage Gain

The circuit yields

$$A_v = \frac{V_o}{V_i} = -\frac{g_m R}{1 + \dfrac{R}{r_p}} \tag{7.11}$$

The negative sign indicates a 180° phase reversal between V_i and V_o.

Input Resistance

The value of R_g is usually made about 1 MΩ and at audio frequencies the input of the triode represents an open circuit.

Output Resistance

We short-circuit V_i, making $V_i = V_{gk} = 0$ and the current source is then open (zero current). As a result, from the 2,2 terminals we see only

$$R_o = r_p \tag{7.12}$$

as the output resistance of the amplifier.

The common-cathode circuit is generally used for medium gain at audio frequencies.

7.7 The Pentode Common-Cathode Amplifier

The pentode common-cathode circuit in Fig. 7.7(a) was generally applied because of greater gain and stable operation at radio frequencies. To maintain stability the reactance of C_s should be only a few hundred ohms at the lowest operating frequency. Capacitors C_s and C and resistors R_g and R_s are parts of the bias circuits and are dropped from the equivalent circuit for the ac signal.

Voltage Gain

The equivalent circuit is simple and the voltage gain can be written as

$$A_v = \frac{V_o}{V_i} = -g_m R \tag{7.13}$$

Loads as high as 100,000 Ω are used and gains of several hundred are obtained.

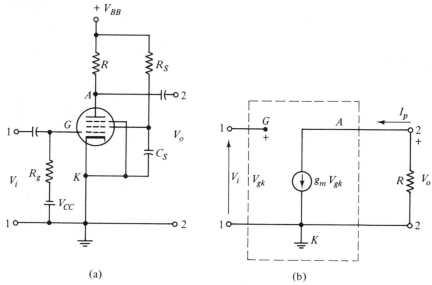

Figure 7.7 (a) Common-cathode pentode amplifier; (b) equivalent circuit for (a).

Input Resistance

With $R_g \cong 1$ MΩ, the input resistance can be assumed to be an open circuit.

Output Resistance

Following the method used for the triode, we have

$$R_o = r_p \tag{7.14}$$

and this is very large.

The pentode common-cathode amplifier is generally used up to frequencies of many megahertz. A disadvantage of the circuit was the need for C_s and R_s in bypassing the screen grid to the cathode. Grid biasing may be done by any of the methods employed with FETs.

7.8 The Cathode Follower

In the common-anode circuit or *cathode follower* in Fig. 7.8, we place the anode at ground for the signal but maintain it positive for acceleration of the electrons in the tube. Resistor R_g is large and dropped from the equivalent circuit. Circuit performance is found to be similar to that obtained from the emitter follower or the source follower. Resistor R_g may approximate 1 MΩ.

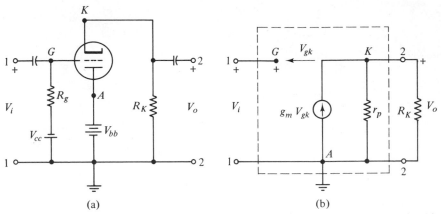

Figure 7.8 (a) Cathode follower circuit; (b) equivalent circuit for (a).

Voltage Gain

From the output circuit we can obtain

$$V_o = g_m V_{gk} \frac{r_p R_k}{r_p + R_k}$$

Around the input loop

$$V_i = V_{gk} + V_o$$

and

$$V_{gh} = V_i - V_o$$

Substituting for V_{gk} and rearranging,

$$V_o \left(1 + \frac{g_m r_p R_k}{r_p + R_k} \right) = \frac{g_m r_p R_k}{r_p + R_k} V_i$$

Clearing the denominator and dividing by r_p, we have

$$V_o \left(1 + \frac{R_k}{r_p} + g_m R_k \right) = g_m R_k V_i$$

But usually $R_k \ll r_p$ and R_k/r_p can be dropped as small compared to unity. Then we can determine the gain as

$$A_v = \frac{V_o}{V_i} = \frac{g_m R_k}{1 + g_m R_k} \tag{7.15}$$

Since $g_m R_k > 1$, the gain is less than but near unity. One side of the output signal is at ground potential and this may be an advantage.

Output Resistance

The output resistance at the 2,2 port is low and approximates

$$R_o \cong \frac{1}{g_m} \quad (\Omega) \tag{7.16}$$

The input resistance is very high.

The cathode follower acts as an impedance transformer, from a high input resistance to a low output resistance, as did the emitter follower and the source follower circuits.

The circuit is usually designed with triodes since with only unity gain the high μ of the pentode provides no advantage and the pentode requires an expensive screen-cathode bypass capacitor. Grid bias is obtained as with the FET.

Example: Find the performance of a cathode follower using a triode with $\mu = 30$ and $r_p = 14,000 \ \Omega$, $R_k = 3000 \ \Omega$.

We have

$$g_m = \frac{\mu}{r_p} = \frac{30}{14,000} = 0.00214 \ \text{mho}$$

With Eq. 7.15 we need

$$g_m R_k = 2.14 \times 10^{-3} \times 3 \times 10^3 = 6.42$$

Therefore

$$A_v = \frac{6.42}{1 + 6.42} = 0.87$$

as the voltage gain. The output resistance is

$$R_o = \frac{1}{g_m} = \frac{1}{0.00214} = 467 \ \Omega$$

7.9 The Grounded-Grid Amplifier

The circuit in Fig. 7.9 is the *grounded-grid amplifier*, similar to the C-B transistor amplifier. Grounding of the grid provides an electrostatic shield and reduces the transfer of energy between output and input circuits. The frequency range of the triode is extended to many megahertz by this circuit but the input resistance is low.

Voltage Gain

The voltage gain is

$$A_v = \frac{V_o}{V_i} \cong \frac{g_m R}{1 + g_m R_s + \dfrac{R}{r_p}} \tag{7.17}$$

and this is relatively low. We have assumed $1 + \mu \cong \mu$.

Output Resistance

The internal resistance factors of Eq. 7.17 lead to

$$R_o = r_p(1 + g_m R_s) \tag{7.18}$$

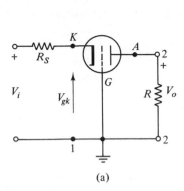
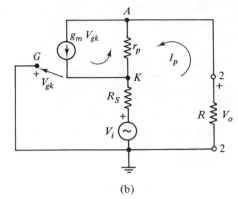

(a)   (b)

Figure 7.9   The grounded-grid amplifier.

Input Resistance

The input resistance is conveniently written as

$$R_i \cong \frac{1}{g_m}\left(1 + \frac{R}{r_p}\right) \tag{7.19}$$

and this is a small resistance.

The circuit provides an impedance transformer of low gain, operating from a low input resistance to a high output resistance. Again, a pentode offers no gain advantage; when used the grids are connected together giving the characteristics of a high-μ triode.

Example: With $g_m = 2500$ μmhos, $r_p = 18,000$ Ω, $R_s = 500$ Ω and $R = 4000$ Ω, find the gain and resistances when this triode is used in a G-G circuit.

By Eq. 7.17,

$$A_v = \frac{2.5 \times 10^{-3} \times 4 \times 10^3}{1 + 2.5 \times 10^{-3} \times 500 + (4 \times 10^3)/(18 \times 10^3)}$$
$$= \frac{10}{2.47} = 4.05$$

By Eq. 7.18,

$$R_o = 18 \times 10^3(1 + 2.5 \times 10^{-3} \times 500)$$
$$= 40,500 \ \Omega$$

By Eq. 7.19,

$$R_i = \frac{1}{2.5 \times 10^{-3}}\left(1 + \frac{4000}{18,000}\right)$$
$$= 489 \ \Omega$$

These values confirm the statements made concerning the relative magnitudes of the input and output resistances.

7.10 The Cathode-Ray Tube

The *cathode-ray tube* is a vacuum-tube device used for television viewing and as a laboratory instrument for visualization of electrical voltages and currents in circuits. It is built in an evacuated glass envelope and employs electron-beam deflection by electric or magnetic fields that vary with the signals applied. The tube includes an electron-emitting cathode and beam-focusing electrodes in an assembly called an *electron gun*, followed by two pairs of mutually perpendicular deflecting plates and a fluorescent viewing screen. When magnetic field deflection of the beam is used, one set of deflecting plates is replaced with a pair of coils, producing a magnetic field. A visible spot of light is produced at the point of impact of the beam on the screen; the color of the light is dependent on the fluorescent material with which the screen is coated on its interior face. A cathode-ray tube is diagrammed in Fig. 7.10(a) and (b).

The accelerating potential V_a increases the beam velocity and brightens the spot of light. The velocity of the electrons is very high; upon impact,

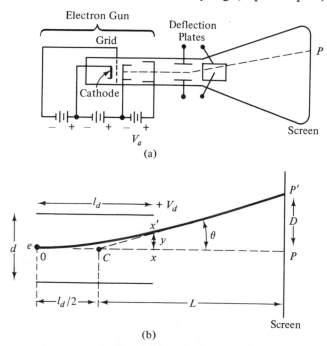

Figure 7.10 (a) A cathode-ray tube; (b) deflection system.

part of the electron energy is converted to visible light and part to heat on the screen. A stationary spot can cause a burn on the screen.

The geometry of the deflecting plate system in Fig. 7.10(b) can be used to determine the deflection D on the screen as

$$D_e = \frac{L l_d V_d}{2_d V_a} \quad (\text{m}) \tag{7.20}$$

With two pairs of plates at right angles, the position of the spot can be moved in both x and y axes on the screen. More usually a *sweep voltage*, which increases linearly with time, is applied to the x deflection plates. A voltage applied to the y deflection plates will then appear as if plotted against time. The complete formation of a sine wave on the screen is shown in Fig. 7.11, for one sweep. By use of a sawtooth form of sweep voltage, as in Fig. 7.12, the figure can be repeated.

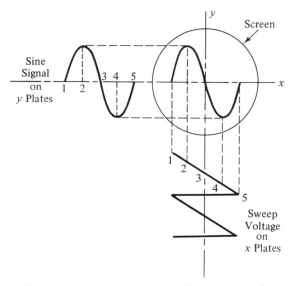

Figure 7.11 Plotting of a sine wave against a linear sweep voltage.

In television, one set of deflection plates is replaced with a pair of coils producing a magnetic flux density B, perpendicular to the electron beam. The deflection is that of a magnetic field on a current and is

$$D_m = \sqrt{\frac{e}{2m}} \frac{L l_m B}{\sqrt{V_a}} \quad (\text{m}) \tag{7.21}$$

For a given deflection we can use higher acceleration voltages and obtain brighter spots with magnetic deflection.

Materials used for the fluorescent screen have varying properties of light persistence after bombardment by the electrons. We have screens with image persistence in microseconds to screens with image persistence measured in minutes. Various colors are also obtainable, with green or blue common in laboratory equipment, and the three basic colors available for color television screens.

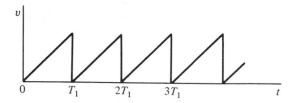

Figure 7.12 A sweep waveform.

7.11 Comments

Because of its broad frequency range and high gain, the pentode was once widely employed as a voltage amplifier. Triode use was confined to audio frequencies or for impedance transformation when used as a cathode follower. The grounded-grid circuit still finds some application in high-power transmitting equipment.

Transistors are now generally used, however, because of the advantages previously listed.

REVIEW QUESTIONS

7.1 Explain what is meant by V_{BB}, V_{CC}, V_b, I_b, I_p, i_b, V_{gk}.

7.2 Explain how the grid controls the passage of electrons through a triode.

7.3 What happens if a vacuum tube is operated with a positive grid?

7.4 What amount of power is dissipated in the triode in Fig. 7.4(b) with a Q point at $V_c = -1.8$ V, $V_b = 150$ V? (Interpolate the curves.)

7.5 If the maximum allowable $P_d = 2.5$ W for a triode, would a Q point at $V_b = 250$ V, $V_c = -1$ V be desirable for the triode in Fig. 7.4(b)?

7.6 What boundaries would you select for the placement of a Q point on the curves in Fig. 7.3(a) for distortionless amplification?

7.7 Define μ, g_m, and r_p. How are they related?

7.8 Triode A has a wide-spaced grid, tube B has a close-spaced grid. Which has the highest μ? Which has the highest r_p? Could you determine g_m?

7.9 What can you say about current amplification in a triode?

7.10 What happens to the anode current if a triode is operated with a grid bias more negative than $-V_{BB}/\mu$?

7.11 Why do we add extra grids to form the pentode?

7.12 Why is a pentode said to be a current source?

7.13 How do we accelerate the electrons through the screen grid?

7.14 Why do we use bypass capacitor C_s with the pentode?

7.15 How large should C_s be? Why is the value of C_s determined at the lowest frequency of interest?

7.16 Why would you select a cathode follower in place of a common-cathode circuit?

7.17 List the similar types of amplifiers, using bipolar transistors, FETs, and vacuum tubes.

7.18 Name two reasons for use of a grounded-grid amplifier.

7.19 What condition must be assumed to justify $-g_m R$ as the gain of a pentode common-cathode amplifier?

7.20 Why does the grounded-grid circuit operate with stable gain at many megahertz?

PROBLEMS

7.1 For the triode in Fig. 7.3, plot curves of anode voltage on the ordinate and grid voltage on the abscissa, for several values of constant anode current. Measure $\Delta v_b/\Delta v_c$; what is this parameter called?

7.2 The triode in Fig. 7.3(b) is biased to a Q point at $I_b = 1.5$ mA, $V_c = -4$ V. What is the peak value of a sine-wave signal that can be used without driving the grid into the positive region? In terms of operating within a small region of the curves, would this sine voltage really be a small signal?

7.3 What is the μ of the triode represented in Fig.7.3(a) in the region near $v_b = 250$ V, $i_b = 1.5$ mA?

7.4 The transconductance of a triode is 4000 μmhos and μ is 8. The grid voltage is kept constant; find the increase in anode current when the anode voltage is changed from 200 to 235 V.

7.5 An equation for the output curves of a triode is

$$i_b = K(8v_c + v_b) \quad \text{(mA)}$$

At $v_b = 200$ V and $v_c = -12$ V, the current is 6 mA.
(a) What is the current at $v_b = 300$ V and $v_c = -8$ V?
(b) What v_c is needed to return i_b to 6 mA at $v_b = 300$ V?

7.6 Taking Δ changes in voltages, determine μ, g_m, and r_p for the triode described by

$$\text{Amp,} \quad i_b = 38 \times 10^{-5}(v_b + 8v_c)$$

near $v_b = 200$ V, $v_c = -12$ V.

7.7 The triode of Problem 7.6 has an anode current which is to increase 0.3 mA when the grid voltage changes from -2 to -3 V; what change in anode voltage must be made simultaneously?

7.8 From the following data taken on a triode, find μ and g_m:

i_b (mA)	v_c	v_b
8.2	0	165
6.0	0	130
6.0	-1	165
4.1	-1	130
4.1	-2	165

7.9 A pentode has $r_p = 650,000\ \Omega$, $g_m = 2500\ \mu$mhos. What is the value of μ? What load R should be used to obtain a voltage gain of 120 in a common-cathode circuit?

7.10 For the pentode in Fig. 7.4(b), determine graphically the value of r_p for $v_c = -1.5$ V and the value of g_m near $v_b = 200$ V, $v_c = -1.5$ V. What is the value of μ?

7.11 For a certain triode with $g_m = 3300\ \mu$mhos, $r_p = 5100\ \Omega$.
 (a) Find the anode current change produced by variation of the grid-cathode voltage from -2 to -6 V, at $v_b = 140$ V.
 (b) What change in anode voltage will bring the anode current back to its original value, with $v_c = -6$ V?

7.12 A common-cathode circuit uses a triode with $r_p = 4000\ \Omega$, $\mu = 8$, and input signal $V_i = 2$ V rms, ac. Find the output rms voltage across a 5000-Ω load.

7.13 Starting with Eq. 7.15, derive a gain expression for the cathode follower in the form

$$A_v = \frac{\mu R_k}{r_p + \mu R_k}$$

7.14 In the cathode follower in Fig. 7.8, $R_k = 1000\ \Omega$, $g_m = 0.003$ mho, and $\mu = 25$. If $V_i = 1$ V rms, find the output rms voltage. Also find the output resistance.

7.15 A pentode has $g_m = 0.0035$ mho, $r_p = 650,000\ \Omega$ and is used in a common-cathode amplifier to provide 35-V output when the input signal $V_i = 0.05$ V rms. What value of R is being used?

7.16 A grounded-grid amplifier uses a triode having $\mu = 70$, $r_p = 40,000\ \Omega$. With $R_s = 300\ \Omega$, $R = 10,000\ \Omega$, find the voltage again and the input and output resistances.

8

Frequency Response
of RC Amplifiers

We have been assuming that the series capacitors in amplifiers are infinite and the shunt capacitances of transistors and tubes are ideally zero. A practical amplifier uses nonideal capacitor values, however, and the effect on the operation is dependent on the reactance of the capacitors that varies with frequency. When we use multistage *cascaded* systems for greater overall gain, the frequency effects are compounded.

The frequency response of amplifiers will be studied here in terms of the very commonly applied resistance-capacitance (*RC*) amplifier.

8.1 Cascaded Amplifiers

Figure 8.1 shows a typical *RC* transistor amplifier of two stages, from A_1 to A_2 and from A_2 to A_3. These stages may be analyzed separately, with the overall gain determined as the product of the individual stage gains.

Output V_{33} is

$$V_{33} = -A_{v_2}V_{22} \tag{8.1}$$

and

$$V_{22} = -A_{v_1}V_{11} \tag{8.2}$$

Then by substitution for V_{22}, we have

$$V_{33} = (-A_{v_2})(-A_{v_1})V_{11}$$

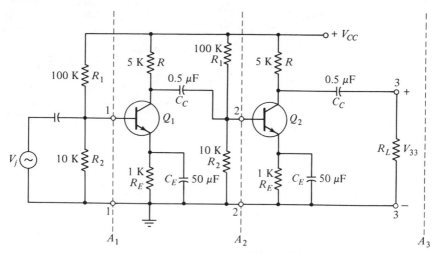

Figure 8.1 Two-stage *RC* amplifier.

and the overall gain between port 1,1 and port 3,3 is

$$A_{v(\text{ov})} = \frac{V_{33}}{V_{11}} = A_{v_1}A_{v_2} \tag{8.3}$$

and in general

$$A_{v(\text{ov})} = A_{v_1}A_{v_2}A_{v_3} \ldots \tag{8.4}$$

We have previously shown that when the individual stage gains are expressed in decibels, then the overall gain is obtained as the sum of the individual decibel gains:

$$\text{dB, } A_{v(\text{ov})} = A_{v_{1dB}} + A_{v_{2dB}} + A_{v_{3dB}} + \cdots \tag{8.5}$$

8.2 The Amplifier Passband

We usually wish the output waveform to be the same as the waveform at the input to the amplifier and that the distortion be negligible. However, our amplifier input signal may consist of many frequencies distributed over a wide frequency band. For instance, a 1-μs pulse signal contains frequency components of which those to 4 MHz are important if we are to obtain an accurate reproduction of the input waveform. To have an undistorted waveform, we must amplify all frequencies equally. If this is not done, we have *frequency distortion* in the amplifier. Amplification of the pulse to 1 MHz will produce a distorted, but recognizable, pulse form.

If we plot the performance of an amplifier as a gain versus frequency curve, ideally an amplifier should have a curve that appears as a horizontal

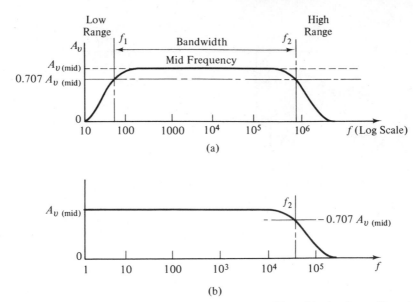

Figure 8.2 (a) Gain-frequency plot for an *RC* amplifier; (b) plot for a direct-coupled amplifier.

straight line over the desired frequency range. More practically, we obtain a curve like that shown in Fig. 8.2(a).

We have been assuming that blocking capacitor C_C will have sufficient capacitance to be a negligible series reactance in the equivalent circuit. Such capacitors must have reasonable cost and size, however, and there will always be some range of frequency from zero upward in which the blocking capacitor is too small to meet the ideal requirement, and represents an appreciable reactance. This is the reason for the fall in gain in the low-frequency range in Fig. 8.2.

Also, we have been assuming the maximum frequency of the signal to be low enough that the very small internal capacitance of a transistor or tube and the stray mounting and wiring capacitance represent such high reactances as to be considered an open circuit to ground across the input or load resistances. There is, however, always some frequency above which these react-ances have an appreciable effect in shunting the load. This accounts for the fall in gain in the high-frequency range in Fig. 8.2.

Only in the mid-frequency range are we able to satisfy the assumptions on both series and shunt reactances and consider them negligible in effect. In this mid-range of frequency we obtain the predicted gain figures of the preceding chapters.

The low-frequency range and the high-frequency range are almost always well separated in frequency and quite distinct, as in Fig. 8.2. We are able to analyze the amplifier response in each region independently and to determine and evaluate the circuit factors responsible for the fall in gain.

We arbitrarily bound the mid-frequency region of uniform gain by choosing limit frequencies, f_1 and f_2, at which the gain has fallen to 0.707 or $1/\sqrt{2}$ of its value in the mid-frequency region. These frequencies are called the *cutoff, band-limit,* or *half-power frequencies.* At f_1 and f_2 the output power is one-half of the mid-range output. That is,

$$V_o = A_{v(\text{mid})}V_i$$

and

$$P_{\text{mid}} = \frac{V_o^2}{R} = \frac{(A_{v(\text{mid})}V_i)^2}{R} \tag{8.6}$$

At f_1 or f_2,

$$V_o' = 0.707A_{v(\text{mid})}$$

and the power is

$$P_{1,2} = \frac{(V_o')^2}{R} = \frac{[0.707A_{v(\text{mid})}V_i]^2}{R} = \frac{0.5[A_{v(\text{mid})}V_i]^2}{R} \tag{8.7}$$

We see that

$$P_{1,2} = 0.5P_{\text{mid}} \tag{8.8}$$

as predicted for the band-limit frequencies.

The amplifier *passband* is defined as the mid-frequency region, with a bandwidth given by

$$\text{bandwidth (BW)} = f_2 - f_1 \quad \text{(Hz)} \tag{8.9}$$

A direct-coupled (*DC*) amplifier omits C_C and its low-frequency response extends to zero frequency. It still has a high-frequency region due to the inherent capacitances of the transistor or tube. The gain-frequency curve appears in Fig. 8.2(b).

In deciding to neglect one resistance or reactance as large or small with respect to another, sufficient accuracy is usually obtained if there is a ratio of 10:1 in the respective impedance magnitudes. For example, a resistor of 1000 Ω is in parallel with another of 10,000 Ω. The combined resistance is

$$R_p = \frac{10^3 \times 10^4}{L \times 10^3 + 10 \times 10^3} = \frac{10^4}{11 \times 10^3} = 910 \ \Omega$$

If we drop the 10,000-Ω resistor from consideration as large with respect to the 1000-Ω resistor, we are saying that we have a resistance of 1000 Ω instead of the actual 910 Ω. This error is less than 10 per cent and can usually be overlooked because of the larger circuit variations introduced by the parameters of the transistors and tubes.

8.3 The Frequency Plot

The frequency axis of a gain-frequency curve is usually plotted on a logarithmic scale. In this way each multiple of 10:1 in frequency is given equal distance on the abscissa and a very large frequency range can be covered,

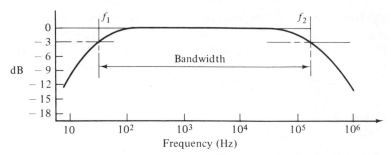

Figure 8.3 A gain plot in decibels.

as shown in Fig. 8.3. Each multiple of 10:1 in frequency is called a *frequency decade*.

The mid-frequency gain does not vary with frequency and can be used as a standard. By comparing the low- and high-frequency gains to the mid-range gain, we can obtain a general gain-frequency curve, said to be *normalized* on the mid-frequency gain. To normalize a gain figure, we divide by the mid-frequency gain, to give terms such as

$$\frac{A_{v(lo)}}{A_{v(mid)}} \quad ; \quad \frac{A_{v(hi)}}{A_{v(mid)}}$$

The value of these ratios in the mid-range is unity for all amplifiers and so the gains are normalized to 1.

The normalized gain figures are ratios and can be converted to decibels, as

$$\frac{A_{v(lo)}}{A_{v(mid)}}, \text{ dB} = 20 \log \left| \frac{A_{v(lo)}}{A_{v(mid)}} \right| \tag{8.10}$$

In the mid-frequency range the ratio is 1 and the logarithm of 1 is 0 so that when stated in decibels the *normalized mid-frequency gain* is 0 dB. At the limit frequencies we have $20 \log (0.707) = -20 \log (1.414) = -3$ dB and so the limit frequencies are correctly called the -3-dB frequencies. A normalized gain-frequency plot in decibels is drawn in Fig. 8.3.

The phase angle of $A_{v(mid)}$ for a single C-E stage is 180°. Normalizing of $A_{v(lo)}$ and $A_{v(hi)}$ against $A_{v(mid)}$ will result in their phase angles being normalized against 180° as a reference angle. The total phase shift will then be $180° + \theta_{lo}$ and $180° - \theta_{hi}$. At low frequencies the total phase shift tends toward $180° + 90° = 270°$ at zero frequency. In the high-frequency range, the total phase shift tends toward $180° - 90° = 90°$ at a very high frequency.

8.4 Low-Frequency Response

The most commonly used amplifier is the *RC*-coupled, common-emitter circuit of which Fig. 8.1 is an example. The purpose of each of the circuit elements should be understood since the four-resistor bias network is employed for each transistor. As drawn, the circuit is too complicated to be readily analyzed and we break it down by stages for better understanding. For this purpose we consider stage 1 as between A_1 and A_2 and redraw that part of the circuit in Fig. 8.4(a).

We assume that C_E is large and represents a negligibly small reactance so that we can drop R_E and C_E from the equivalent circuit. The effect of an unbypassed R_E will be considered in Sec. 8.7. We replace the transistor Q_1 with its equivalent circuit, the input resistance h_{ie}, and the current generator $g_m V_{be}$.

We previously showed that the effect of R_1 and R_2 could be replaced by a single resistance R_B, where

$$R_B = \frac{R_1 R_2}{R_1 + R_2}$$

and R_B appears in the circuit in Fig. 8.4(b). By proper bias circuit design, however, we can choose R_1 and R_2 so that $R_B \gg R_{i_2}$ where R_{i_2} is the input resistance of the transistor, usually h_{ie}. The equivalent circuit in Fig. 8.4(c) follows after dropping R_B.

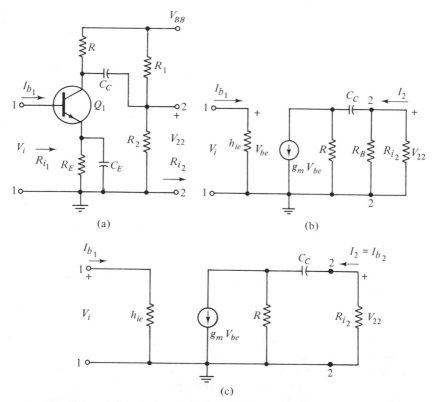

Figure 8.4 (a) One stage of the *RC* amplifier; (b) partially simplified; (c) equivalent circuit at low frequency.

We first write the mid-range gain by assuming the reactance of C_C negligible compared to R_{i_2}; this is really the definition of a mid-range frequency, with capacitor C_C represented as a short circuit. Then

$$V_{22} = -g_m V_{be} \frac{R R_{i_2}}{R + R_{i_2}} \qquad (8.11)$$

Then the *mid-range gain* for the first stage is

$$A_{v(\text{mid})} = \frac{V_{22}}{V_{be}} = -g_m \frac{R R_{i_2}}{R + R_{i_2}} \qquad (8.12)$$

We now reinsert C_C and derive the *low-frequency gain* for the circuit in Fig. 8.4(c). The output V_{22} is

$$V_{22} = -I_2 R_{i_2}$$

and we need to find I_2 as a portion of the transistor current $g_m V_{be}$.

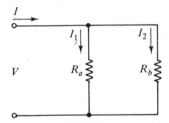

Figure 8.5 The current divider example.

This question can be resolved by use of the *current-division factor*. In Fig. 8.5 the voltage across the parallel branches is equal and so we can write the equations

$$V = R_a I_1 = \frac{R_a R_b}{R_a + R_b} I \qquad (8.13)$$

$$V = R_b I_2 = \frac{R_a R_b}{R_a + R_b} I \qquad (8.14)$$

In Eq. 8.13 we cancel R_a on each side and the branch current is

$$I_1 = \frac{R_b}{R_a + R_b} I \qquad (8.15)$$

Likewise, with Eq. 8.14 we obtain

$$I_2 = \frac{R_a}{R_a + R_b} I \qquad (8.16)$$

The current I_1 divides out of I in proportion to R_b, the resistance of the *other* path divided by the sum of the resistances $R_a + R_b$.

Similarly, I_2 divides out of I in proportion to R_a, the resistance of the *other* path, divided by the sum of the two resistance paths $R_a + R_b$.

Equations 8.15 and 8.16 demonstrate the use of current-division factors.

We now write the reactance of C_C as

$$X_C = \frac{1}{\omega C_C} = \frac{1}{2\pi f C_C} \qquad (8.17)$$

using $\omega = 2\pi f$. The impedance of the I_2 path is then

$$R_{i_2} + X_C \angle -90° = R_{i_2} - \frac{j}{\omega C_C} \qquad (8.18)$$

Accordingly, with R as the resistance of the other path and $R + R_{i_2} - (j/\omega C_C)$ as the impedance of both paths, we have current I_2 as a fraction of $g_m V_{be}$:

$$I_2 = g_m V_{be} \frac{R}{R + R_{i_2} - (j/\omega C_C)}$$

Dividing out $R + R_{i_2}$, we have

$$I_2 = g_m V_{be} \frac{R}{R + R_{i_2}} \left[\frac{1}{1 - j\dfrac{1}{\omega C_C(R + R_{i_2})}} \right] \qquad (8.19)$$

As a magnitude, this is

$$I_2 = g_m V_{be} \frac{R}{R + R_{i_2}} \frac{1}{\sqrt{1 + \left[\dfrac{1}{\omega C_C(R + R_{i_2})} \right]^2}} \qquad (8.20)$$

The output voltage is

$$V_{22} = -I_2 R_{i_2} = -g_m V_{be} \frac{R}{R + R_{i_2}} \frac{1}{\sqrt{1 + \left[\dfrac{1}{\omega C_C(R + R_{i_2})} \right]^2}}$$

Dividing out V_{be}, we find the low-frequency gain as

$$A_{v(\text{lo})} = \frac{V_{22}}{V_{be}} = -g_m \frac{R R_{i_2}}{R + R_{i_2}} \frac{1}{\sqrt{1 + \left[\dfrac{1}{\omega C_C(R + R_{i_2})} \right]^2}} \qquad (8.21)$$

Comparison of the multiplier term with Eq. 8.12 shows that the multiplier is $A_{v(\text{mid})}$. Then the gain ratio is

$$\frac{A_{v(\text{lo})}}{A_{v(\text{mid})}} = \frac{1}{\sqrt{1 + \left[\dfrac{1}{\omega C_C(R + R_{i_2})} \right]^2}} \qquad (8.22)$$

There is an associated phase angle:

$$\theta = \tan^{-1} \frac{1}{\omega C_C(R + R_{i_2})} \qquad (8.23)$$

in addition to the 180° phase shift in $A_{v(\text{mid})}$.

The effect of frequency on the gain is shown by the radical in the denominator of Eq. 8.22. If $f = 0$, $\omega = 2\pi f = 0$, and the second term in the radical becomes infinite, the gain is zero, with a phase angle of 180° + 90°

$= 270°$. As frequency increases, the second term in the radical decreases with respect to 1 and the gain ratio increases toward the mid-frequency value of unity.

8.5 The Low-Frequency Limit

Let us examine the situation when the terms in the radical of Eq. 8.22 become equal; that is, at f_1

$$1 = \frac{1}{2\pi f_1 C_C(R + R_{i_s})} \tag{8.24}$$

Then we have

$$\frac{A_{v(\text{lo})}}{A_{v(\text{mid})}} = \frac{1}{\sqrt{1+1}} = \frac{1}{\sqrt{2}} = 0.707 \tag{8.25}$$

We see that the frequency defined in Eq. 8.24 is the lower limit frequency of the amplifier. Frequency f_1 is associated with the circuit elements chosen by the designer as

$$f_1 = \frac{1}{2\pi C_C(R + R_{i_s})} \tag{8.26}$$

The lower limit frequency is dependent on the size of the blocking capacitor and the resistances through which it charges. These are factors that the designer can select to place the cutoff of the mid-frequency region where cost of parts and size considerations permit.

We can substitute this value of $\omega_1 = 2\pi f_1$ into Eq. 8.22 and 8.23 to obtain a general expression

$$\frac{A_{v(\text{lo})}}{A_{v(\text{mid})}} = \frac{1}{\sqrt{1 + (f_1/f)^2}} \tag{8.27}$$

with a phase angle

$$\theta = \tan^{-1}\frac{f_1}{f} \tag{8.28}$$

These expressions are important results. Any amplifiers having the same *product* of blocking capacitance and charging resistances will have the same frequency response. That is, a small capacitance associated with large resistances or a large capacitance with small resistance values can lead to the same limit frequency.

A general curve for the low-frequency response of all *RC* amplifiers is plotted in Fig. 8.6, from the decibel values of Table 8.1, which is calculated from Eq. 8.27 and 8.28. To use the curve, one need merely know the f_1 frequency for a particular amplifier and the response at other frequencies is readily found. In order to show the gain increasing with frequency as a normal low-frequency response, the curve is plotted in terms of f/f_1. Figure 8.7 is the general phase response.

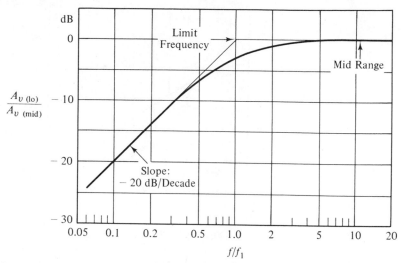

Figure 8.6 Low-frequency response in decibels.

TABLE 8.1 Low-Frequency Response from Eqs. 8.27 and 8.28

f_1/f	f/f_1	dB	Phase Angle (degrees)	Total Phase[a] Angle (degrees)
10	0.1	−20	+84°	264°
5	0.2	−14	79°	259°
2	0.5	−7	63°	244°
1	1.0	−3	45°	225°
0.5	2	−1	27°	207°
0.2	5	−0.2	11°	191°
0.1	10	−0.04	6°	186°

[a]Including the inherent 180° phase shift of a C-E amplifier at mid-frequency.

The table and curve show that for a 1 : 10 change in frequency (1 decade), at low frequencies, the gain changes by 20 dB. It is hardly necessary to plot the curve below $0.1 f/f_1$ because we know that the gain varies at 20 dB per decade and

f/f_1	dB
0.1	−20
0.01	−40
0.001	−60

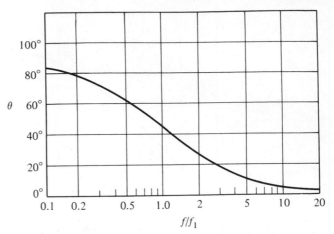

Figure 8.7 Phase angle variation at low frequencies.

The expression of Eq. 8.27 can be plotted for any amplifier by drawing a horizontal asymptote at 0 dB for frequencies above $f = f_1$ and a sloping asymptote falling 20 dB per decade below $f = f_1$. At $f = f_1$, the limit frequency gain is plotted at -3 dB. The response can then be sketched through the -3-dB point to the asymptotes.

Example: The first-stage circuit in Fig. 8.1, between A_1 and A_2, has $C_C = 0.5\ \mu\text{F}$, $R = 5\ \text{k}\Omega$, and a transistor with $h_{ie} = 900\ \Omega$. Since

$$R_B = \frac{R_1 R_2}{R_1 + R_2} = \frac{100 \times 10^3 \times 10 \times 10^3}{110 \times 10^3} = 9100\ \Omega$$

this resistance can be neglected with respect to h_{ie} at $900\ \Omega$, and $R_{i_2} = h_{ie}$

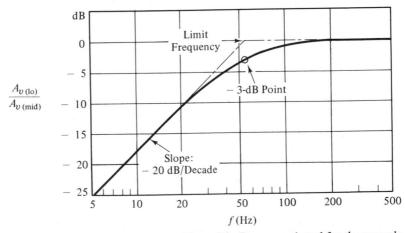

Figure 8.8 Response plotted for the example.

$= 900 \, \Omega$. Then

$$f_1 = \frac{1}{2\pi C_c(R + R_{i_s})} = \frac{1}{6.28 \times 0.5 \times 10^{-6} \times (5000 + 900)}$$

$$= \frac{1}{0.0185} = 54 \text{ Hz}$$

This is the band-limit frequency. The result is plotted in Fig. 8.8, by use of the asymptotes and the -3-dB point at f_1.

8.6 Low-Frequency Response for the FET and Vacuum-Tube Amplifiers

In Fig. 8.9(a) we show an FET amplifier using a four-resistor bias circuit. We assume that C_S is large and adequately bypasses R_S so that the combination can be dropped from the equivalent circuit in Fig. 8.9(b). We again have

$$R_B = \frac{R_1 R_2}{R_1 + R_2}$$

Comparison of the equivalent circuit of the FET amplifier with that for the bipolar transistor in Fig. 8.4(c) shows the circuits identical, except for some of the resistor designations. Therefore the low-frequency analysis of Sec. 8.4 and 8.5 applies to the FET. We need note that the limit frequency for the mid-frequency region of the FET amplifier is written

$$f_1 = \frac{1}{2\pi C_c(R + R_B)} \tag{8.29}$$

As a rule $R + R_B$ is much larger than for the bipolar transistor and the needed value of C_C will be reduced for the same f_1 frequency.

Figure 8.10 shows a pentode amplifier with its equivalent circuit. Once more we compare the equivalent circuit with Fig. 8.4(c) and see that it is identical to that analyzed for the low-frequency response of the bipolar transistor. For the pentode the limit frequency will be written

$$f_1 = \frac{1}{2\pi C_c(R + R_g)} \tag{8.30}$$

and all the universal gain and phase curves apply with f_1 as the limit frequency for the mid-range of the pentode amplifier.

Example: Typical values for a pentode amplifier would be $R = 50{,}000 \, \Omega$, $R_g = 500{,}000 \, \Omega$, and $g_m = 0.0035$ mho. Find the value needed for C_C to make $f_1 = 54$ Hz as in the example of Sec. 8.5 and find the mid-frequency and f_1 gains.

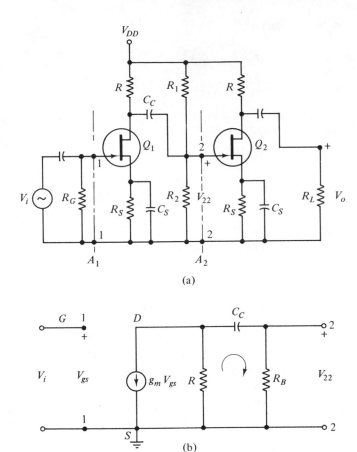

(a)

(b)

Figure 8.9 (a) FET amplifier; (b) equivalent circuit.

We have

$$f_1 = \frac{1}{2\pi C_C(R + R_g)}$$

and $R + R_g = 50{,}000 + 500{,}000 = 0.55 \times 10^6 \ \Omega$. Then with $f_1 = 54$ Hz, we have

$$
\begin{aligned}
C_C &= \frac{1}{2\pi \times 54 \times 0.55 \times 10^6} \\
&= 5.36 \times 10^{-9} = 0.00536 \times 10^{-6} \ \text{F} \\
&= 0.005 \ \mu\text{F}
\end{aligned}
$$

This is much smaller than the value of $C_C = 0.5 \ \mu\text{F}$ needed to give the same limit frequency for the bipolar transistor. The voltage rating required for the $0.005 \ \mu\text{F}$ capacitor in the pentode circuit may be 300 V and that for the bipolar transistor may be 15 V, however, so that costs and size may not differ appreciably.

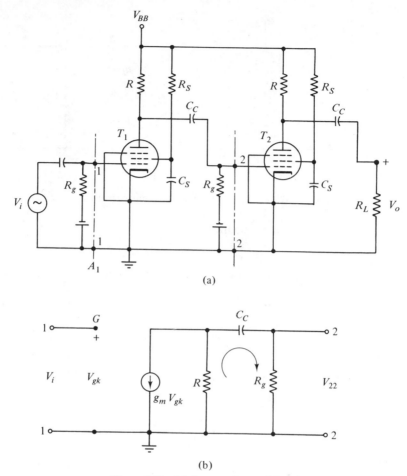

Figure 8.10 (a) A pentode amplifier; (b) equivalent circuit.

The gain at mid-frequencies is

$$A_{v(\text{mid})} = \frac{g_m R R_g}{R + R_g}$$
$$= \frac{0.0035 \times 5 \times 10^4 \times 0.5 \times 10^6}{0.55 \times 10^6}$$
$$= 159$$

The gain at $f_1 = 0.707 \times A_{v(\text{mid})} = 0.707 \times 159 = 112$.

8.7 The Unbypassed Emitter Resistor

While we have assumed the emitter bypass capacitor C_E to be of sufficient capacity to bypass R_E adequately, the capacitor is not often used, for reasons of space and cost. Then we have the C-E amplifier input circuit as in Fig.

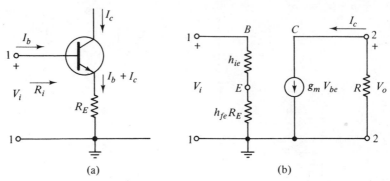

Figure 8.11 (a) Unbypassed emitter resistor; (b) equivalent circuit.

8.11(a). The input circuit equation is

$$V_i = h_{ie}I_b + (I_b + I_c)R_E \qquad\qquad (8.31)$$

Since $I_c = h_{fe}I_b$, we have the input resistance R_i as

$$R_i = \frac{V_i}{I_b} = h_{ie} + (1 + h_{fe})R_E \qquad\qquad (8.32)$$
$$\cong h_{ie} + h_{fe}R_E$$

The input circuit is drawn in Fig. 8.11(b), as part of the usual transistor equivalent circuit, and consists of h_{ie} in series with a large resistor $h_{fe}R_E$. Actually, this is the input circuit of a C-C amplifier.

We have

$$I_b = \frac{V_i}{h_{ie} + h_{fe}R_E}$$

and

$$V_{be} = h_{ie}I_b = \frac{h_{ie}V_i}{h_{ie} + h_{fe}R_E}$$
$$= \frac{V_i}{1 + \dfrac{h_{fe}}{h_{ie}}R_E} = \frac{V_i}{1 + g_m R_E} \qquad\qquad (8.33)$$

The load voltage is $V_o = -g_m R V_{be}$ and substitution of Eq. 8.33 for V_{be} gives

$$V_o = -g_m R \frac{1}{1 + g_m R_E} V_i$$

But the term $-g_m R$ is recognizable as $A_{v(\text{mid})}$ and so with R_E in the circuit we have a gain designated as A'_v:

$$A'_v = \frac{V_o}{V_i} = A_{v(\text{mid})} \frac{1}{1 + g_m R_E} \qquad\qquad (8.34)$$

The gain is reduced by the presence of R_E.

Similar expressions are obtained for the gain with an unbypassed source resistor R_S for an FET or for an unbypassed cathode resistor R_K for a tube, used in the circuits in Fig. 8.12.

When these bias resistors are not bypassed, the gain expressions of the preceding sections are multiplied by the factor $1/(1 + g_m R_E)$, or equivalent, derived above. Using appropriate g_m and resistance values, this factor is the same for all three active devices.

The gain reduction is often made up by additional gain elsewhere in transistor circuits since the cost and space requirements for the bypass capacitor are excessive. With the relatively expensive and bulky vacuum tube, the extra gain was more costly and the bypass capacitor was commonly employed.

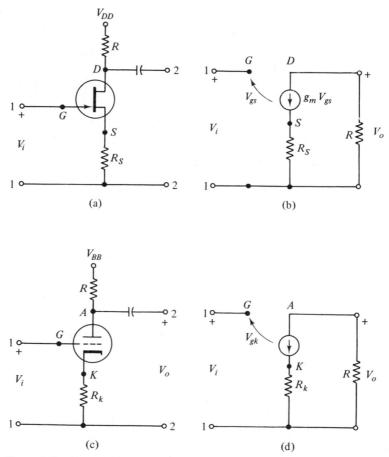

Figure 8.12 (a) and (b) FET with unbypassed source resistance; (c) and (d) triode with unbypassed cathode resistor.

8.8 High-Frequency Equivalent Circuits; the Miller Effect

The internal shunt capacitances of transistor and tube become small reactances that cannot be neglected at some high frequency. A modified equivalent circuit for the three devices is the result.

For the junction transistor, the effect of the capacitances is illustrated by use of the *hybrid-π equivalent circuit* shown in Fig. 8.13. Capacitance C_{be} holds the charge stored in the base and is of many picofarads. Capacitance C_{bc} is that of the depletion region at the reverse-biased base-collector junction and is only a few picofarads. Resistance $r_{oe} = 1/h_{oe}$ is large and eliminated in the equivalent circuit as we have done before.

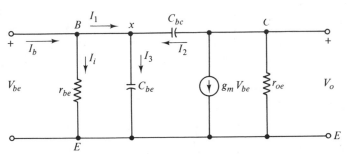

Figure 8.13 Hybrid-π transistor circuit.

Figure 8.13 shows the capacities in the input circuit. The currents add at X as

$$I_1 + I_2 = I_3 \tag{8.35}$$

where

$$I_3 = \frac{V_{be}}{1/\omega C_{be}} = \omega C_{be} V_{be}$$

$$I_2 = \frac{V_o - V_{be}}{1/\omega C_{bc}} = \omega C_{bc}(V_o - V_{be})$$

But $V_o = A_v V_{be}$ and we substitute, giving

$$I_2 = \omega C_{bc}(A_v V_{be} - V_{be}) = \omega C_{bc} V_{be}(A_v - 1) \tag{8.36}$$

Substitution of I_2 and I_3 into Eq. 8.35 yields

$$I_1 = I_3 - I_2 = \omega C_{be} V_{be} - \omega C_{bc} V_{be}(A_v - 1)$$

Since $A_v = -g_m R$ for the C-E circuit, the terms can be combined to give

$$I_1 = \omega V_{be}[C_{be} + (1 + g_m R)C_{bc}] \tag{8.37}$$

Current I_1 passes into an apparent input capacitance

$$C_{ie} = C_{be} + (1 + g_m R)C_{bc} \tag{8.38}$$

This capacitance is the result of a multiplied value of C_{bc}, moved into the input circuit in parallel with C_{be}, between B and E. The effective value of C_{ie} is much larger than either C_{bc} or C_{be} and this multiplying of capacitance is called the *Miller effect*.

We then have the two branches of the input circuit in Fig. 8.14, which is the *high-frequency* g_m *model* of the transistor. For a given I_b the voltage V_{be} will fall as the reactance of C_{ie} falls with frequency. Therefore, the input capacitance C_{ie} is responsible for the reduction of gain of the transistor amplifier at high frequencies.

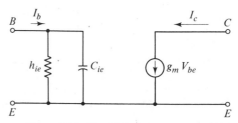

Figure 8.14 The high-frequency g_m model.

The FET has a capacitance C_{gs} from the gate to the source through the insulating layer of a MOSFET or the depletion region of a JFET. The FET also has a capacitance C_{gd} between gate and drain which includes the capacitance to the mounting of the transistor. These capacitances are usually in the range of 1 to 5 pF. The input circuit of the FET is drawn in Fig. 8.15(a) and is seen to be identical to the branched capacitances of the junction transistor in Fig. 8.13. The FET, of course, has no equivalent of h_{ie}.

We could perform an analysis similar to that of Eq. 8.35 to 8.38 and would find that the capacitances of the FET can be represented by a single shunt capacitance

$$C_{if} = C_{gs} + (1 + g_m R)C_{gd} \qquad (8.39)$$

much larger than C_{gd} alone. The Miller effect also appears with the FET and the *high-frequency model* for the FET is that shown in Fig. 8.15(b),

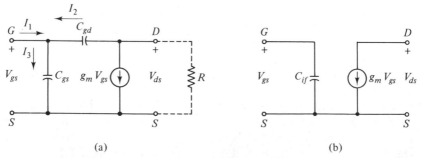

(a)   (b)

Figure 8.15 (a) Capacitances for the FET; (b) high-frequency FET model.

with C_{if} in shunt to the input terminals. At high frequencies the input impedance of the FET is not infinite but falls as frequency increases due to C_{if}. With a driving voltage source having an internal resistance, the current taken by C_{if} reduces the value of V_{gs} with increasing frequency and the gain falls.

We might carry through the same reasoning process with the triode or pentode in Fig. 8.16. We find a Miller-effect capacitance across the grid-cathode terminals as

$$C_{it} = C_{gk} + (1 + g_m R)C_{gp} \qquad\qquad (8.40)$$

We have a high-frequency model for the vacuum tube as in Fig. 8.16(c). The capacitance C_{it} is responsible for the fall of gain with increasing frequency as for the FET.

Again, our three active devices operate in an equivalent manner.

Example: A junction transistor has $C_{bc} = 4\,\text{pF}$, $C_{be} = 60\,\text{pF}$, and $g_m = 4\,\text{mA/V} = 0.004\,\text{mho}$ and the amplifier load is $10,000\,\Omega$. Find the input capacitance C_{ie}.

First

$$g_m R = 0.004 \times 10^4 = 40$$

Using the relation of Eq. 8.38,

$$C_{ie} = C_{be} + (1 + g_m R)C_{bc}$$
$$C_{ie} = 60 + (1 + 40) \times 4 = 220\,\text{pF}$$

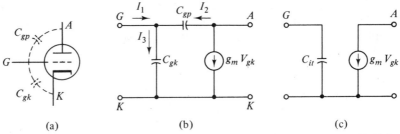

Figure 8.16 (a) Capacities in the triode; (b) circuit of capacitances; (c) high-frequency vacuum-tube model.

8.9 High-Frequency Response

In the high-frequency range of an amplifier the series blocking capacitors represent zero reactance; this is also true for C_E bypasses for the emitter resistors. But the internal C_i capacitances of the active devices appear in shunt and we draw the first stage, A_1 to A_2, from Fig. 8.1 in the high-

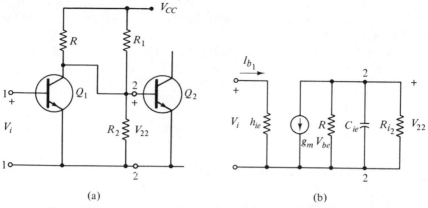

Figure 8.17 (a) Stage 1 of the amplifier of Fig. 8.1; (b) high-frequency equivalent circuit.

frequency g_m model in Fig. 8.17. Resistance R_{i_2} is the input resistance of Q_2, normally expected to be h_{ie}, and C_{ie} is the Miller-effect capacitance of Q_2.

The gain of the C-E amplifier is

$$A_v = -g_m Z_p$$

where Z_p is the parallel impedance of R, and R_{i_2} and the reactance of C_{ie}. This parallel impedance has a magnitude

$$|Z_p| = \frac{R R_{i_2}}{\sqrt{(R + R_{i_2})^2 + (\omega C_{ie} R R_{i_2})^2}}$$

$$= \frac{R R_{i_2}}{R + R_{i_2}} \frac{1}{\sqrt{1 + \left(\dfrac{\omega C_{ie} R R_{i_2}}{R + R_{i_2}}\right)^2}} \tag{8.41}$$

The voltage gain in the high-frequency range of the amplifier is

$$A_{v(\text{hi})} = -\frac{g_m R R_{i_2}}{R + R_{i_2}} \frac{1}{\sqrt{1 + \left(\dfrac{\omega C_{ie} R R_{i_2}}{R + R_{i_2}}\right)^2}} \tag{8.42}$$

which has a phase angle

$$\theta = \tan^{-1}\left(-\frac{\omega C_{ie} R R_{i_2}}{R + R_{i_2}}\right) \tag{8.43}$$

The first term of Eq. 8.42 is the mid-frequency gain of the amplifier, $A_{v(\text{mid})}$, however, and so

$$A_{v(\text{hi})} = A_{v(\text{mid})} \frac{1}{\sqrt{1 + \left(\dfrac{\omega C_{ie} R R_{i_2}}{R + R_{i_2}}\right)^2}} \tag{8.44}$$

The gain ratio follows as

$$\frac{A_{v(\text{hi})}}{A_{v(\text{mid})}} = \frac{1}{\sqrt{1 + \left(\dfrac{\omega C_{ie} R R_{i_2}}{R + R_{i_2}}\right)^2}} \tag{8.45}$$

The term in the parentheses in the radical increases with frequency f and the high-frequency gain falls.

When the terms in the radical are equal,

$$\frac{2\pi f_2 C_{ie} RR_{i_2}}{R + R_{i_2}} = 1$$

we find the *upper limit frequency* f_2 as

$$f_2 = \frac{1}{\dfrac{2\pi C_{ie} RR_{i_2}}{R + R_{i_2}}} \quad \text{(Hz)} \tag{8.46}$$

The upper band-limit frequency is dependent on the *product* of C_{ie} and the parallel value of the associated resistances. The frequency response of the amplifier is established as soon as the values of C_{ie}, R, and R_{i_2} are chosen by the circuit designer.

Using f_2 from Eq. 8.46, we can write the gain ratio in the more general form

$$\frac{A_{v(\text{hi})}}{A_{v(\text{mid})}} = \frac{1}{\sqrt{1 + (f/f_2)^2}} \tag{8.47}$$

with a phase angle

$$\theta = \tan^{-1}\left(-\frac{f}{f_2}\right) \tag{8.48}$$

The frequency response in Fig. 8.18 is the general response curve for Eq. 8.47 in decibels; the curve in Fig. 8.19 shows the phase response from Eq. 8.48. Knowing the component values that produce a given f_2 frequency, the response curve can be plotted by use of a few points selected from the curve or values from Table 8.2.

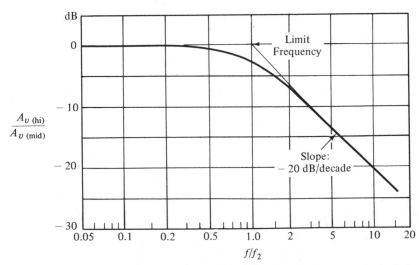

Figure 8.18 High-frequency response in decibels.

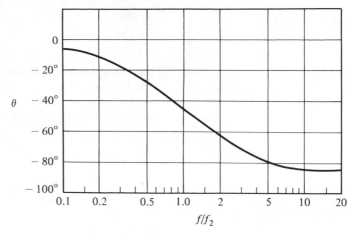

Figure 8.19 Phase angle variation at high frequencies.

TABLE 8.2 High-Frequency Responses from Eqs. 8.47 and 8.48

f/f_2	Decibels (gain)	Phase Angle (degrees)	Total Phase Angle* (degrees)
0.1	−0.04	−6°	174°
0.2	−0.2	−11°	169°
0.5	−1	−27°	153°
1	−3	−45°	135°
2	−7	−63°	117°
5	−14	−79°	101°
10	−20	−84°	96°

*Including the inherent 180° phase shift of a C-E amplifier at mid-frequency.

The behavior of the amplifier at high frequencies is the inverse of the behavior at low frequencies when frequency ratios are used; that is, at $0.5f_1$ the gain is the same as the gain at $2f_2$, using the mid-frequency gain as a reference.

At higher frequency ratios the gain continues to fall at the rate of −20 dB per frequency decade.

8.10 The Frequency Limit of the Transistor

Common-emitter current gain h_{fe} falls at high frequencies, primarily because of the internal capacitances represented by C_{ie}. If we use h_{feo} as the usual gain figure at low or mid-frequencies, then h_{fe} at any frequency is given by the relation

$$h_{fe} = \frac{h_{feo}}{\sqrt{1 + (f/f_\beta)^2}} \qquad (8.49)$$

When compared with Eq. 8.47, the form of this equation indicates that $f = f_\beta$ is a limit frequency or -3-dB frequency for the transistor, at which $h_{fe} = h_{feo}/\sqrt{2}$.

By definition, h_{fe} is the current gain for a short-circuit load, as in Fig. 8.20. With $R = 0$, the Miller-effect equation is

$$\begin{aligned} C_{ie} &= C_{be} + (1 + g_m \times 0)C_{bc} \\ &= C_{be} + C_{bc} \end{aligned} \qquad (8.50)$$

This equation also comes from the circuit, Fig. 8.20, since C_{be} and C_{bc} are placed in parallel by the short-circuit load.

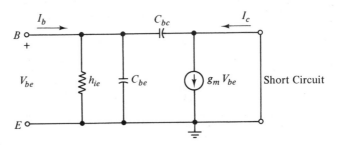

Figure 8.20 The transistor at high frequencies, under short-circuit load.

With short-circuit load there are no circuit elements external to the transistor to downgrade the performance. Therefore, with Eq. 8.50 as the least possible input capacitance, the *short-circuit limit frequency* for the transistor is defined as

$$f_\beta = \frac{1}{2\pi(C_{be} + C_{bc})h_{ie}} \qquad (8.51)$$

A transistor is operable at frequencies above f_β but at values of h_{fe} below h_{feo}, as shown in the plot of h_{fe} against f/f_β in Fig. 8.21. The value of f_β is used as a frequency *figure of merit* to compare transistors.

With frequency f well above f_β, the value of $f/f_\beta \gg 1$ and the denominator of Eq. 8.49 simplifies to give

$$h_{fe} = \frac{h_{feo}}{f/f_\beta} \qquad (8.52)$$

The h_{fe} curve at large f/f_β values falls at -20 dB per frequency decade and if the curve is extended to a frequency $f = f_T$, where h_{fe} has fallen to unity or 0 dB, we have another frequency limit of the transistor, f_T. Then from Eq. 8.52

$$1 = \frac{h_{feo}}{f_T/f_\beta}$$

$$f_T = f_\beta h_{feo} \qquad (8.53)$$

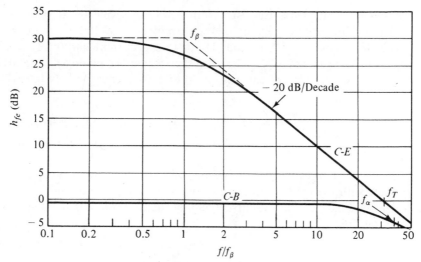

Figure 8.21 High-frequency performance of a transistor.

Substituting f_β from Eq. 8.51 gives

$$f_T = \frac{h_{feo}}{2\pi(C_{be} + C_{bc})h_{ie}} = \frac{g_{mo}}{2\pi(C_{be} + C_{bc})} \tag{8.54}$$

since $g_{mo} = h_{feo}/h_{ie}$.

The transistor bandwidth is $f_2 - f_1$ but, for a transistor, f_1 is at zero frequency and $f_2 = f_\beta$ so that

$$\text{BW} = f_\beta \tag{8.55}$$

for the transistor. Equations 8.53 and 8.54 represent a product of gain and bandwidth. This *gain-bandwidth product*, f_T, is dependent only on transistor parameters and is a constant for a given transistor. Values of f_β or f_T are given by the manufacturers and both serve as figures of merit, useful in selection of a transistor for a given frequency range.

When connected in a practical circuit the frequency limit becomes f_2 because of the presence of a load R and the Miller-effect capacitance. The first step in selection of a transistor is to determine that it has a needed value of h_{fe} and the second step is to determine that f_β is well above the expected highest operating frequency. If f_T is given in the transistor specifications, Eq. 8.53 may be used to find f_β.

We have a similar gain-bandwidth figure of merit for the FET:

$$\text{G-BW} = f_T = \frac{g_{mo}}{2\pi(C_{gs} + C_{gd})} \tag{8.56}$$

and for a triode or pentode vacuum tube:

$$\text{G-BW} = \frac{g_{mo}}{2\pi(C_{gk} + C_{gp})} \tag{8.57}$$

Example: Consider a transistor with f_T rating given as 20 MHz. This gain-bandwidth product tells us that a current gain of 100 is theoretically possible with a bandwidth of 200 kHz or that a current gain of 10 is obtainable to a frequency of 2 MHz. Another transistor having $f_T = 5$ MHz would have a possible current gain of 10 to a frequency of only 0.5 MHz.

8.11 The Common-Base Connection at High Frequencies

By use of the figures of merit, we can compare the frequency range of a transistor with the emitter common to one with the base common. At mid-frequency we have α_{mid} as the current gain of the common-base configuration and at any frequency

$$\alpha = \frac{\alpha_{\text{mid}}}{\sqrt{1 + (f/f_\alpha)^2}} \tag{8.58}$$

The frequency f_α is identified as the upper band limit or -3-dB frequency of the transistor alone, in the short-circuit common-base connection in Fig. 8.22.

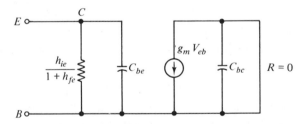

Figure 8.22 Transistor in the C-B circuit with a short-circuit load.

The short circuit is across C_{bc} and eliminates that capacitance from the circuit so that

$$C_{ie} = C_{be} \tag{8.59}$$

Also

$$h_{ib} = \frac{h_{ie}}{1 + h_{fe}} \simeq \frac{h_{ie}}{h_{fe}} \tag{8.60}$$

and the input resistance is much less than that of the transistor with emitter common. The reactance of C_{be} remains large with respect to its shunting resistance to a higher frequency with base common. Using C_{ie} and h_{ib}, we have the limit frequency f_α defined as

$$f_\alpha = \frac{1}{2\pi C_{be}(h_{ie}/h_{fe})} = \frac{h_{fe}}{2\pi C_{be} h_{ie}} = \frac{g_m}{2\pi C_{be}} \tag{8.61}$$

Comparison of Eq. 8.61 and 8.52 shows that because of the reduction in input capacitance

$$f_\alpha > h_{fe}f_\beta \tag{8.62}$$

and by reference to Eq. 8.54

$$f_\alpha > f_T \tag{8.63}$$

Comparison curves for the current gains of a transistor connected with emitter common and with base common are shown in Fig. 8.21. The common-base connection has a greater gain-bandwidth product than does the connection with emitter common.

The parameter f_α is a *figure of merit* for a transistor with base common to input and output.

8.12 Bandwidth of Cascaded Amplifiers

The overall high-frequency gain for a number of stages of amplification may be written in decibels as

$$\left|\frac{A_{v(\text{hi})}}{A_{v(\text{mid})}}\right|, \text{dB} = 20 \log \frac{1}{[1 + (f/f_{2a})^2]^{1/2}} + 20 \log \frac{1}{[1 + (f/f_{2b})^2]^{1/2}}$$
$$+ 20 \log \frac{1}{[1 + (f/f_{2c})^2]^{1/2}} + \cdots$$

where $f_{2a}, f_{2b}, f_{2c}, \ldots$ are the limit frequencies of the respective stages. We invert the ratios, use a negative sign, and transfer the square root operation outside the logarithm:

$$\left|\frac{A_{v(\text{hi})}}{A_{v(\text{mid})}}\right|, \text{dB} = -10 \log\left[1 + \left(\frac{f}{f_{2a}}\right)^2\right] - 10 \log\left[1 + \left(\frac{f}{f_{2b}}\right)^2\right]$$
$$- 10 \log\left[1 + \left(\frac{f}{f_{2c}}\right)^2\right] - \cdots \tag{8.64}$$

A similar expression involving $f_{1a}/f, f_{1b}/f, f_{1c}/f, \ldots$ can be written for the low-frequency response. The highest f_1 and the lowest f_2 frequency would primarily determine the limit frequencies of the complete amplifier. Therefore it does not pay to overdesign the separate stages.

Consequently we often use amplifiers of n identical stages in cascade. We would have the gain ratio

$$\text{overall,} \left|\frac{A_{v(\text{hi})}}{A_{v(\text{mid})}}\right| = \left[\frac{1}{\sqrt{1 + (f/f_2)^2}}\right]^n \tag{8.65}$$

for n stages, with f_2 the same for all stages.

We define f_2' as the limit frequency of the overall amplifier; that is, at $f = f_2'$ the overall gain ratio is $1/\sqrt{2}$. Then

$$\text{overall,} \frac{A_{v(\text{hi})}}{A_{v(\text{mid})}} = \frac{1}{\sqrt{2}} = \left[\frac{1}{1 + (f_2'/f_2)^2}\right]^{n/2} \tag{8.66}$$

Squaring and taking the reciprocal,

$$2 = \left[1 + \left(\frac{f_2'}{f_2} \right)^2 \right]^n$$

$$2^{1/n} = 1 + \left(\frac{f_2'}{f_2} \right)^2$$

$$\frac{f_2'}{f_2} = \sqrt{2^{1/n} - 1}$$

$$f_2' = f_2 \sqrt{2^{1/n} - 1} \quad \text{(Hz)} \tag{8.67}$$

where f_2 is the limit frequency of one stage.

Similarly we derive an expression for f_1', the low-frequency limit for the overall amplifier, as

$$f_1' = \frac{f_1}{\sqrt{2^{1/n} - 1}} \quad \text{(Hz)} \tag{8.68}$$

where f_1 is the limit frequency for one stage.

With gains plotted in decibels, the high- and low-frequency asymptotes for the gain curves have slopes of $20n$ dB per decade and $-20n$ dB per decade, as shown for the high-frequency region in Fig. 8.23.

Table 8.3 and Fig. 8.23 indicate the bandwidth narrowing that occurs as we cascade identical amplifier stages to obtain increased overall gain. For instance, with three stages the low-frequency cutoff or limit frequency is raised by a factor of almost 2 and the upper limit frequency is reduced to almost $\frac{1}{2}$. Thus by cascading three stages we have reduced the overall bandwidth to about 50 per cent of that of each stage.

TABLE 8.3 **Bandwidth Limits for *n* Identical Stages**

$f_1', f_2' = $ overall limit frequency
$f_1, f_2 = $ limit frequencies per stage

n	f_1'/f_1	n	f_2'/f_2
1	1	1	1
2	1.56	2	0.65
3	1.96	3	0.51
4	2.30	4	0.43
5	2.59	5	0.39

To restore bandwidth, it is necessary to overdesign each stage, in this example by doubling its f_2 frequency and halving its f_1 frequency.

If we have stages with differing values of f_2, as $f_a, f_b, \ldots$, an approximate value for f_2', the overall frequency limit of the amplifier, can be found by

$$\frac{1}{f_2'} = 1.1 \sqrt{\frac{1}{f_a^2} + \frac{1}{f_b^2} + \cdots}$$

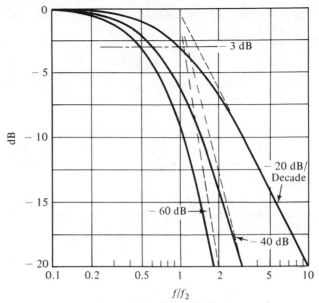

Figure 8.23 Generalized *n*-stage gain curves.

8.13 Frequency Emphasis and De-Emphasis

The variation of output with frequency obtained with *RC* circuits is applied in other ways, including audio system tone controls, pre-emphasis and de-emphasis circuits in FM transmitters and receivers, and in tape recording. In such applications the limit frequencies are usually called *turnover frequencies*. The several curves can be drawn by the asymptote techniques of Sec. 8.5 and 8.9.

Figure 8.24 shows a circuit for *bass boost* in an audio amplifier. It operates with a gain of *A* and the circuit reduces the middle and high frequencies. By

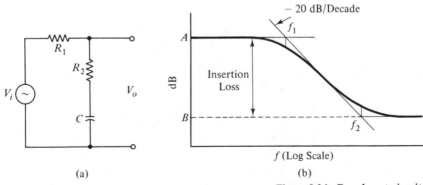

(a)   (b)

Figure 8.24 Bass-boost circuit.

reducing these frequencies the circuit appears to boost the unchanged low frequencies. The attenuation begins at

$$f_1 = \frac{1}{2\pi(R_1 + R_2)C} \tag{8.69}$$

with f_1 usually chosen in the neighborhood of 1000 Hz. The transition ends at

$$f_2 = \frac{1}{2\pi R_2 C} \tag{8.70}$$

The level of differentiation between low and high frequencies is the *insertion loss*, obtainable from Table 8.4 as a function of the respective turnover frequencies.

The circuit in Fig. 8.25 illustrates a *treble-boost* circuit, which actually reduces the mid- and low-frequencies by the amount of the insertion loss. The effect appears to boost the level of the high frequencies with respect to the mid- and low-frequency regions.

TABLE 8.4   **Insertion Loss for Figs. 8.24 and 8.25**

f_2/f_1	*Decibels*	f_2/f_1	*Decibels*
2	−6	20	−26
3	−9.5	30	−29.5
4	−12	40	−32
5	−14	50	−34
6	−15.5	60	−35.5
7	−17	70	−37
8	−18	80	−38
9	−19	90	−39
10	−20	100	−40

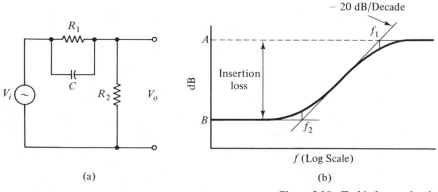

Figure 8.25   Treble-boost circuit.

The treble-boost insertion loss is found from Table 8.4 when the turnover frequencies are chosen as

$$f_1 = \frac{1}{2\pi \dfrac{R_1 R_2}{R_1 + R_2} C} \tag{8.71}$$

$$f_2 = \frac{1}{2\pi R_1 C} \tag{8.72}$$

By using a variable resistor for R_2 of the bass-boost circuit, the value of f_2 can be continuously varied. The insertion loss varies and the bass boost appears to be changed. Similar action is obtained in the treble-boost circuit by variation of resistor R_1.

In frequency modulation radio systems, the high frequencies of the signal are pre-emphasized by the circuit in Fig. 8.26(a), giving a result that is the inverse of the curve in Fig. 8.26(c). The values of R and C are determined from the time constant $RC = 75 \times 10^{-6}$s, resulting in a turnover frequency $f_2 = 1/RC = 1/(75 \times 10^{-6}) = 2123$ Hz. At the receiver the de-emphasis circuit of Fig. 8.26(b) is used with the same turnover frequency, giving a fall at high frequencies as in Fig. 8.26(c). All signal frequencies above the turnover are dropped by the amount of the pre-emphasis and their original level is restored; however, the noise frequencies originating in the transmission path and in the receiver are reduced by the amounts of the curve in Fig. 8.26(c). The signal comes out at its original level, while the noise is reduced in level.

In tape recording the high frequencies do not record well and accordingly they are boosted in recording. At the playback the high frequencies are reduced to normal using a circuit with the elements in Fig. 8.24. The resulting reproducer curve is drawn in Fig. 8.27.

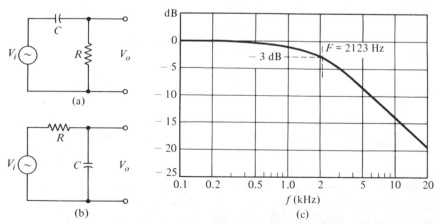

Figure 8.26 (a) Transmitter pre-emphasis circuit, $RC = 75 \times 10^{-6}$ s; (b) FM receiver de-emphasis circuit; (c) de-emphasis of the high frequencies at the receiver.

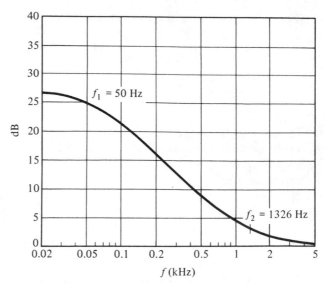

Figure 8.27 Standard tape playback curve, 3.75 ips (9.5 cm/s).

8.14 Review

The frequency range covered by an *RC*-coupled amplifier may be divided into three parts: the low-frequency region, the high-frequency region, and the frequencies between, called the mid-frequency region. In the last region, all reactances in the circuit are negligible and the gain is constant with frequency. We use this region as a gain reference, or 0 dB. If the gain is not constant across all signal frequencies, then we have an amplifier with frequency distortion.

In the low-frequency region the gain is affected by the series blocking capacitor and the gain starts at a low value, rising at 20 dB per decade up to f_1 at which the gain is -3 dB. The frequency f_1 is determined by the blocking capacitance C_C and the series value of the load R and h_{ie} of the following transistor.

The high-frequency region is bounded by a limit frequency f_2 at -3 dB from the mid-frequency gain level; above f_2 the gain falls, ultimately reaching a rate of -20 dB per decade. The frequency f_2 is inversely proportional to C_{ie}, the Miller capacitance, and to the parallel value of R and h_{ie}.

Each type of device has its own ultimate figure of merit as the gain-bandwidth product. This is a frequency at which the gain is unity, or 0 dB, and comparison of frequency performance is possible by using the G-BW figures of the devices.

REVIEW QUESTIONS

8.1 Why do we need to consider the frequency response of an amplifier?

8.2 Define frequency distortion.

8.3 Why are we able to consider the three frequency regions separately?

8.4 What type of distortion is often present in *RC* amplifiers?

8.5 For a rectangular 1-μs pulse, repeated 60 times per second, there are harmonics every 60 Hz. How many harmonics must be reproduced by an amplifier for perfect reproduction of the pulse? For an approximate reproduction?

8.6 What function does the blocking capacitor serve?

8.7 The series reactive element determines what frequency limit? What frequency limit is fixed by the shunt reactances?

8.8 What function does R_1 and R_2 serve in a transistor amplifier?

8.9 What function does R_G in an FET amplifier, and R_g in a vacuum-tube amplifier, serve?

8.10 Draw the small-signal high-frequency model of an FET and label the elements.

8.11 What is the effect if the emitter resistor R_E is not bypassed?

8.12 How would you choose the capacitance value of a blocking capacitor?

8.13 What frequencies bound the mid-frequency region?

8.14 What factors determine f_1 for the low-frequency range?

8.15 What factors determine f_2 for the high-frequency range?

8.16 A transistor amplifier has $f_1 = 150$ Hz; what is the gain at 300 Hz, with mid-frequency as the reference gain?

8.17 A transistor has $h_{ie} = 800\ \Omega$; what maximum reactance may C_C have in a well-designed amplifier at the lowest frequency used?

8.18 In Question 8.17, why are we interested in the reactance of C_C at the lowest frequency?

8.19 An amplifier has a mid-frequency gain of 37 dB; its f_2 value is 15,000 Hz. What is the gain in decibels at a frequency of 45,000 Hz?

8.20 What is the physical origin of the two internal capacitances in the high-frequency model of the junction transistor?

8.21 Repeat Question 8.20 for the FET.

8.22 Repeat Question 8.20 for the triode tube.

8.23 What is the Miller effect?

8.24 Is it a capacitance or a gain that is responsible for the Miller effect?

8.25 Why do we use the short-circuit load to find the frequency limit of a transistor?

8.26 Define f_β; define f_T; what is the relationship between f_β and f_T?

8.27 Why does h_{fe} of a transistor decrease at high frequencies?

8.28 How should f_2 of an amplifier be related to f_T of the transistor?

8.29 Show that variation of R_1 in Fig. 8.25 will vary the high-frequency boost of an amplifier with a tone control.

8.30 Why is a de-emphasis circuit used in an FM radio receiver?

PROBLEMS

8.1 In the circuit of Fig. 8.28(a) with transistors having $h_{ie} = 2000\ \Omega$ and $h_{fe} = 40$, find the value of C that will reduce the gain at 100 Hz by 3 dB below the mid-range region.

8.2 An *RC*-coupled amplifier with a single stage has $f_1 = 82$ Hz, $f_2 = 12,500$ Hz. Sketch the complete frequency-response curve, using a log scale on the frequency axis and gain in decibels.

8.3 In the circuit of Fig. 8.28(b) the transistors have $g_m = 0.015$ mho and $h_{ie} = 900\ \Omega$. What is the decibel mid-frequency gain from port 1,1 to port 2,2? What is f_1?

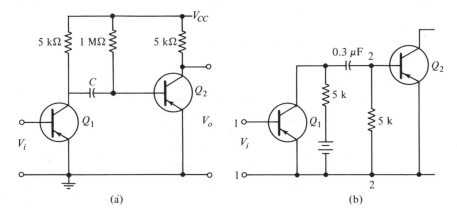

Figure 8.28

8.4 An FET with $g_m = 0.004$ mho is used in the circuit of Fig. 8.29(a), with $C_{gs} = 2$ pF and $C_{gd} = 4$ pF. What are the upper and lower half-power frequencies? What I_i will produce 10 V at V_o in the mid-frequency range?

8.5 A triode amplifier has $C_C = 0.005\ \mu$F, $R_g = 0.25$ MΩ, $r_p = 27,700\ \Omega$, $g_m = 0.0026$ mho, and $R = 0.10$ MΩ. Find $A_{v(\text{mid})}$; also find A_v at 100 Hz. What is f_1?

8.6 Using a transistor with $h_{ie} = 850\ \Omega$, $h_{fe} = 60$, $R = 5000\ \Omega$, $C_{be} = 100$ pF, and $C_{bc} = 3$ pF, what is the transistor input capacitance?

8.7 The transistors in the amplifier of Fig. 8.29(b) have $h_{ie} = 1600\ \Omega$, $h_{fe} = 40$, $C_{bc} = 4$ pF, and $C_{be} = 100$ pF.
(a) Find f_1 and f_2 for the amplifier.
(b) What is the mid-frequency gain?

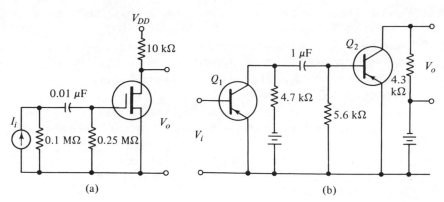

Figure 8.29

8.8 A transistor Q_1 in the circuit of Fig. 8.30(a) has $h_{ie} = 1200 \, \Omega$, $g_m = 0.005$ mho, and $C = 0.05 \, \mu F$. Find f_1 and the gain at 30 Hz.

8.9 The triode of Fig. 8.30(b) has $r_p = 12,000 \, \Omega$, $g_m = 2500 \, \mu mhos$, $C_{gk} = 3 \, pF$, and $C_{pk} = 3 \, pF$. Find the bandwidth in hertz for the gain between ports 1,1 and 2,2, with $R = 40,000 \, \Omega$ for T_2.

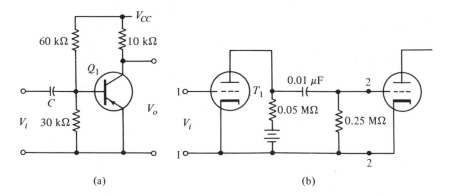

Figure 8.30

8.10 Given the following transistor measurements at low frequency:

$$h_{ie} = 600 \, \Omega \qquad C_{bc} = 3 \, pF$$
$$h_{fe} = 30 \qquad C_{be} = 100 \, pF$$

Find f_β, f_T, for the transistor.

8.11 When the transistor of Problem 8.10 is used in the C-E amplifier of Fig. 8.30(a), find f_2. Also find the decibel gain at $f = 2.5f_2$.

8.12 Plot the high-frequency gain in decibels for Problem 8.11. Prove that your gain figure is correct at $2.5f_2$.

8.13 An amplifier has three identical stages with each having $f_1 = 10$ Hz and $f_2 = 25$ kHz. What are the limit frequencies of the amplifier?

8.14 With $h_{fe} = 80$, $C_{bc} = 1$ pF, $C_{be} = 75$ pF, and $h_{ie} = 600\ \Omega$, find f_β and f_T of the transistor. What gain is possible with BW $= 100$ MHz?

8.15 Specify the f_1 and f_2 design frequencies for each stage of a four-identical-stage amplifier if the overall bandwidth required is from 100 to 450,000 Hz.

8.16 Show that a rate of fall of 20 dB per decade is equivalent to 6 dB per octave (double frequency).

8.17 You wish the gain to fall at a rate of -60 dB per decade at a frequency that is high with respect to your operating frequency. How will you design your amplifier?

8.18 An amplifier has $f_1 = 80$ Hz. Sketch the frequency-response curve on a log frequency scale from 10 to 500 Hz. What is the rate of gain fall at 0.1 Hz? Use a decibel gain scale.

8.19 A transistor has $f_\beta = 10$ MHz and $h_{feo} = 100$ at a mid-frequency. Find h_{fe} at 15 MHz.

8.20 A transistor has $f_T = 250$ MHz and $h_{feo} = 64$. Find h_{fe} at 5.0 MHz.

8.21 An amplifier has three nonidentical stages with gains of $A_1 = 18$ dB, $A_2 = 20$ dB, and $A_3 = 14$ dB at mid-frequency. What is the overall gain, expressed as $A_v = V_o/V_i$?

8.22 A two-nonidentical stage amplifier has a first stage $f_1 = 40$ Hz and a second stage $f_1 = 15$ Hz. Find f_1 for the overall amplifier.

8.23 Plot the high-frequency region gain on a log frequency scale for a six-stage amplifier, with $f_2 = 1$ MHz for each stage. Use a decibel gain scale.

8.24 The G-BW product for a transistor is 120 MHz. With mid-frequency $g_{mo} = 0.010$ mho and $R = 2000\ \Omega$ for the load, $h_{ie} = 2200\ \Omega$. Determine the gain to be expected at a band limit of 4.5 MHz.

8.25 A given amplifier has an f_2 value of 30 kHz. At what frequency is the amplifier gain down only 0.1 dB from its mid-range value?

8.26 For the amplifier of Problem 8.24, we find the gain is 1.5 dB below the mid-range gain. What is the frequency?

8.27 A transistor has a mid-range $\alpha = 0.97$ and $f_\alpha = 0.55$ MHz. Find the magnitude of α at 0.75 MHz.

8.28 A one-stage amplifier has $A_v = 30$ and $f_2 = 400$ kHz. The f_2 value is to be increased to 600 kHz. What change can you make in the amplifier? What will the new A_v become?

8.29 An amplifier for an electronic voltmeter must have A_i constant within 5 per cent up to 100,000 Hz. What f_2 value must the amplifier have? If composed of three identical stages, what f_2 value must be specified for each stage?

9

Negative Feedback
in Amplifiers

In electronic circuit design we are often forced to sacrifice performance in one area to achieve better performance elsewhere; this is called a *trade-off*. In the use of *negative feedback*, we build extra gain into the original design and trade off this excess gain to obtain reduced distortion, stable gain, greater bandwidth, and changed amplifier input and output resistances. When we use solid-state devices and integrated circuits, we find that the price paid for the additional gain is relatively low. Negative feedback makes our amplifiers nearly precise and ideal.

9.1 The Black Box with Feedback

Negative feedback is an old and fundamental process, used in many fields to make the output response of a system more nearly correspond to the input signal. We compare the output with the input and utilize any difference as a corrective signal. The principle is apparent in such a simple activity as placing a pen on paper; without optical feedback of the difference between actual (output) and desired (input) locations of the pen, we could not write. The children's game of "pin the tail on the donkey" is another example of our inability to perform as usual without feedback in some form.

In electronic amplifier feedback we compare a sample of the output waveform against the input waveform. Any difference between the two signals gives an error voltage that is applied to the amplifier so as to change the output waveform and reduce the difference toward zero. Thus we force the output toward equality with the input and somewhat unprecise amplifiers become almost ideal gain elements that give an output like the input.

The C-E amplifier in Fig. 9.1 has *negative feedback* applied through resistor R_f and the blocking capacitor C_f. A portion of V_o is transmitted back to the input circuit through R_f. Because of the 180° phase inversion in the C-E amplifier, the voltage fed back as V_f is opposite in phase or negative to V_i and the feedback voltage is compared with and subtracted from the input voltage. The difference becomes the amplifier input.

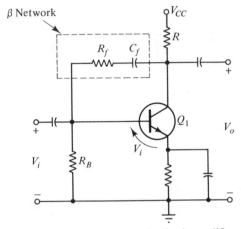

Figure 9.1 A practical feedback amplifier.

Figure 9.2 shows an amplifier system of gain

$$A' = \frac{V'_o}{V_s} \tag{9.1}$$

from port 1 to port 3. The *internal gain* of the black box from port 2 to port 3 is

$$A = \frac{V'_o}{V'_i} \tag{9.2}$$

The output is sampled by the β network and the result V_f is subtracted from the input at the mixing point Σ (sigma), as shown by the indicated polarities.

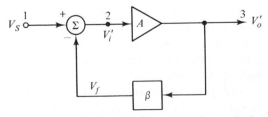

Figure 9.2 Feedback applied to a black box amplifier.

We have

$$V_f = \beta V'_o$$

where β is less than unity and the β circuit is designed to produce zero phase shift in the operating frequency range. A resistance voltage divider serves well, as the R_f, R_b divider in Fig. 9.1.

At the amplifier input,

$$V'_i = V_s - V_f = V_s - \beta V'_o \qquad (9.3)$$

From Eq. 9.2 we have $AV'_i = V'_o$ and substitution of Eq. 9.3 gives

$$A(V_s - \beta V'_o) = V'_o$$
$$V'_o(1 + A\beta) = AV_s \qquad (9.4)$$

We find the gain from port 1 to port 3 of the feedback system by use of Eq. 9.1 as

$$A' = \frac{V'_o}{V_s} = \frac{A}{1 + A\beta} \qquad (9.5)$$

This is the fundamental equation of feedback, expressing the *closed-loop gain A'* as dependent on the *internal gain A* and on the *feedback factor β*.

In Eq. 9.3 the feedback voltage V_f is presented to the input circuit in subtractive fashion. The denominator $|1 + A\beta| > 1$, and the *feedback is negative*. Equation 9.5 then shows that $|A'| < |A|$ and the gain of the system with feedback is less than the internal amplifier gain. Thus gain is sacrificed with negative feedback.

If A is negative, as is usual in C-E amplifiers, we reverse V_f from the β network, resulting in a positive $A\beta$ term in Eq. 9.5 and so retain the negative feedback.

If the phase of V_f reverses, as may happen with nonresistive β networks, the feedback voltage V_f becomes additive to V_s in Eq. 9.3 and the denominator of Eq. 9.5 shows that $|1 + A\beta| < 1$ and the *feedback is positive*. The closed-loop gain is $|A'| > |A|$ and the gain of the feedback system is greater than the internal gain. This is a condition of gain instability since it includes the case at $A\beta = -1$ where the gain becomes infinite. The amplifier then becomes a generator of signals and when such action is wanted, we call the circuit an *oscillator;* these circuits will be studied in Chapter 13. The condition of positive feedback is avoided in amplifiers.

A measure of the amount of negative feedback introduced into an amplifier is given by the gain change in decibels as

$$\text{dB of feedback} = 20 \log \frac{A}{A'} \qquad (9.6)$$

The latter expression can be translated into another that is useful. Look at

$$\frac{A}{A'} = \frac{A}{\dfrac{A}{1 + A\beta}} = 1 + A\beta \qquad (9.7)$$

so that the decibels of feedback can also be written as

$$\text{dB of feedback} = 20 \log (1 + A\beta) \tag{9.8}$$

Because of the reduction of system gain by feedback, the input at port 1, 1 must be greater than the amplifier input at port 2, 2. By our definitions

$$V'_o = A'V_s$$
$$V'_o = AV'_i$$

From these relations, for equal output

$$V_s = \frac{A}{A'}V'_i = \frac{A}{\dfrac{A}{1 + A\beta}}V'_i = (1 + A\beta)V'_i \tag{9.9}$$

The system input must be $(1 + A\beta)$ greater than the amplifier input at port 2, 2.

Example: An amplifier has a gain $A = 70$ and a normal input signal of 0.1 V. Feedback with $\beta = 0.1$ is added, giving $A\beta = 7.0$. The closed-loop gain is

$$A' = \frac{A}{1 + A\beta} = \frac{70}{1 + 7.0} = 8.75$$

An input voltage

$$V_s = (1 + A\beta)V'_i = (1 + 7) \times 0.1 = 0.8 \text{ V}$$

will be needed.

The output voltage of the amplifier is

$$V'_o = A'V_s = 8.75 \times 0.8 = 7.0 \text{ V}$$

or

$$V'_o = AV'_i = 70 \times 0.1 = 7.0 \text{ V}$$

These relations apply to the system and to the internal amplifier, at ports 1, 1 and 2, 2, respectively.

With feedback the amplifier input voltage is

$$V'_i = \frac{V'_o}{A} = \frac{7.0}{70} = 0.1 \text{ V}$$

The external signal requirement is raised from 0.1 to 0.8 V by the addition of negative feedback, but inside the loop the amplifier is operating with an input of 0.1 V and an output of 7.0 V, with or without feedback.

9.2 Stabilization of Gain by Negative Feedback

When amplifiers are used in calibrated electronic instruments, such as voltmeters, internal gain changes will affect the instrument accuracy. Changes can be expected, due to supply voltage shifts, aging, and particularly operating temperature. Negative feedback is used to make the amplifier gain independent of these variables.

For a percentage change in A, $\Delta A/A$, we find a resultant percentage change in the system gain as $\Delta A'/A'$. That is,

$$\frac{\Delta A'}{A'} = \frac{1}{1 + A\beta}\frac{\Delta A}{A} \qquad \textbf{(9.10)}$$

Thus the feedback gain change is much less than the change in the internal amplifier gain.

To reduce the sensitivity to internal gain change still further, we can make $A\beta$ very large so that Eq. 9.5 reduces to

$$A' \cong \frac{1}{\beta} \qquad \textbf{(9.11)}$$

With $A\beta \gg 1$, we no longer are concerned with maintenance of an exact value of A because the overall system gain is dependent only on the elements of the β network. When constructed of precision resistors, a precise gain is retained over long periods of time. The price paid for this stability, of course, is a reduction in feedback system gain.

Example: A range of ± 10 per cent is allowed for the internal gain of an amplifier, with $A = 100$. How can this gain change be reduced to ± 1 per cent? What is the resultant gain?

We have $\Delta A/A = 0.10$ and desire $\Delta A'/A'$ to be 0.01. Then

$$0.01 = \frac{1}{1 + A\beta} \times 0.10$$
$$1 + A\beta = 10$$

Since $A = 100$,

$$100\beta = 10 - 1 = 9$$
$$\beta = 0.09$$

We then have $A\beta = 9$ for the feedback system.

With $1 + A\beta = 10$, the feedback system gain is

$$A' = \frac{A}{1 + A\beta} = \frac{100}{10} = 10$$

and we have reduced the gain by a factor of 10 in stabilizing the gain by a factor of 10.

If we design our amplifier for a gain of 1000 ± 10 per cent, however, we have

$$0.01 = \frac{1}{1 + A\beta} \times 0.10$$
$$1 + A\beta = 10$$

and with $A = 1000$

$$1000\beta = 9$$
$$\beta = 0.009$$

and we have $A\beta = 9$ again. But the feedback system gain is now

$$A' = \frac{1000}{1 + 9} = 100$$

This is the desired gain but it is now stabilized to ± 1 per cent.

We have had to buy an amplifier with a gain of 1000 to obtain 1 per cent stability at a gain of 100. This, in certain applications, may be a small price to pay.

Example: An amplifier is designed with $A = 4000$. Choosing $\beta = 0.04$, we have

$$A\beta = 4000 \times 0.04 = 160$$

Then

$$A' = \frac{A}{1 + A\beta} = \frac{4000}{1 + 160} = 24.84$$

Suppose the internal gain doubles to 8000. Then

$$A\beta = 8000 \times 0.04 = 320$$
$$A' = \frac{8000}{1 + 320} = 24.92$$

Gain A might drop to 2000, in which case

$$A\beta = 2000 \times 0.04 = 80$$
$$A' = \frac{2000}{1 + 80} = 24.69$$

We have demonstrated that with A large, Eq. 9.11 holds and the gain can be maintained close to

$$A' = \frac{1}{\beta} = \frac{1}{0.04} = 25$$

regardless of large variations of the internal amplifier gain A.

9.3 Bandwidth Improvement with Negative Feedback

In Chapter 8 we showed that the high-frequency gain of an RC-coupled amplifier was

$$A_{v(\text{hi})} = \frac{A_{v(\text{mid})}}{1 + \dfrac{jf}{f_2}} \tag{9.12}$$

Using a resistive β network, the phase angle of β can be held constant over the frequency band. For a feedback system, we could write

$$A' = \frac{A_{v(\text{hi})}}{1 + \beta A_{v(\text{hi})}} = \frac{\dfrac{A_{v(\text{mid})}}{1 + (jf/f_2)}}{1 + \dfrac{\beta A_{v(\text{mid})}}{1 + (jf/f_2)}} = \frac{A_{v(\text{mid})}}{[1 + \beta A_{v(\text{mid})}] + \dfrac{jf}{f_2}} \qquad (9.13)$$

We learned that we are at a limit frequency when the terms in the denominator are equal so

$$\frac{f'_2}{f_2} = 1 + \beta A_{v(\text{mid})}$$

and our upper limit frequency with feedback, designated f'_2, is

$$f'_2 = [1 + \beta A_{v(\text{mid})}]f_2 \qquad (9.14)$$

The upper frequency limit of the system has been raised by $(1 + A\beta)$.
  Similarly, we can show

$$f'_1 = \frac{f_1}{1 + \beta A_{v(\text{mid})}} \qquad (9.15)$$

and the low-frequency limit is reduced by feedback. The bandwidth has been materially increased.

  Using f'_2 as the bandwidth or assuming $f'_2 \gg f'_1$, the gain-bandwidth product with feedback is

$$\text{G-BW}' = \frac{A_{v(\text{mid})}}{1 + \beta A_{v(\text{mid})}}[1 + \beta A_{v(\text{mid})}]f_2 = A_{v(\text{mid})}f_2 \qquad (9.16)$$

and the G-BW figure is seen to be independent of the feedback β. The curves in Fig. 9.3 illustrate this, being reduced in gain as the frequency band widens. It is apparent that we have traded gain for bandwidth.

  Physically large and expensive blocking capacitors can often be avoided when the mid-frequency bandwidth is expanded to lower frequencies, by use of negative feedback.

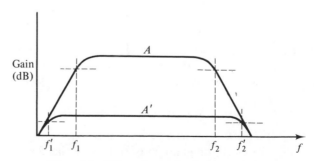

Figure 9.3 Effect of negative feedback on bandwidth.

9.4 Reduction of Nonlinear Distortion

Large-signal amplifiers, forced to operate in wide excursions across the device volt-ampere characteristics, will generate harmonics in the output signal. Negative feedback reduces such internally generated distortion of the the waveform.

With the feedback loop open in Fig. 9.4, we have

$$V_o = AV'_i + V_h \tag{9.17}$$

where V_h is the generator representing the internally generated distortion waveform.

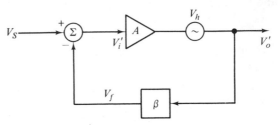

Figure 9.4 Distortion inside the feedback loop.

If we add negative feedback, the input signal is

$$V'_i = V_s - V_f \tag{9.18}$$

and the output signal is

$$V'_o = AV'_1 + V_h \tag{9.19}$$

Substituting

$$V'_o = A(V_s - V_f) + V_h$$

and using Eq. 9.9 for V_s with feedback, as well as $V_f = \beta V'_o$, we have

$$V'_o = A[(1 + A\beta)V'_i - \beta V'_o] + V_h$$

Sorting the terms, we have

$$V'_o(1 + A\beta) = A(1 + A\beta)V'_i + V_h$$

$$V'_o = AV'_i + \frac{V_h}{1 + A\beta} \tag{9.20}$$

Making outputs equal, or $V_o = V'_o$, we compare Eq. 9.17 without feedback and Eq. 9.20 with feedback. We see that the use of negative feedback has reduced the harmonic distortion by the factor $1/(1 + A\beta)$, or

$$D' = \frac{D}{1 + A\beta} \tag{9.21}$$

where D' and D represent per cent distortion.

This result is of great importance in the design of high-power audio amplifiers. But with large $A\beta$ values the bandwidth is increased also and the high-frequency limit of audio amplifiers is frequently raised above 100 kHz. This occurs even though audio signals present no components over 20 kHz. Should $A\beta$ shift in phase angle in this extreme frequency range, there may be positive feedback and gain instability can be created.

Example: An audio amplifier of 500 voltage gain produces 11 per cent harmonic distortion at full output. It was designed, considering C_{ie} of the transistors, to yield an upper frequency limit of 8 kHz.

Tolerable distortion is considered to be 1 per cent. What value of β is needed to reduce the distortion and what is the bandwidth extension?

From the distortion relation

$$D' = \frac{D}{1 + A\beta}$$

$$0.01 = \frac{0.11}{1 + A\beta}$$

$$1 + A\beta = 11$$

and $A\beta = 10$. With $A = 500$ we have

$$\beta = \frac{10}{500} = 0.02$$

The f_2 limit of the amplifier is extended by $(1 + A\beta) = 11$ so that the f_2' limit with feedback is $11 \times 8 \text{ kHz} = 88 \text{ kHz}$. To improve the distortion situation, it is seen that we have extended the frequency range far beyond the needs of the audio signal.

9.5 Control of Amplifier Output and Input Resistances

For a constant input signal, the feeding back of a voltage sample V_f, proportional to the output voltage of an amplifier, tends to maintain the output voltage at a constant value. This is done in Fig. 9.5(a) and constitutes *voltage feedback*. The amplifier appears to have a low output resistance.

Similarly, the feedback of a voltage V_f proportional to the load current tends to control the load current at a constant magnitude, independent of load resistance. This is the property of the circuit in Fig. 9.5(b) and represents *current feedback*. The amplifier appears to have a high output resistance.

It is often desired to alter amplifier output resistances so as to supply a needed output current easily or to power match a load such as a loudspeaker. The two methods of obtaining feedback voltage provide the circuit designer with means for lowering or raising the output resistance of an amplifier.

Considering voltage feedback first, as in Fig. 9.5(a), the output resistance R_o' may be found by short-circuiting the independent signal source V_s. Then

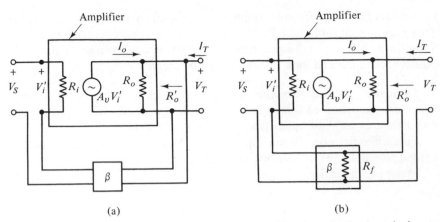

(a)   (b)

Figure 9.5   (a) Voltage feedback, series input; (b) current feedback, series input.

we apply a voltage V_T at the output and $R'_o = V_T/I_T$. With a short at V_s, the input is $V'_i = -\beta V_T$ with the negative sign as a result of the subtraction of V_f. From the currents in R_o,

$$V_T = \left(\frac{-A_v\beta V_T}{R_o} + I_T\right)R_o$$

and it follows that

$$R'_o = \frac{V_T}{I_T} = \frac{R_o}{1 + A_v\beta} \tag{9.22}$$

The output resistance of the amplifier is *reduced* by the introduction of negative voltage feedback.

Referring to the current feedback circuit in Fig. 9.5(b), we short V_s and have $V'_i = R_f I_T$. Then

$$V_T = (A_v R_f + R_o + R_f)I_T \tag{9.23}$$

With the feedback factor being

$$\beta = \frac{R_f}{R_f + R_o} \tag{9.24}$$

we have

$$V_T = \left(\frac{A_v R_f}{R_f + R_o} + 1\right)(R_f + R_o)I_T$$

Since $R_f \ll R_o$,

$$V_T = (A_v\beta + 1)R_o I_T \tag{9.25}$$

$$R'_o = \frac{V_T}{I_T} = R_o(1 + A_v\beta) \tag{9.26}$$

which shows that the output resistance *increases* by use of negative current feedback.

The manner in which the feedback voltage is introduced into the input circuit can alter the input resistance of an amplifier. With the feedback volt-

age V_f introduced in *series*, shown in Fig. 9.5, the input resistance with feedback is

$$R_i' = R_i(1 + A_v\beta) \tag{9.27}$$

and increases with β.

When V_f is introduced in *shunt* as in Fig. 9.6, the input resistance is reduced as

$$R_i' = \frac{R_i}{1 + A_v\beta} \tag{9.28}$$

The circuit designer can choose β and the method by which he derives the
voltage V_f to adjust the output resistance to a desired value. He can choose
β and the method by which V_f is inserted in the input circuit to alter the
input resistance of the amplifier. These results are tabulated in Table 9.1.

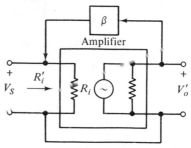

Figure 9.6 Shunt input of the feedback voltage.

TABLE 9.1 **Effect of Feedback on Amplifier Resistances**

Voltage derived:	R_o decreases
Current derived:	R_o increases
Series input:	R_i increases
Shunt input:	R_i decreases

9.6 A Current Series-Feedback Circuit

In Fig. 9.7(a) the unbypassed emitter resistor R_E provides a feedback proportional to load current; therefore it is *current feedback*. The voltage across
R_E is V_f and this is introduced in series with the input signal so that we have
a *current-series feedback circuit*. That is,

$$V_{be} = V_s - V_f \tag{9.29}$$

From the equivalent circuit

$$V_s = V_{be} + V_f = h_{ie}I_b + (I_b + I_c)R_E$$
$$= [h_{ie} + (1 + h_{fe})R_E]I_b \tag{9.30}$$

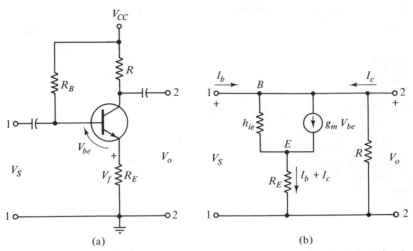

Figure 9.7 (a) Current feedback, series input; (b) equivalent circuit.

Since $V_o = -RI_c = -h_{fe}RI_b$, the gain can be obtained by use of Eq. 9.30 as

$$A_{v(\text{mid})} = \frac{V_o}{V_s} = \frac{-h_{fe}R}{h_{ie} + (1 + h_{fe})R_E} \qquad (9.31)$$

By dividing out the h_{ie} term and neglecting $1 \ll h_{fe}$, we have

$$A_{v(\text{mid})} \cong \frac{-g_mR}{1 + g_mR_E} \qquad (9.32)$$

It is of interest to find that the feedback theory previously developed does apply to this practical circuit.

The feedback factor is

$$\beta = \frac{\text{voltage feedback}}{\text{load voltage}} = \frac{(1 + h_{fe})\,R_EI_b}{-h_{fe}RI_b} \cong -\frac{R_E}{R} \qquad (9.33)$$

Without feedback ($R_E = 0$), the gain of the amplifier is

$$A = -g_mR \qquad (9.34)$$

We showed that with feedback the gain is

$$A' = \frac{A}{1 + A\beta}$$

Using Eq. 9.33 and 9.34 we can form the feedback gain:

$$A' = \frac{-g_mR}{1 + (-g_mR)\left(-\dfrac{R_E}{R}\right)} = \frac{-g_mR}{1 + g_mR_E} \qquad (9.35)$$

which is Eq. 9.32 written from the feedback equation. The negative sign on β

appears because A_v is negative for the one-stage C-E amplifier; this situation was previously discussed.

Thus our feedback theory is confirmed.

The input resistance can be obtained directly from Eq. 9.30 as V_s/I_b, where

$$R_i = \frac{V_s}{I_b} = h_{ie} + (1 + h_{fe})R_E$$

$$\cong h_{ie} + h_{fe}R_E = h_{ie}(1 + g_m R_E) \tag{9.36}$$

The series input of the feedback voltage has raised the input resistance from the h_{ie} value, present without feedback. Our theory says

$$R_i' = R_i(1 + A\beta) \tag{9.37}$$

The input resistance is h_{ie} with no feedback ($R_E = 0$ in Eq. 9.36). Using the A and β relations,

$$R_i' = h_{ie}\left[1 + (-g_m R)\left(-\frac{R_E}{R}\right)\right]$$

$$= h_{ie}(1 + g_m R_E) \tag{9.38}$$

which is Eq. 9.36 derived from the circuit.

To obtain the output resistance, we shall merely use Eq. 9.26:

$$R_o' = \frac{1 + A\beta}{R}$$

since R is the output resistance at the 2, 2 port with no feedback. Using our A and β relations,

$$R_o' = R\left[1 + (-g_m R)\left(-\frac{R_E}{R}\right)\right]$$

$$= R(1 + g_m R_E) \tag{9.39}$$

with feedback. This is increased by reason of the current feedback.

Example: For the circuit in Fig. 9.7, we use $R_E = 1.5\,\text{k}\Omega$, $R = 10\,\text{k}\Omega$, $h_{ie} = 2\,\text{k}\Omega$, $h_{fe} = 50$ and $h_{oe} = 10^{-4}$ mho. Find the gain and input and output resistances, without and with feedback.

We have

$$g_m = \frac{h_{fe}}{h_{ie}} = \frac{50}{2000} = 0.025\ \text{mho}$$

Without feedback,

$$A = -g_m R = -25 \times 10^{-3} \times 10 \times 10^3 = -250$$
$$R_i = h_{ie} = 2000\ \Omega$$
$$R_o = R = 10,000\ \Omega \qquad (1/h_{oe}\ \text{neglected})$$

With feedback,

$$A' = \frac{-g_m R}{1 + g_m R_E} = \frac{-250}{1 + (25 \times 10^{-3} \times 1.5 \times 10^3)} = -6.5$$

To confirm the gain, we calculate β as

$$\beta = -\frac{R_E}{R} = -\frac{1500}{10,000} = -0.15$$

and using the feedback gain expression

$$A' = \frac{A}{1 + A\beta} = \frac{-250}{1 + (-250)(-0.15)} = -6.5$$

$$R_i' = h_{ie}(1 + g_m R_E) = 2000(1 + 25 \times 10^{-3} \times 1.5 \times 10^3)$$
$$= 77,000 \ \Omega$$

$$R_o' = \frac{1 + g_m R_E}{h_{oe}} = 10^4(1 + 25 \times 10^{-3} \times 1.5 \times 10^3)$$
$$= 380,000 \ \Omega$$

9.7 Voltage-Shunt Feedback Circuit

The circuit in Fig. 9.8 gives shunt input for the voltage-derived feedback; therefore it is a *voltage-shunt feedback circuit*.

A current summation at the base yields

$$I_s + I_f = I_b$$

and we can write

$$I_s = \frac{V_s - V_{be}}{R_s} \cong \frac{V_s}{R_s} \tag{9.40}$$

$$I_f = \frac{V_o - V_{be}}{R_f} \cong \frac{V_o}{R_f} \tag{9.41}$$

$$I_b = \frac{I_c}{h_{fe}} \cong -\frac{V_o}{h_{fe}R} \tag{9.42}$$

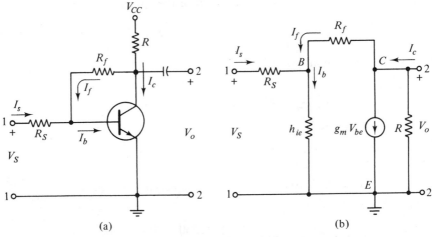

(a)   (b)

Figure 9.8 (a) Voltage feedback, shunt input; (b) equivalent circuit.

In writing Eq. 9.40 and 9.41 we have assumed that the internal gain A is large and V_{be} is therefore small. Likewise, in using $I_c = -V_o/R$ in Eq. 9.42, we have said that $I_f \ll I_c$ or that $R_f \gg R$.

Summing the currents,

$$\frac{V_s}{R_s} + \frac{V_o}{R_f} = -\frac{V_o}{h_{fe}R} \tag{9.43}$$

$$-V_o\left(\frac{1}{h_{fe}R} + \frac{1}{R_f}\right) = \frac{V_s}{R_s}$$

Then

$$\frac{V_o}{V_s} = -\frac{1}{R_s}\left(\frac{1}{\dfrac{1}{h_{fe}R} + \dfrac{1}{R_f}}\right) \tag{9.44}$$

and the gain is

$$A' = \frac{V_o}{V_s} = -\frac{R_f}{R_s}\frac{h_{fe}R}{R_f + h_{fe}R}$$

For $h_{fe}R \gg R_f$, the gain is determined solely by the feedback resistors R_f and R_s, as

$$A' \cong -\frac{R_f}{R_s} \tag{9.45}$$

giving a very stable gain. This is equivalent to the result of Eq. 9.11 and so we see that

$$\beta = -\frac{R_s}{R_f} \tag{9.46}$$

for the shunt-feedback circuit.

The result of Eq. 9.45 will be further discussed in Chapter 10.

9.8 Voltage Feedback with the FET

In the FET amplifier in Fig. 9.9, a portion of the output voltage is provided by the R_1, R_2 voltage divider and inserted in series with the input. A similar circuit is useful with the vacuum tube. This is a *voltage-series feedback* application.

By opening R_1 the feedback is removed from the circuit and the gain is

$$A_v = -g_m R_D \tag{9.47}$$

The sum $R_1 + R_2$ is made large with respect to R_D so that $I_1 \ll I_d$. The current in the voltage divider is

$$I_1 = -\frac{V_o}{R_1 + R_2}$$

and the voltage feedback V_f is

$$V_f = I_1 R_2 = -\frac{R_2}{R_1 + R_2}V_o \tag{9.48}$$

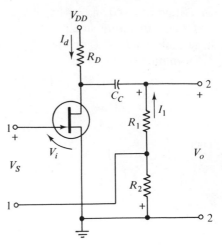

Figure 9.9 An FET with voltage-series feedback.

This feedback signal is introduced as

$$V_i = V_s - V_f \tag{9.49}$$

and the feedback is negative. The feedback factor can be obtained as

$$\beta = \frac{V_f}{V_o} = \frac{-R_2}{R_1 + R_2} \tag{9.50}$$

This is an accurate result because there is no gate current to consider.
The gain with negative feedback is then

$$A' = \frac{A}{1 + A\beta} = \frac{-g_m R_D}{1 + (-g_m R_D)\left(-\dfrac{R_2}{R_1 + R_2}\right)}$$

$$= \frac{-g_m R_D}{1 + g_m \dfrac{R_D R_2}{R_1 + R_2}} \tag{9.51}$$

Example: An FET amplifier has $g_m = 0.004$ mho, $R_D = 10,000\ \Omega$, $R_1 + R_2 = 100\ \text{k}\Omega$, and $\beta = -0.10$. Find the gain with and without feedback.
Without feedback,

$$A_v = -g_m R_D = -4 \times 10^{-3} \times 10^4 = -40$$

The feedback resistor R_2 is found from

$$\beta = -\frac{R_2}{R_1 + R_2}$$
$$R_2 = -\beta(R_1 + R_2) = 0.10 \times 10^5 = 10^4\ \Omega$$

The gain with feedback is

$$A'_v = \frac{-g_m R_D}{1 + g_m \dfrac{R_D R_2}{R_1 + R_2}} = \frac{-40}{1 + 4 \times 10^{-3} \dfrac{10^4 \times 10^4}{10^5}}$$

$$= \frac{-40}{5} = -8$$

Using Eq. 9.5 as a check,

$$A'_v = \frac{A}{1 + A\beta} = \frac{-40}{1 + (-40)(-0.10)} = -8$$

9.9 The Emitter Follower as a Feedback Amplifier

From the emitter follower circuit in Fig. 9.10, we have

$$V_{be} = V_s - V_o$$

which is equivalent to

$$V'_i = V_s - V_f \tag{9.52}$$

for our general feedback amplifier. The result of Eq. 9.52 tells us that $V_f = V_o$ and the entire output voltage is being fed back to the input. This means that

$$\beta = \frac{V_f}{V_o} = \frac{V_o}{V_o} = 1 \tag{9.53}$$

Since the output voltage is being fed back, the circuit gives *voltage feedback* with $\beta = 1$ and is extremely stable.

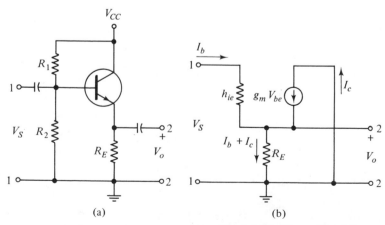

Figure 9.10 The emitter follower.

Writing

$$V_o = (I_b + I_c)R_E \cong h_{fe}R_E I_b$$

and

$$V_s = h_{ie}I_b + h_{fe}R_E I_b = (h_{ie} + h_{fe}R_E)I_b$$

we find the gain as

$$A_v = \frac{h_{fe}R_E}{h_{ie} + h_{fe}R_E} = \frac{g_m R_E}{1 + g_m R_E} \cong 1 \qquad (9.54)$$

which is the result obtained for the C-C circuit. The input resistance is increased by the series feedback and the output resistance is reduced to $\approx 1/g_m$, as for the C-C circuit, which the emitter follower is.

Similar results are obtained for the FET source follower and the triode cathode follower.

9.10 Multiple-Stage Feedback

When we cascade C-E amplifiers, each stage adds a phase shift of 180°, or $n \times 180°$ is the phase shift for n stages. The polarity of the V_f voltage must be maintained negative to V_s for negative feedback and $A\beta$ must be positive, either as a result of $(+A)(+\beta)$ or $(-A)(-\beta)$.

With a single stage of C-E amplification, the output V_o is at 180° to the input signal and voltage feedback V_f is introduced into the base, where it is negative to the signal as required. This is the method in Fig. 9.8.

Because of the 360° phase shift of V_o with respect to V_s in the two-stage C-E circuit of Fig. 9.11, the voltage feedback V_f must be introduced into the emitter. This amounts to reversal of the V_f voltage and gives $(+A)(+\beta) = A\beta$. Thus we have methods for insertion of the feedback voltage with n odd or even.

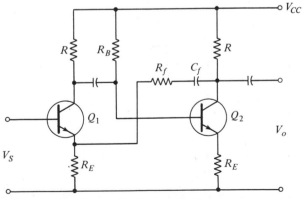

Figure 9.11   Voltage feedback, series input, over two stages.

In combining C-E, C-B, or C-C stages or FET or tube equivalents inside the feedback loop, we must determine if the overall phase shift is an odd or even multiple of 180°. The proper entry point for the feedback voltage V_f is then known.

Feedback over multiple stages gives a larger value of A, resulting in a larger $A\beta$ term; therefore, we can have greater reduction of distortion. There is some risk in amplifier stability because with increased numbers of capacitors inside the feedback loop, the phase angle of A may not be exactly $180n°$ at all frequencies in the response range. Conditions for positive feedback may be approached at the limits of frequency response, resulting in amplifier instability.

The major feedback in Fig. 9.11 is voltage derived but a small amount of current feedback is added in each stage by the unbypassed emitter resistors. Feedback of both types is additive in an amplifier and the effective β is the sum of the voltage and current feedback β values.

Figure 9.12 shows a two-stage C-E current feedback amplifier. With a 360° phase shift, the current supplied from R_E opposes the signal current in R_s so that the feedback is negative or subtractive to the input signal.

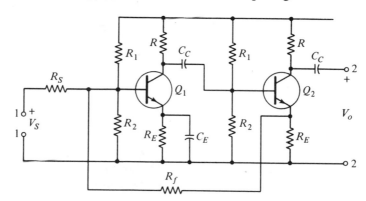

Figure 9.12 Current-shunt negative feedback.

9.11 Amplifier Gain Stability with Feedback

We have used ideal feedback conditions as they occur in the mid-frequency range of an amplifier, requiring the feedback voltage V_f to be opposite or at 180° to the input signal voltage. We can design amplifiers to meet this condition satisfactorily over a specified frequency range.

We also encounter situations, however, in which the required phase angle is not present at extremely high or low frequencies, at which the angle of A_v approaches $90n°$ in RC amplifiers of n stages. At frequencies above f_2, or below f_1, the voltage V_f may not directly subtract from V_s, and V_i' increases. As a result the amplifier output is larger and the gain increases at frequency extremes, as shown in Fig. 9.13, and we have the condition of positive feedback known as *regeneration*.

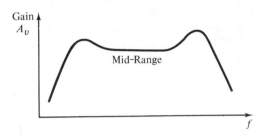

Figure 9.13 A feedback amplifier gain curve with regeneration at high and low frequencies.

The dividing line between negative feedback and positive feedback appears at

$$|1 + A\beta| = 1 \tag{9.55}$$

and with

$$|1 + A\beta| < 1 \tag{9.56}$$

the gain is unstable as a result of positive feedback. When the $A\beta$ value leads to

$$|1 + A\beta| = 0 \tag{9.57}$$

or $A\beta = -1$, the gain A' becomes infinite according to

$$A' = \frac{A}{1 + A\beta} = \frac{A}{0} = \infty$$

and this is the extreme condition of gain instability known as *oscillation*.

Using $A\beta$ as a stability criterion, we can write

$$\begin{aligned}
\text{Stability:} && |A\beta| &> 0 \\
\text{Instability:} && 0 > |A\beta| &> -1 \\
\text{Oscillation:} && A\beta &= -1
\end{aligned}$$

A Nyquist plot may be used to study the action of $A\beta$ and is a polar plot of $A\beta$ at all frequencies from zero to infinity. Table 9.2 shows $A\beta$ magnitudes and phase angles for an RC amplifier, for $A = 40$, $\beta = 0.1$ in Eqs. 8.27, 8.28, 8.47, 8.48. Values from the table are plotted in Fig. 9.14. The frequency point moves clockwise around the plot; at $\theta = +45°$ we have f_1, at $\theta = -45°$ we have f_2. The mid-frequency range appears on the x axis at $A\beta = 4.0$, since the phase angle is zero in the mid-range. Since $A\beta$ never becomes less than 0, the one-stage RC amplifier is unconditionally stable.

Suppose that two such stages are used in cascade. At the frequency extremes the amplifier phase angles approach $\pm 180°$, and the Nyquist plot

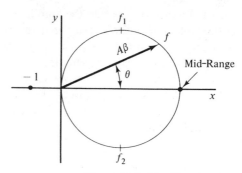

Figure 9.14 The Nyquist plot.

TABLE 9.2 *Aβ* Plot of *RC* Amplifier

f/f_1	$A\beta$	θ (degrees)	f/f_2	$A\beta$	θ (degrees)
0	0	+90	Mid-range	4.0	0
0.1	0.4	84	0.1	≈ 4.0	−6
0.25	1.0	76	0.25	3.9	−14
0.5	1.8	63	0.5	3.6	−27
1.0	2.8	45	1.0	2.8	−45
2.0	3.6	27	2.0	1.8	−63
4.0	3.9	14	4.0	1.0	−76
Mid-range	4.0	0	∞	0	−90

swings into the negative amplitude region. It may approach or surround the critical point at $A\beta = -1$. Such situations are shown in Fig. 9.15.

The *Nyquist criterion* states *an amplifier is unstable if the Nyquist curve encircles the* −1 *point and is stable otherwise.*

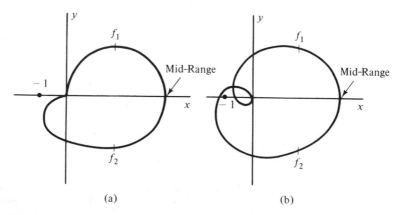

(a)   (b)

Figure 9.15 (a) Nyquist plot for a regenerative amplifier; (b) an unstable amplifier.

By this rule we see that the amplifier response diagrammed in Fig. 9.15(b) is unstable and will oscillate; that in Fig. 9.15(a) is regenerative due to the near approach to the -1 point. For this amplifier the gain-frequency curve will have a high-frequency hump.

In general, instability can be corrected by change of the $A\beta$ phase angle with a parallel R, C circuit in the line supplying V_f to the input; or by reduction of A through use of a series R, C circuit across a load, thus reducing gain at f_2 and above.

9.12 Gain and Phase Margin

The Nyquist diagram requires much labor but simple frequency plots of $A\beta$, in decibel magnitude and in phase angle, can be used for determination of the safety margin below the instability levels at $A\beta = 0$ dB and $\theta = 180°$.

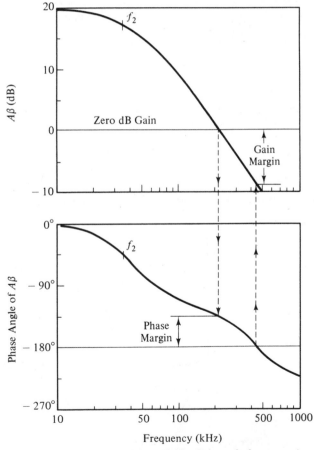

Figure 9.16   Gain and phase margin.

The *gain margin* is the decibel value of $A\beta$ at the frequency at which the phase angle reaches 180°. If negative, the amplifier is stable and can tolerate a theoretical gain increase equal to the margin, without regeneration. The *phase margin* is the angle of $A\beta$ at the frequency at which $A\beta$ reaches the 0-dB level or the unity magnitude ratio.

Figure 9.16 shows about -9 dB of gain margin and 45° of phase margin for a particular audio amplifier. Usual design limits are considered to be -10 dB of gain margin and 30° of phase. The frequencies at which these margins are measured are far beyond the normal mid-range of audio use; for instance, $f_2 = 34$ kHz but the gain margin is measured at 440 kHz. Again, this shows the necessity for phase control and gain limitation by RC compensating circuits at frequencies well beyond the operating band in feedback amplifiers. It indicates that reduced distortion has been purchased at the expense of problems associated with stability in greater bandwidths.

9.13 Comments

Negative feedback is primarily used to reduce distortion in amplifiers and to make gain figures precise. We design excess gain into the amplifiers and sacrifice this gain in using feedback in order to achieve desired results. With a large excess of internal gain, a large amount of negative feedback can be used and we obtain a precise overall gain without concern for variations of the internal gain figure. Widened bandwidth is an incidental result of using large amounts of feedback to reduce harmonic distortion.

Summarizing the advantages obtained by the negative feedback process,

1. Reduction of nonlinear distortion.
2. Stabilized gain figures.
3. Improved frequency response.
4. Voltage feedback—lower output resistance.
5. Current feedback—higher output resistance.
6. Lower or higher input resistance.

REVIEW QUESTIONS

9.1 Define negative feedback; regeneration.

9.2 What is the feedback factor?

9.3 What is meant by closed-loop gain?

9.4 Define voltage feedback; current feedback.

9.5 What is meant by series-input feedback; shunt-input feedback?

9.6 What conditions lead to increased gain with feedback?

9.7 What conditions lead to decreased gain with feedback?

9.8 In negative feedback, what is the phase relation of the voltage fed back to the input voltage?

9.9 Equation 9.3 tells you whether the feedback is negative or positive. What do you look for?

9.10 How does Eq. 9.52 tell us that the emitter follower has negative feedback?

9.11 Under what conditions does feedback reduce distortion?

9.12 What form of feedback increases the output resistance of an amplifier?

9.13 What form of feedback reduces the output resistance of an amplifier?

9.14 What is the circuit connection used to increase the input resistance of an amplifier? To decrease it?

9.15 What is the effect of negative feedback on bandwidth?

9.16 What is the definition of β?

9.17 When must β be positive and when negative for negative feedback?

9.18 How does the connection of the feedback circuit of an amplifier differ with an odd number of C-E stages from that with an even number of stages?

9.19 Why do audio amplifiers often have excessive bandwidth?

9.20 How does feedback affect the stability of amplifier gain?

9.21 How do you describe the form of feedback in an emitter follower?

9.22 How is current feedback obtained in a triode circuit?

9.23 Why does an unbypassed emitter resistor reduce the gain of a C-E amplifier?

9.24 Trace out the feedback loop in Fig. 9.11.

9.25 Trace out the feedback loop in Fig. 9.12.

9.26 What is a Nyquist diagram?

9.27 Why should $A\beta$ avoid the value -1?

9.28 Why is the mid-frequency range concentrated at one point in a Nyquist diagram?

9.29 What is the Nyquist criterion for stability?

9.30 Plot the $A\beta$ values of Table 9.1; show that a circle is obtained.

9.31 What is gain margin?

9.32 What is the phase margin?

9.33 Name six advantages provided by negative feedback.

9.34 How do we pay for the advantages of negative feedback?

PROBLEMS

9.1 For the black box feedback system in Fig. 9.17, if $V_s = 0.2$ V, $A = 20$, and $V'_o = 1$ V, find β, V_f, V'_i, and A'.

9.2 For the circuit in Fig. 9.17, we have $A = 50$, $\beta = 0.03$, and $V'_o = 5$ V. Find V_s, V'_i, and A'.

9.3 For the circuit in Fig. 9.17, we supply $V_s = 5$ V, with $A = -20$ and $\beta = -1.0$. Find the output V'_o, V_f, V'_i, and A'.

9.4 A 1-V input signal for V_s is used in Fig. 9.17 with $A = 60$, $\beta = 0.07$. What is V'_o, V_f, and the gain A'?

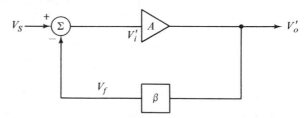

Figure 9.17

9.5 An amplifier of three stages with voltage gains of -50, -15, and -10 has overall feedback applied with $\beta = -0.01$. What is the overall gain with feedback?

9.6 What value of β is needed to reduce the gain of the amplifier of Problem 9.5 to -100?

9.7 In the circuit in Fig. 9.7(a), $R = 10,000\ \Omega$, $R_E = 1000\ \Omega$, $h_{ie} = 1700\ \Omega$, and $g_m = 0.007$ mho. Find the voltage gain; find R'_i; find R'_o.

9.8 In the circuit in Fig. 9.8(a) we wish to incorporate 10 dB of negative feedback. Find R_f if $R_s = 50,000\ \Omega$, $R = 4000\ \Omega$, and $A = -50$.

9.9 An amplifier has $A_{v(\text{mid})} = 200$ and $f_2 = 50$ kHz. When we add negative feedback with $\beta = 0.10$, what is the mid-frequency gain and what value of high-frequency band limit do we obtain?

9.10 In the circuit in Fig. 9.9, we have $R_D = 10,000\ \Omega$, $g_m = 0.004$ mho. With $R_1 + R_2 = 100$ kΩ, find R_2 to give a gain of -5.

9.11 In the emitter follower of Fig. 9.10(a) we have $R_E = 5000\ \Omega$, $g_m = 0.0025$ mho, and $h_{ie} = 1200\ \Omega$. Find the value of V_{be} for $V_s = 1.0$ V. What is the gain?

9.12 An amplifier has a mid-frequency gain of 300 and $f_2 = 500$ kHz. We wish to raise the upper frequency limit to 5 MHz by the use of negative feedback. What gain will remain? What can you say about the gain-bandwidth product?

9.13 An amplifier has $A = -100$ and $R_i = 5000\ \Omega$. What value of β should be used to increase the input resistance to $50,000\ \Omega$? How would you suggest that the feedback circuit be arranged? What is the gain with feedback?

9.14 A transistor with $h_{ie} = 1000\ \Omega$, $h_{fe} = 60$ is used in a C-E circuit with $R = 2000\ \Omega$, R_E (unbypassed) $= 700\ \Omega$. Find β and the gain with the feedback present.

9.15 Find the input impedance of the amplifier of Problem 9.14, with and without R_E present.

9.16 An amplifier without feedback has $A_v = -2700$. With feedback, the gain A'_v

is -97. What is the value of β being used? How much feedback in decibels is used?

9.17 Feedback of 15 dB is added to an amplifier with internal gain $A = 250$. What is the value of β required, and what is the gain with feedback? For the same output voltage, what will be the change in input voltage?

9.18 An amplifier has $A_v = -80$ and $V_o' = 100$ V with 8 per cent harmonic distortion. We wish to reduce the distortion to 0.5 per cent. What value of β should be used, and what input voltage is needed to give the same output as before?

9.19 The gain of an amplifier without feedback is 100. The output resistance without feedback was 1500 Ω. Plot a curve of output resistance with negative voltage feedback as a function of β over the range $\beta = 0.005$ to $\beta = 0.15$.

9.20 An amplifier has an internal gain of 37 dB and at 50-V output has 11 per cent distortion. Feedback is to be used to reduce the distortion to 1 per cent.
(a) What gain will be obtained?
(b) What input signal must be supplied for the same output?

9.21 An amplifier has an internal voltage gain $A = 512$ and an output of 12 V. Feedback is added until 0.95 V is required as input to give the same output. Find the β being used.

9.22 An amplifier in an electronic voltmeter must have a voltage gain of 100 within 0.5 per cent but the internal gain may change as much as 12 per cent due to component changes. Determine the value of β needed to meet the guarantee and the needed internal gain.

9.23 With negative feedback an amplifier gives an output of 10.5 V with an input of 1.12 V. When feedback is removed, it requires 0.15-V input for the same output. Find the value of β and of the gain without feedback.

9.24 An *RC* amplifier has three identical stages, with $f_1 = 48$ Hz, $f_2 = 140$ kHz for each stage. The overall internal gain is -450 and $\beta = -0.05$ is applied. Determine f_1' and f_2' for the amplifier with feedback.

10
Integrated
Amplifiers

Circuits that eliminate the series blocking capacitor are said to be *direct-coupled*. The mid-frequency range is thereby extended down to zero frequency. More importantly, by removal of the bulky blocking capacitor the amplifier size is reduced and it becomes possible to integrate an amplifier circuit on a silicon wafer along with its transistors and diodes. Such monolithic construction leads to small size, high reliability, reduced cost, and offsetting of temperature effects. Problems arise because of bias voltage needs but these are met by use of dual power supplies or by special circuit design.

By adding negative feedback the operational amplifier or so-called "op amp" has evolved. With this device we closely approach an ideal element having constant and controlled gain. In addition, high input resistance and low output resistance are achieved. Such gain elements can be cascaded without much concern for the effects of variable loads at the output.

10.1 The Integrated Amplifier

In Fig. 10.1 we have the circuit of an integrated direct-coupled amplifier. With negative feedback added externally, the circuit becomes a stable general-purpose high-gain amplifier.

The complete circuit contains 10 transistors, 2 diodes, and 16 resistors and

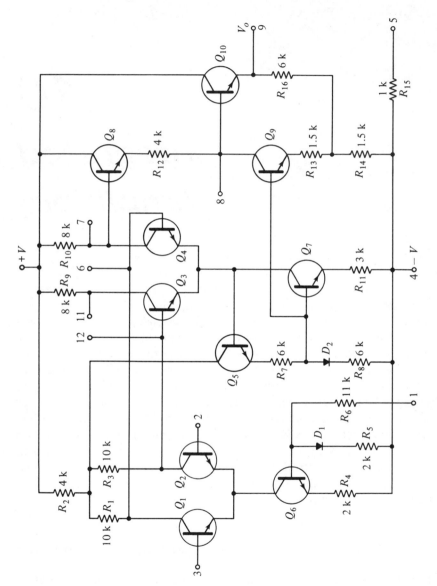

Figure 10.1 CA 3008, 60-dB gain to 0.1 MHz, no feedback.

is built on a chip of silicon approximately 1 mm^2. This chip is enclosed in a 12-terminal package about 1 cm in diameter and $\frac{1}{4}$ cm high.

While the amplifier appears to be complex, the circuit functions can be readily explained. Transistors Q_1 and Q_2 constitute an emitter-coupled differential amplifier stage. Input to terminal 3 gives an output in phase with the input and A is positive; input to terminal 2 inverts the signal or gives a reversed phase output and negative overall gain, $-A$. Terminal 4 is at $-V_{CC}$ and serves as the ground connection.

The outputs from Q_1 and Q_2 drive a second emitter-coupled differential pair at Q_3 and Q_4. The output is taken from Q_4 and drives Q_8, which supplies the base input for Q_{10}. This transistor furnishes the amplifier output at terminal 9 as an emitter follower, used to ensure a low output resistance.

Transistor Q_5, through its action on the currents of Q_3 and Q_4, cancels shifts in dc level that would occur with changes in dc supply voltage. A decrease in V_{CC} causes a decrease in the voltage at the emitters of Q_3 and Q_4. This negative-going change in voltage acts on transistor Q_5 with its emitter output going to transistors Q_7 and Q_9, and less current passes these transistors. But less current in Q_3, Q_4, and Q_8 raises the collector voltages of these transistors, canceling most of the original downward shift from V_{CC}. Transistor Q_6 is a constant-current control for the first differential pair.

The output circuit of Q_9 and Q_{10} shifts the dc level at the output so that it is substantially equal to the zero level at 2 or 3 with zero signal.

All transistors are manufactured in the same operations and while Q_5, Q_6, Q_7, and Q_9 could be eliminated without reduction in the overall gain of the block, their presence adds little to the cost and much to the overall stability of output with varying supply voltages. Made as a monolithic element on silicon, the transistor characteristics are more nearly the same than when discrete units are assembled. The resistors are formed on the chip as well and the connection costs are not increased by the circuit complexity.

The two transistors of each differential pair will cancel common variations. Since the two transistors are formed on the same chip with only a few thousandths of a millimeter separating them, temperature differentials between transistors are negligible and the effects of temperature changes on transistor characteristics are canceled.

There is a great variety of such integrated circuits available. In general, they consist of four stages and Fig. 10.1 illustrates these. The first stage is the differential amplifier of Q_1 and Q_2, with differential output; the second stage is a cascaded differential amplifier with single-ended output, composed of Q_3 and Q_4. Third comes an emitter follower Q_8 to lower the dc voltage level back toward that of the input, and the fourth stage is the output at Q_{10}.

A study of the elemental circuits employed and some of the applications

of these "gain blocks" are the objectives of this chapter. Additional information on integrated circuit processing will be provided at the end of the
chapter.

10.2 The Differential Amplifier

As can be concluded from the previous description, the *differential amplifier*
is the basic element of the integrated amplifier. Drawn in Fig. 10.2, the
differential amplifier circuit is symmetrical about the vertical dashed line
a-a' and identical parameter changes on each side are balanced out.

Direct-coupled amplifiers are unable to distinguish between changes in
the dc value of a signal and temperature-caused changes in v_{BE}, h_{FE}, and the
reverse saturation current and so the output current will drift with temperature. In the differential amplifier the two transistors and the load resistors
form a balanced resistance bridge at zero signal. A simultaneous change in
v_{BE} of the two transistors will change the collector currents; the voltages at
A and *B* will change identically but the difference $A - B$ will not be affected.
Temperature drifts can be limited to effective input values of 3 μV per °C
(2 μV per °F).

The circuit is well suited to integrated unit production because the
simultaneously produced transistors and resistors will be matched closely.
Equality of resistors is an important factor that can be easily achieved,
whereas to produce resistors of an exact magnitude is difficult in processing.

If a *differential input* voltage is applied, the inputs V_1 and V_2 will be equal
in magnitude but of opposite polarity. With equal transistor parameters, one
collector current will increase and the other will decrease. Voltages at *A* and
B change up and down so that there is a voltage V_o between *A* and *B*. The
changes in the respective collector currents are illustrated in the *transfer curves*
in Fig. 10.3. Since the sum of the two currents remains constant, there is no
signal voltage change across R_E.

If a signal is introduced to both transistors in a *common mode*, with both
inputs being equally positive or in phase as an example, the collector currents
increase identically and the bridge remains balanced with equal voltages at
A and *B*. The output V_o remains zero. The current in R_E does change with a
common-mode input, however, and a common-mode signal appears across
R_E.

The equivalent circuit is drawn in Fig. 10.2(b), in a form that emphasizes
the bridge action of the amplifier. With identical transistor parameters we can
write the circuit relations as

$$V_1 = (R_s + h_{ie})I_{b_1} + R_E I_e \qquad (10.1)$$
$$V_2 = (R_s + h_{ie})I_{b_2} + R_E I_e \qquad (10.2)$$
$$I_e = I_{e_1} + I_{e_2} \cong h_{fe}I_{b_1} + h_{fe}I_{b_2} \qquad (10.3)$$

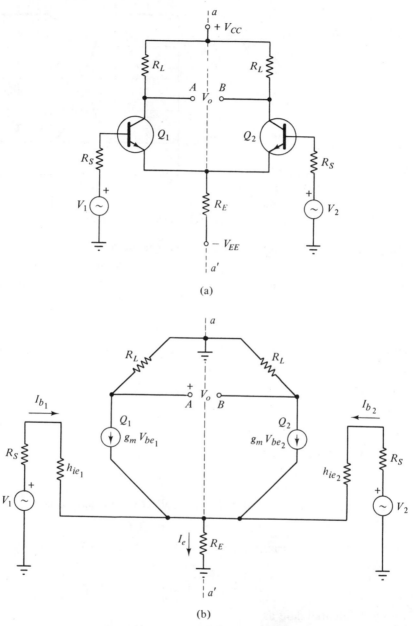

Figure 10.2 (a) The differential amplifier; (b) equivalent circuit.

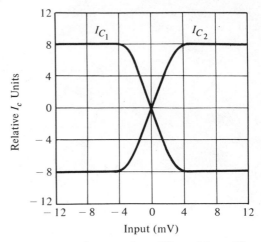

Figure 10.3 Transfer curve for a differential amplifier.

Using Eq. 10.3 in Eq. 10.1 and 10.2, we arrange the results as

$$V_1 = (R_s + h_{ie})I_{b_1} + h_{fe}R_E I_{b_1} + h_{fe}R_E I_{b_2} \qquad (10.4)$$
$$V_2 = (R_s + h_{ie})I_{b_2} + h_{fe}R_E I_{b_2} + h_{fe}R_E I_{b_1} \qquad (10.5)$$

Subtracting, we have

$$V_1 - V_2 = (R_s + h_{ie})(I_{b_1} - I_{b_2})$$

However, with balanced or matched transistors a change in I_{b_1} is a negative change in I_{b_2}, or $I_{b_1} = -I_{b_2}$, and so

$$V_1 - V_2 = 2(R_s + h_{ie})I_{b_1}$$

from which

$$I_{b_1} = \frac{V_1 - V_2}{2(R_s + h_{ie})} \qquad (10.6)$$

Similarly we can find that

$$I_{b_2} = \frac{V_2 - V_1}{2(R_s + h_{ie})} \qquad (10.7)$$

The voltage between A and B at the collector connections is

$$V_o = h_{fe}R_L I_{b_1} - h_{fe}R_L I_{b_2}$$

Substitution of the current values from Eqs. 10.6 and 10.7 yields

$$V_o = \frac{g_m R_L (V_1 - V_2)}{1 + \dfrac{R_s}{h_{ie}}}$$

with a differential gain as

$$A_{vd} = \frac{V_o}{V_1 - V_2} = \frac{g_m R_L}{1 + \dfrac{R_s}{h_{ie}}} \qquad (10.8)$$

With V_1 and V_2 opposite in polarity, $V_1 = -V_2$, we have an output V_o. If V_1 and V_2 are equal in magnitude and positive, the output is zero.

We have stated that no signal voltage appears across R_E with differential input voltage; confirming this is the absence of R_E in the output voltage expression.

It is evident from either Eq. 10.6 or Eq. 10.7 that the input resistance to either base is

$$R_i = \frac{V_1}{I_{b_1}} = \frac{V_2}{I_{b_2}} = 2(R_s + h_{ie}) \qquad (10.9)$$

This represents the series resistance of the path through the transistors and V_1 and V_2; the signal currents bypass the emitter resistor and it could be eliminated from the equivalent circuit because no signal voltage appears across it.

When the output is taken from one collector to ground, the operation is said to be *single-ended* and the gain is one-half of the differential gain, as

$$V_o \text{ (single-ended)} = \frac{-g_m R_L (V_1 - V_2)}{2\left(1 + \dfrac{R_s}{h_{ie}}\right)} \qquad (10.10)$$

The differential amplifier circuit becomes an *emitter-coupled phase inverter* when the signal is applied to one input and the second base is grounded. The output voltages at 2 and 2′ are of equal magnitude and 180° in phase. This is shown in Fig. 10.4.

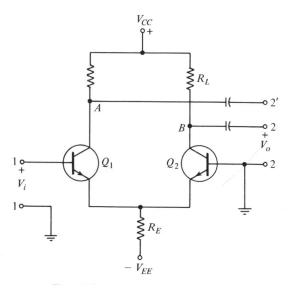

Figure 10.4 An emitter-coupled phase inverter.

10.3 Rejection of Common-Mode Signals

The differential amplifier tends to cancel effects common to its two sides and this action extends to signal voltages that are equally introduced in phase (not oppositely, as with differential plus and minus signals). Such *common-mode signals* are frequently caused by ac variations in the supply voltage or by voltages from stray magnetic fields in the ground or signal leads. The common-mode signal is introduced equally to transistors Q_1 and Q_2 as shown in Fig. 10.5 and these signals are usually unwanted in the amplifier output.

In Fig. 10.5 we have the two inputs to the amplifier as V_d for the differential input and V_c for the common-mode input. With differential input the transistor voltages are equal but plus and minus, or $V_{i_1} = -V_{i_2}$. For the common-mode signal, $V_{i_1} = V_{i_2}$ and

$$V_c = V_{i_1}$$
$$V_c = V_{i_2}$$

By adding,

$$V_c = \frac{V_{i_1} + V_{i_2}}{2} \tag{10.11}$$

With a pure differential signal and $V_{i_1} = -V_{i_2}$, the common signal is zero.

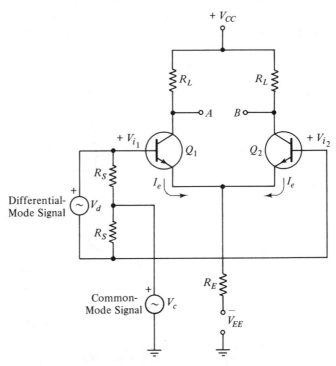

Figure 10.5 Showing a common-mode signal.

As long as we have a perfectly symmetrical circuit, the common-mode signal will be canceled but we want to know how to maximize the rejection of common-mode signals when circuit unbalances occur. If a positive common-mode signal causes the transistor currents to rise, as would be expected with the *npn* units in Fig. 10.5, the voltage drop across R_E also rises. This increased emitter voltage subtracts from the input signal and causes negative feedback, reducing the gain for the common signal. Differential-mode signals are not passed through R_E and no differential voltage appears there. We usually want the common-mode signal to be rejected so that a high value of R_E for feedback is indicated.

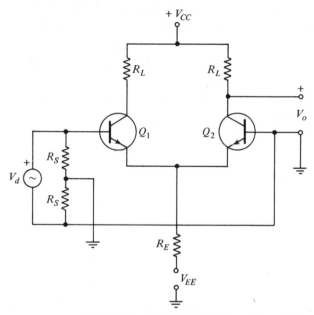

Figure 10.6 Single-ended differential amplifier.

We often wish to have one side of the output at ground potential and use a single-ended output circuit, as in Fig. 10.6. Because of the loss of circuit symmetry, a common-mode signal will produce some output, although there will be a substantial reduction in gain for the common-mode signal. A measure of the rejection of the common-mode signal in the output is given by the *common-mode rejection ratio*, defined as

$$\text{CMRR} = \frac{\text{differential-mode gain}}{\text{common-mode gain}}$$
$$= \frac{A_{vd}}{A_{vc}} \tag{10.12}$$

The output consists of differential output and common output so that

$$V_o = A_{vd}V_d + A_{vc}V_c \tag{10.13}$$

We can develop a useful relation as

$$V_o = A_{vd} V_d \left(1 + \frac{A_{vc} V_c}{A_{vd} V_d} \right)$$

$$= A_{vd} V_d \left(1 + \frac{1}{\text{CMRR}} \frac{V_c}{V_d} \right) \tag{10.14}$$

Since $A_{vd} V_d$ is the desired output, we readily determine the effect at the output by the additive term $(1/\text{CMRR})(V_c/V_d)$.

To obtain the CMRR for the circuit in Fig. 10.4, we split the circuit down the a-a' line; since R_E appears in each half, it is shown as $2R_E$ in Fig. 10.7. It becomes R_E when paralleled in the actual circuit. Each half is a C-E transistor circuit with an emitter resistor $2R_E$ and the gain is

$$A_{vc} \cong \frac{-h_{fe} R_L}{h_{ie} + R_s + 2h_{fe} R_E} = \frac{-g_m R_L}{1 + \dfrac{R_s}{h_{ie}} + 2g_m R_E} \tag{10.15}$$

With Eq. 10.8 for the differential gain we can calculate the CMRR value for the circuit as

$$\text{CMRR} \cong 2g_m R_E \tag{10.16}$$

This result confirms the assertion that R_E should be large for a large value of CMRR.

Example 1: With a transistor having $g_m = 0.0025$ mho, what value must R_E have to obtain a CMRR value of 40 dB?

To obtain CMRR as a ratio,

$$40 = 20 \log \text{CMRR}$$

$$\log \text{CMRR} = \frac{40}{20} = 2$$

$$\text{CMRR} = 10^2 = 100 = 2g_m R_E$$

from which

$$R_E = \frac{100}{2 \times 2.5 \times 10^{-3}} = \frac{10^5}{5} = 20{,}000 \ \Omega$$

Example 2: We have input voltages of $V_{i_1} = 50 \ \mu\text{V}$ and $V_{i_2} = -50 \ \mu\text{V}$. The difference-mode gain of the amplifier is $A_{vd} = 1000$ and the CMRR value is (a) 100 (40 dB); (b) 10,000 (80 dB). Calculate the output.

We find

$$V_d = V_{i_1} - V_{i_2} = 50 - (-50) = 100 \ \mu\text{V} = 0.1 \ \text{mV}$$

Then

$$V_c = \frac{V_{i_1} + V_{i_2}}{2} = \frac{50 - 50}{2} = 0$$

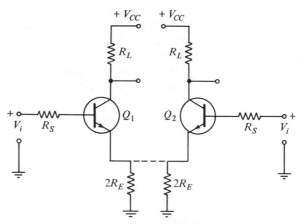

Figure 10.7 Circuit analysis for CMRR.

The differential signal output is

$$V_o = A_{vd}V_d = 10^3 \times 0.1 \times 10^{-3} = 100 \text{ mV}$$

and we have zero common-mode output.

Example 3: For the same amplifier we now have signals $V_{i_1} = 1.00 \text{ mV}$ and $V_{i_2} = 0.90 \text{ mV}$.
  We find

$$V_d = V_{i_1} - V_{i_2} = 1.00 - 0.90 = 0.10 \text{ mV}$$

as the differential signal.
  The common-mode signal is

$$V_c = \frac{V_{i_1} + V_{i_2}}{2} = \frac{1.00 + 0.90}{2} = 0.95 \text{ mV}$$

(a) With CMRR = 100, we use Eq. 10.14 to find

$$V_o = A_{vd}V_d \left(1 + \frac{1}{\text{CMRR}} \frac{V_c}{V_d}\right)$$
$$= 10^3 \times 0.10 \left(1 + \frac{1}{100} \frac{0.95}{0.1}\right) = 109.5 \text{ mV}$$

For the same differential input as in Example 1, the output has been increased 9.5 per cent by the presence of the common-mode signal.
  (b) With CMRR = 10^4, we again use Eq. 10.14 to find

$$V_o = 10^4 \times 0.10 \left(1 + \frac{1}{10^4} \frac{0.95}{0.1}\right) = 100.095 \text{ mV}$$

With the larger CMRR the error in the output signal caused by the common-mode voltage is only 0.095 mV, or about 0.1 per cent.

10.4 A Constant-Current Circuit for R_E

For differential-mode signals we expect the current in R_E to be constant. We would also like R_E to be very large to give us a high value of CMRR. We can achieve these results by use of transistor Q_3 in the *constant-current circuit* in Fig. 10.8.

Looking at the I_3 series circuit,

$$(R_2 + R_3)I_3 + V_D - V_{EE} = 0$$

after neglecting I_B as small. If $V_{EE} \gg V_D$, then we have for I_3

$$I_3 \cong \frac{V_{EE}}{R_2 + R_3} \tag{10.17}$$

Around the base-emitter circuit of Q_3 we can write

$$R_1 I_o + V_{BE_3} = R_2 I_3 + V_D$$

and using I_3 from Eq. 10.17

$$R_1 I_o = \frac{R_2 V_{EE}}{R_2 + R_3} + V_D - V_{BE_3} \tag{10.18}$$

If we select diode D to have a voltage characteristic equivalent to that of the base-emitter diode of Q_3, then $V_D = V_{BE_3}$ and

$$I_o \cong \frac{R_2 V_{EE}}{R_1(R_2 + R_3)} \tag{10.19}$$

This expression for transistor current contains no transistor parameters and is a constant.

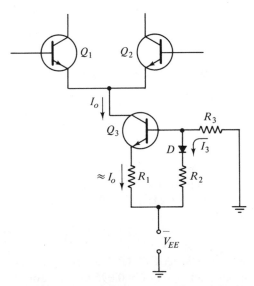

Figure 10.8 A constant-current bias transistor.

Transistor Q_3 is a current feedback amplifier and so has a large output resistance at the collector. This may approximate several megohms. This gives a very large effective R_E value in Eq. 10.16 and a very large CMRR value.

The value of V_{BE} decreases about 2.5 mV per °C (1.5 mV per °F) but the diode can be chosen to vary similarly and the cancellation of V_D and V_{BE} in Eq. 10.18 can be achieved at all usual temperatures. As a result, I_o is independent of temperature.

These circuits appear at Q_6 and Q_7 in the emitter leads of the differential amplifiers in Fig. 10.1.

A simpler but less accurate constant-current circuit is formed by a transistor with a fixed voltage reference applied to the base. The collector current is then $I_C = h_{FE}I_B$ and constant. The accuracy of current control is dependent on constant v_{BE} and h_{FE} values. This is the function of Q_9 in the amplifier in Fig. 10.1.

10.5 Voltage References

We have just discussed an application in which a transistor was biased by the forward-voltage drop of a diode. The forward-biased silicon diode may be assumed as presenting a constant voltage, about 0.7 V, regardless of current. To obtain greater voltages, several diodes may be connected in series. By choice of the temperature characteristic of the diode, temperature compensation can be added to the function of a voltage reference diode, as shown in Sec. 10.4.

These drops are smaller than can be obtained by use of a Zener diode and, more importantly, a diode such as D in Fig. 10.8 can be processed in the same steps and with the same materials as are required for the emitter-base junction of a transistor on the silicon chip.

10.6 The DC Level Shifter

With each stage deriving its base input voltage from the preceding collector, the dc voltage level of each base rises with respect to ground, as one progresses through a direct-coupled amplifier. It is therefore necessary to shift the voltage level back down to obtain an output at which zero output voltage corresponds to zero input voltage. Using resistors and transistors, the *dc level shifter* in Fig. 10.9 is well suited to integrated circuit production methods.

Input is supplied to Q_1, which operates as an emitter follower. Signal output is taken from Q_3, also an emitter follower, to provide a low output resistance.

Transistor Q_2 is a constant-current device, with its base supplied by a fixed reference source; therefore $I_C = h_{FE}I_B$ and is constant. With current I_E

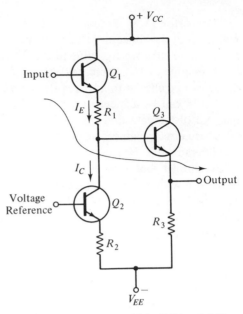

Figure 10.9 DC level shifter.

also constant, since $I_E \cong I_C$, the drop in R_1 is constant. Along the signal path shown we have

$$V_{BE_1} + R_1 I_E + V_{BE_2} = K \qquad \text{constant volts}$$

We call ΔV_{in} a change at the input and ΔV_{out} the resultant change at the output. Voltages with no signal at input and output are V_{in} and V_{out}. Then we can write through the signal path

$$\Delta V_{in} + V_{in} - K \text{ volts} = \Delta V_{out} + V_{out}$$

A *change* in the input is transferred directly to the output as

$$\Delta V_{in} = \Delta V_{out}$$

but the no-signal voltage level at the output is lower than that at the input because

$$V_{in} - K = V_{out} \tag{10.20}$$

A level shifter of this kind appears in Fig. 10.1, using Q_8 and Q_9 with Q_{10} for output.

10.7 The Operational Amplifier

When negative feedback is added to an integrated dc amplifier of large internal gain, we have an *operational amplifier* or op amp. It was originally given the operational name because of its use in performing the mathematical operations of addition, integration, and differentiation but its applications are now much more general. The basic circuit is that in Fig. 10.10(a).

The amplifier may have a number of stages and resistor R_f, or some other circuit element, supplies voltage-shunt feedback around the gain element, stabilizes the gain, and lowers the output resistance. We define the internal gain as

$$A = -\frac{V_o}{V_i'} \tag{10.21}$$

We shall require that gain A be very large as a condition of amplifier design, perhaps $A > 10^4$. This permits us to say that $V_i' \cong 0$ at the amplifier terminal at 1. To justify this statement, let $A = 10^4$ and $V_o = 10$ V; then $V_i' = 1$ mV. With large A, we are also saying that $A\beta \gg 1$ in the usual feedback amplifier expression.

With $V_i' \cong 0$, however, the current to the amplifier must be negligible as well, or $I_i' = 0$ at terminal 1. Then a current summation there gives

$$I_1 = -I_2 \tag{10.22}$$

With the voltage V_i' as zero, these currents can be stated

$$I_1 = \frac{V_i}{R_1}$$

$$I_2 = \frac{V_o}{R_f}$$

Using Eq. 10.22

$$\frac{V_i}{R_1} = -\frac{V_o}{R_f}$$

and the gain with feedback is

$$A' = \frac{V_o}{V_i} = -\frac{R_f}{R_1} \tag{10.23}$$

This is the basic gain relation of the inverting operational amplifier.

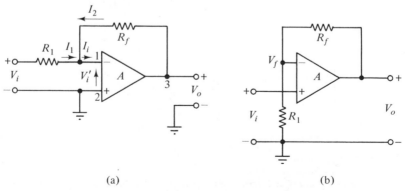

(a)　　　　　　　　　　　　(b)

Figure 10.10　(a) The operational amplifier; (b) the noninverting form.

The stability of the gain with feedback is dependent only on two resistances, or impedances in general, and the value of the gain can be easily adjusted. Note that the gain with feedback is independent of the internal gain, *provided that the internal gain is made large.*

Equation 10.23 is a result of A being large, as was Eq. 9.11:

$$A' = \frac{1}{\beta}$$

so we see that for the inverting operational amplifier

$$\beta = -\frac{R_1}{R_f} \tag{10.24}$$

Using an amplifier integrated on a silicon chip and adding two resistors and a power supply, we have a stable package of gain. The package will have high input resistance, low output resistance, and a wide and controllable bandwidth. The operational amplifier employs the circuits of the preceding sections in functions somewhat as shown by the typical arrangement in Fig. 10.11.

Because of the differential amplifier at the input, there are two input connections available: one resulting in an inverted output or a gain $-A'$ and the other giving a noninverted output and gain $+A'$. This adds further flexibility to the operational amplifier. When it is used, the second input is

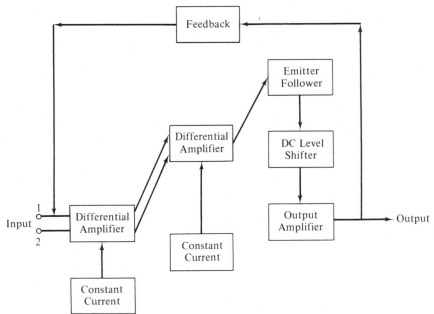

Figure 10.11 Functions of an integrated operational amplifier, illustrating the dc levels.

applied to the positive gain input terminal of the amplifier. The feedback is returned to the negative input terminal so that

$$V_f = \frac{R_1}{R_1 + R_f} V_o = \beta V_o$$

to ground. There is a differential input $V_i - V_f$ and the output of the amplifier is

$$V_o = A(V_i - V_f) = A\left(V_i - \frac{R_1}{R_1 + R_f} V_o\right) \tag{10.25}$$

$$V_o\left(1 + \frac{AR_1}{R_1 + R_f}\right) = AV_i$$

and the gain is

$$A' = \frac{V_o}{V_i} = \frac{A}{1 + \dfrac{AR_1}{R_1 + R_f}} \tag{10.26}$$

If $A \gg 1$ as we have required, then the gain with feedback reduces to

$$A' = \frac{R_1 + R_f}{R_1} \tag{10.27}$$

This approximates a positive equivalent to Eq. 10.23 and becomes equal when $R_f \gg R_1$.

Example 1: The circuit in Fig. 10.10(a) is used for signal inversion, with $A = -100,000$, $R_1 = 1000\ \Omega$, and $R_f = 10,000\ \Omega$. The gain is

$$A' = -\frac{R_f}{R_1} = -\frac{10,000}{1000} = -10$$

Since $\beta = -R_1/R_f = -0.1$, we could use the accurate feedback expression

$$A' = \frac{-A}{1 + A\beta} = \frac{-10^5}{1 + 10^5 \times 0.1} = \frac{-10^5}{1 + 10^4} = -9.9990$$

which shows that when $A = -10^5$, we can certainly neglect $1 \ll A\beta$. This is equivalent to saying that $V_i' = 0$.

Example 2: We use the circuit of Fig. 10.10(b) in the noninverting connections. The gain is

$$A' = \frac{R_1 + R_f}{R_1} = \frac{1000 + 10,000}{1000} = 11.0$$

Since $\beta = R_1/(R_1 + R_f) = 0.0911$, then the accurate feedback expression gives

$$A' = \frac{A}{1 + A\beta} = \frac{10^5}{1 + 9090} = 10.999$$

which again confirms our requirement that A be large.

10.8 The Unity-Gain Isolator

One of the simplest applications of the operational amplifier is the *unity-gain* circuit in Fig. 10.12, useful in isolating one circuit from variations that may occur in a load circuit.

From the circuit,

$$V_i + V_i' = V_o \qquad (10.28)$$

But $V_i' \cong 0$ so that

$$V_i = V_o \qquad (10.29)$$

and the output voltage follows the input signal, without a phase reversal. Since the output resistance is low, needed output currents can be obtained without loading an input circuit at 1,1, and the circuit is a buffer.

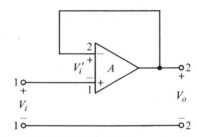

Figure 10.12   Unity-gain isolator circuit.

10.9 The Summing Operation

Several input voltages may be simultaneously operated upon in the operational amplifier in Fig. 10.13. Current I_1 represents the sum of the three current components through R_a, R_b, and R_c. That is, $-I_2 = I_1$ and

$$-\frac{V_o}{R_f} = \frac{V_a}{R_a} + \frac{V_b}{R_b} + \frac{V_c}{R_c}$$

and so

$$V_o = -\left(\frac{R_f}{R_a} V_a + \frac{R_f}{R_b} V_b + \frac{R_f}{R_c} V_c\right) \qquad (10.30)$$

and the output voltage is equal to the weighted negative sum of the several inputs.

If $R_a = R_b = R_c$, then

$$V_o = -\frac{R_f}{R_a}(V_a + V_b + V_c) \qquad (10.31)$$

If $R_f = R_a$, the output represents a negative summation of the input voltages.

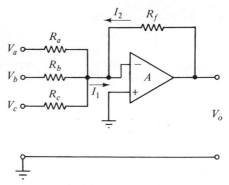

Figure 10.13 A summing amplifier.

10.10 The Integration Operation

By use of a capacitor in place of R_f, the operational amplifier will perform an integration of the input voltage. An integrator circuit is shown in Fig. 10.14.

The result of integration can be understood if we recall that the current-voltage relation for a capacitance, connected between V_o and ground potential at 1, is

$$V_o = \frac{1}{C} \int_0^t I_2 \, dt \qquad (10.32)$$

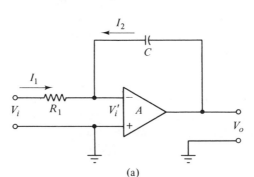

(a)

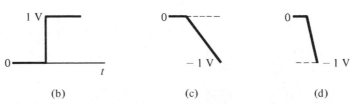

| (b) | (c) | (d) |

Figure 10.14 (a) Operational amplifier as an integrator; (b) input rectangular pulse; (c) output integral, $1/RC = 1$; (d) output integral, $1/RC = 0.1$.

Now we have previously shown that

$$I_1 = -I_2 \qquad \qquad (10.33)$$

because $I_i' \cong 0$. Since $V_i' = 0$, we have said

$$I_1 = \frac{V_i}{R_1}$$

Because of Eq. 10.33,

$$I_2 = -\frac{V_i}{R_1} \qquad \qquad (10.34)$$

Substitution of this result in Eq. 10.32 gives

$$V_o = -\frac{1}{R_1 C} \int_0^t V_i \, dt \qquad \qquad (10.35)$$

as the output-input relation for the operational amplifier. The output voltage is the integral of the input, with a scale factor of $-1/R_1 C$.

If we make $R_1 = 1 \text{ M}\Omega$ and $C = 1 \text{ }\mu\text{F}$, we have

$$R_1 C = 10^6 \times 10^{-6} = 1 \text{ s}$$

and the scale factor is -1. Since RC has units of time, in seconds, we can use the RC scale factor to scale problems using a time variable.

The integration operation is demonstrated on the step input in Fig. 10.14(b), the result being of ramp voltage form as the capacitor charges. If we had made $R = 0.1 \text{ M}\Omega$, then $1/RC = 10$ and the result would plot toward -10, reaching -1 V in one-tenth of the time, so that time is scaled by changing RC.

Figure 10.15 shows a multiple input integrator with different scaling factors, as might be used in an analog computer.

The differentiating operation, opposite to that of integration, can be performed by using C for R_1 and a resistance as R_f. We encounter noise problems in differentiation, however, and the operation is avoided.

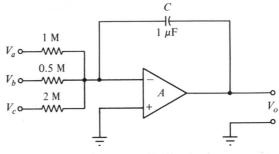

Figure 10.15 Use of different scale factors in integration of several signals.

10.11 The Comparator

One useful application of the differential amplifier is that of a *comparator* of the magnitudes of a signal voltage and a reference voltage. The voltage comparator in Fig. 10.16 is basic in digital computer applications. Its action is demonstrated by the transfer curve of the differential amplifier in Fig. 10.3. When the input signal is slightly greater than the reference voltage, the output swings to saturation; when the input signal is slightly less than the reference voltage, the output swings to saturation on the other side and the output voltage reverses.

A comparator is used in digital voltmeters, where the input voltage is compared with an internally generated ramp voltage. The cycles of an oscillator are counted from turn-on, with the count being stopped with a signal from the comparator at equality between the unknown and the ramp voltage. If the ramp wave rises at 10 mV per millisecond and the oscillator is at 10,000 Hz, the display will read 1500 counts for an input voltage of 1.5 V. Placement of the decimal point will cause the meter to read 1.500 V.

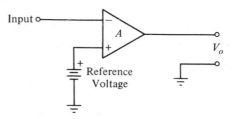

Figure 10.16 Basic comparator circuit.

10.12 A Millivoltmeter

In Fig. 10.17 another application of an operational amplifier yields a *millivoltmeter*. The instrument M may be one of 1-mA full-scale deflection.

Since the operational amplifier gain expression yields

$$V_o = -\frac{R_f}{R_1} V_i \tag{10.36}$$

$$\frac{V_o}{R_o} = I_o = -\frac{R_f}{R_1 R_o} V_i$$

So we have

$$\frac{I_o}{V_i} = -\frac{R_f}{R_1 R_o} \tag{10.37}$$

This is the basic equation for the circuit and states output current per volt of

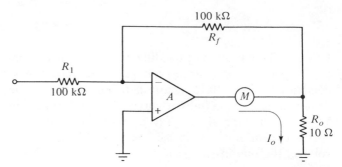

Figure 10.17 A millivoltmeter using an operational amplifier.

input. If we measure I_o in milliamperes, then V_i must be measured in millivolts so we have units of milliamperes per millivolt.

With the circuit values shown in Fig. 10.17, we have

$$\frac{I_o}{V_o} = -\frac{10^5}{10^5 \times 10} = -\frac{1}{10}$$

so that we shall have a reading of 1 mA, or full scale on the instrument, for 10-mV input to the amplifier.

The full-scale value in millivolts can be readily changed to other millivolt values by changing R_1 or R_f.

10.13 Frequency Compensation

As a feedback amplifier of many stages, the operational amplifier can have gain instability and oscillation at the high frequencies. The several stages have input capacitances for the transistors and the gain falls off or "rolls off" above limit frequencies fixed by these capacitances and associated resistances. Some amplifiers are internally compensated for stability but others require external R and C components so that the designer can choose his own bandwidth and stability margins.

A typical uncompensated gain curve for an operational amplifier is shown in Fig. 10.18, without feedback. The curve is drawn by use of the asymptotes of the frequency response curves, for simplicity in sketching. Uncompensated, the amplifier shows 60 dB of gain to the first limit frequency f_2 at about 200 kHz. Two more limit frequencies are determined by the RC combinations of the several stages at f_2' and f_2''. Each combination causes a rate of gain fall of -20 dB per frequency decade and these rates are additive, reaching -60 dB per decade above f_2'' at about 20 MHz.

We want to obtain 20 dB of closed-loop or feedback gain, thus using 40 dB of feedback $= (1 + A\beta)$ dB. We draw the horizontal line at 20 dB and discover the gain to be limited at about 8 MHz and that the amplifier curve is

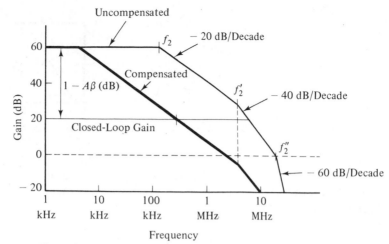

Figure 10.18 Uncompensated and compensated amplifier performance.

falling at a rate of -40 dB per decade. From our study of feedback amplifiers we know that a rate of -20 dB is associated with a phase shift of $\pm 90°$ and is stable and that a rate of -40 dB per decade represents a phase angle of $\pm 180°$ with potential instability. From the discussion of gain and phase margin we know that our gain curve should cross the 0-dB gain line before the phase angle reaches $180°$. For an amplifier to be stable, the gain curve should cross the 0-dB gain line at a slope not greater than -20 dB per decade.

We must compensate the gain curve by introducing a series RC circuit across appropriate terminals designated by the amplifier manufacturer. The roll-off of a simple RC circuit is -20 dB per decade and we design the compensating circuit to have a corner frequency of about 7 kHz, sufficiently low that the -20-dB slope extends to 0 dB before we reach the f_2' frequency. This is shown as the heavy line in the figure. The gain is less than 0 dB before the phase angle of the gain reaches $180°$ at f_2' and the amplifier will be stable.

Instead of reaching a bandwidth of 8 MHz, which might have been hoped, we now have a bandwidth of only about 400 kHz; however, this is still slightly wider than that of the amplifier alone. This is the price we must pay for stability.

10.14 The Slew Rate

Due to the charging time of the input capacitances of the transistors, the rate of rise of voltage is limited when a step input is applied to an operational amplifier. This maximum rate of voltage rise is called the *slew rate*.

Consider the basic amplifier circuit in Fig. 10.10. The full input voltage step appears at input 1 because the output feedback from V_o cannot respond

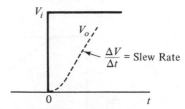

Figure 10.19   Effect of the slew rate on rise of output voltage.

instantly. This large value of input voltage drives one or more stages of the amplifier into saturation and the transistor capacitors are charged. As shown in Fig. 10.19, output V_o can rise only as fast as the internal capacitances can discharge and so we have the slew rate, given by

$$S = \frac{\Delta V_o}{\Delta t} \quad (\text{V}/\mu\text{s}) \tag{10.38}$$

This is a specification supplied by the manufacturer of the amplifier.

The maximum rate of change in a sine wave is a function of frequency, and the amplifier slew rate may limit the amplifier in following a large signal at high frequency. The maximum rate of change of a sine wave occurs at the zero crossing and the slew rate should be greater, or

$$S = \frac{\Delta V_o}{\Delta t} \geq 2\pi f V_{\max} \tag{10.39}$$

where f is the frequency in megahertz, $V_{\max}$ is the peak output voltage in volts, and the slew rate is in volts per microsecond.

The slew rate is primarily of concern in setting a limit on the time in which the amplifier may be switched from on to off.

Example: Given an operational amplifier with a stated slew rate of 10 V per μs and with a supply voltage of $V_{CC} = 15$ V. In switching, the output will go from ≈ 0 to ≈ 15 V as on to off conditions. The time for this operation is determined by the slew rate as

$$T_s = \frac{V}{S} = \frac{15\,\text{V}}{10\,\text{V}/\mu\text{s}} = 1.5 \ \mu\text{s}$$

With a sine wave of $V_o = 15$ V peak value, the slew rate will limit the response of this amplifier to

$$f = \frac{S}{2\pi V_{o(\max)}} = \frac{10}{2\pi \times 15} = 0.106 \ \text{MHz}$$

If the peak of the sine wave is only 1 V, then the expression above shows that the response band would extend to 1.59 MHz.

10.15 Offset Voltage and Current

With both input terminals grounded, a practical operational amplifier will develop a small output voltage, due to inherent imbalance in the circuits. The *input offset voltage* is defined as that voltage that must be supplied to one of the inputs to reduce the output voltage to zero. The value of the input offset voltage V_{is} can be determined from the output offset as

$$V_{is} = \frac{R_1}{R_1 + R_f} V_{os} \tag{10.40}$$

as in Fig. 10.20(a). The offset voltage V_{is} is typically a few millivolts; hence amplifiers with appropriately small values should be selected.

A difference in base bias currents at the input of an operational amplifier is called *bias current offset*. The effect can be reduced by insertion of a resistor R_2 in the circuit having excess current as in Fig. 10.20(b). The resistor should be

$$R_2 = \frac{R_1 R_f}{R_1 + R_f} \tag{10.41}$$

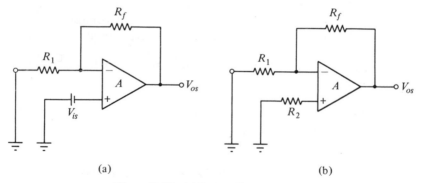

(a)   (b)

Figure 10.20   (a) Input offset voltage; (b) input offset current.

10.16 Definition of Terms

There are a number of terms used in amplifier specifications that need to be understood, and we gather some of them here:

Bandwidth. The frequency at which the voltage gain of the amplifier is 3 dB below the voltage gain at mid-frequency.

Common-mode Voltage Gain. The ratio of the ac voltage between the two output terminals to the ac voltage applied to the two input terminals connected in parallel for ac.

Differential-mode Voltage Gain. The ratio of the voltage change between two output terminals to the change in voltage between the two input terminals.

Common-mode Rejection Ratio (CMRR). The ratio of the differential-mode voltage gain to the common-mode voltage gain.

Differential Voltage Gain—Single-ended Output. The ratio of the change in output voltage with respect to ground at either output terminal to the change in voltage between the two input terminals.

DC Dissipation. The total power consumed by the device under zero input signal and zero output conditions.

Input Offset Current. The difference in the currents at the two input terminals for equal applied voltages.

Input Offset Voltage. The dc voltage that must be applied between the input terminals to obtain equal quiescent voltages at the output terminals.

Maximum Output Voltage. The maximum output voltage swing that can be achieved without distortion of the output waveform at the peak.

Single-ended Output. An amplifier with output taken from only one of the two input terminals.

Slew Rate. The slew rate is the maximum rate of change of output voltage with large signal inputs.

10.17 The Darlington Compound Transistor

Two transistors can be assembled on a common chip in the manner of an integrated circuit, giving us a device called the *Darlington compound transistor*, shown in Fig. 10.21.

By the usual transistor current relations, I_1 and I_2 are

$$I_1 = h_{fe_1} I_i$$
$$I_2 = h_{fe_2} I_{b_2}$$
(10.42)

But I_{b_2} is the emitter current of Q_1 so that

$$I_{b_2} = (1 + h_{fe_1}) I_i$$

and so

$$I_2 = h_{fe_2}(1 + h_{fe_1}) I_i$$
(10.43)

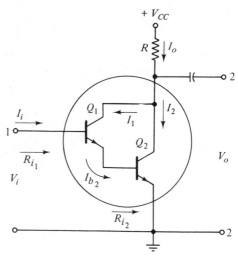

Figure 10.21 The Darlington compound transistor.

Addition of Eq. 10.42 and 10.43 gives for the load current

$$I_o = [h_{fe_1} + h_{fe_2}(1 + h_{fe_1})]I_i \tag{10.44}$$

The current gain is available from this expression as

$$A_i = \frac{I_o}{I_i} \cong h_{fe_1} + h_{fe_1}h_{fe_2} \cong h_{fe_1}h_{fe_2} \tag{10.45}$$

The last expression is simple and provides a fair approximation.

The base-emitter circuit of Q_2 acts as the emitter resistance of Q_1 and by use of the C-C relation we have

$$R_{i_1} = h_{ie_1} + (1 + h_{fe_1})h_{ie_2} \tag{10.46}$$

Transistor Q_2 provides negative current feedback to Q_1 and raises the input resistance at 1,1.

Using the relation between current gain and voltage gain as

$$A_v = -A_i \frac{R_{\text{load}}}{R_{\text{input}}}$$

and Eq. 10.46, the voltage gain of the Darlington connection is

$$A_v \cong -h_{fe_1}h_{fe_2} \frac{R}{h_{ie_1} + h_{fe_1}h_{ie_2}}$$
$$\cong -\frac{h_{fe_2}}{h_{ie_2}}R = -g_{m_2}R \tag{10.47}$$

which is just that of a single transistor in the C-E circuit.

By use of the Darlington connection internally made in the transistor, however, we have a single unit that gives the voltage gain of a single transistor, but with a materially increased input resistance. The Darlington com-

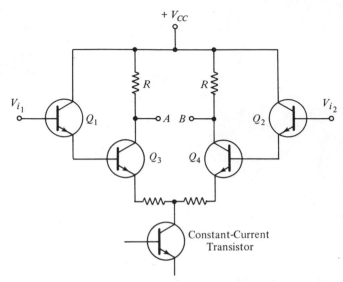

Figure 10.22 Darlington amplifiers added at op-amp inputs.

pound transistor is frequently added at the differential inputs of an operational amplifier to raise the input resistance. This is indicated in Fig. 10.22.

10.18 The Cascode Amplifier

As an example of the design of special circuits in integrated form, we have the *cascode amplifier* in Fig. 10.23. As simplified in Fig. 10.23(b), the signal circuit of Q_1 and Q_3 consists of a C-E transistor followed by a C-B stage. Transistor Q_2 is employed to vary the gain of Q_3, through control voltage applied to terminal 10.

The load of Q_1 is that of the emitter circuits of Q_3 and Q_2. That is,

$$R_{i_3} = \frac{h_{ie_3}}{2(1 + h_{fe_3})} \cong \frac{1}{2g_{m_3}} \tag{10.48}$$

and is therefore small. The voltage gain $-g_m R_{i_3}$ of Q_1 is quite low. With a very low gain, however, the Miller effect cannot appreciably affect the capacitance at the input of Q_1 and so that transistor has a very large f_2 limit frequency and a large bandwidth.

For the C-E stage the gain is approximately

$$A_{v_1} = -g_{m_1} R_{i_3} = \frac{-g_{m_1}}{2g_{m_3}}$$

The gain of Q_3, however, is that of a C-B amplifier

$$A_{v_3} = g_{m_3} R \tag{10.49}$$

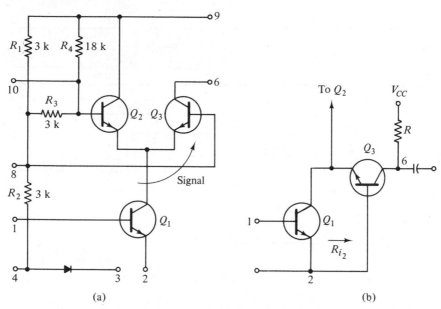

Figure 10.23 (a) The MC 1550 integrated circuit; (b) simplified cascode circuit.

The frequency range of a C-B amplifier is also wide, with $f_\alpha = h_{fe} f_\beta$, and so we have a circuit with three transistors on the same chip, in one package, with the input resistance of a C-E transistor but with a bandwidth of $f_\alpha = f_T$ of the transistor.

The complete package in Fig. 10.23(a) is used as a small-signal high-frequency amplifier.

10.19 Silicon Integrated Circuits

The techniques of impurity diffusion from a gas, masking, oxidation, and metal deposition have made possible the production of complete circuits integrated into a chip of silicon. Reliability is high because interconnections are formed with the circuit elements. Size is drastically reduced and the response can be controlled because lead lengths and stray capacitances are constant.

Figure 10.24 illustrates how the basic manufacturing techniques are applied in a simple integrated circuit. A *p*-silicon wafer, large enough for several hundred circuits, is the starting point. An *n* impurity is diffused through a photoresistant mask to form the *n* pockets in Fig. 10.24(b). The *pn* diodes to the wafer will be back-biased and their resistances will isolate the circuit elements. Additional masking and etching provide holes in another coat of photoresist through which the *p*-base areas in Fig. 10.24(c) are

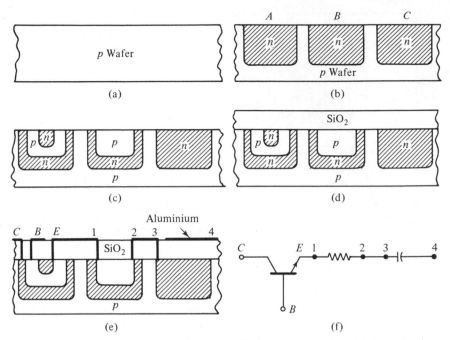

Figure 10.24 Steps in integrated circuit production.

diffused. In the next step an *n* emitter is diffused into the base for a transistor at *A*. The result appears in Fig. 10.24(c).

Finally the surface is oxidized, followed by etching of holes for the connections, and aluminum is deposited through a mask to form the leads as in Fig. 10.24(e). The electrical circuit is shown in Fig. 10.24(f). An *npn* transistor has been formed at *A*; a series resistor of *p* material is at *B*, with a capacitance at *C* using the oxide layer as a dielectric.

The steps in production are basically the same as those required to make a transistor and this is a production benefit. All transistors produced are virtually identical, although there may variations from one batch to the next. Diodes, resistors, and capacitors can also be formed and junctions given alternate usages as capacitances under back bias and resistances under forward bias, as well as serving as voltage references.

When using discrete components, the active device, transistor, FET, or vacuum tube is considered a high cost item and resistors and capacitors are used in preference. In the integrated circuit, cost is a function of circuit area and independent of the number of components per unit area and a transistor can be produced as cheaply as a passive element. New circuit designs become possible, using transistors to perform functions that might otherwise be carried out by passive circuit elements. The use of the constant-current transistor instead of a resistor in the emitter circuit of the operational amplifier is an example of this. The constant-current circuit acts as an almost infinite resistance and greatly increases the common-mode rejection. An actual large R_E would require an excessively large value of emitter supply voltage V_{EE} and would waste power.

10.20 Passive Elements

A circuit capacitance can be developed between collector and base elements formed in the integrated circuit. There is also a capacitance in the reverse-biased diode to the wafer. The depletion layer in these junctions widens with increase in voltage; the capacitance decreases. Thus we have a voltage-variable capacitance. Capacitances as high as 1000 pF per mm² are possible through the use of lightly doped collector materials.

A thin film capacitor can employ the n emitter area as the under plate, the silicon dioxide coating as the dielectric, and aluminum or other metallization as the upper plate of the capacitor. There will also be a stray capacitance to the p layer, in the reverse-biased isolating junction. The silicon dioxide capacitance can develop 300 pF per mm². Higher values can be made by use of tantalum oxide as the insulating film.

In Fig. 10.24 we used the bulk resistivity of one of the diffused areas as a resistor. A thin film resistor can also be vapor deposited on the silicon dioxide insulating layer, using nichrome or tin oxide. The surface is given another layer of silicon dioxide for resistance protection.

The resistance of thin films is given by

$$R = \frac{\rho L}{A} = \frac{\rho L}{Wy} \tag{10.50}$$

where L is the length, W is the width, and y is the depth of the resistance stripe. If $L = W$ or, for a square,

$$R_s = \frac{\rho}{y} \quad (\Omega \text{ per square}). \tag{10.51}$$

The resistance R_s is called the *sheet resistance* and has units of ohms per square. In general, where L and W are not equal, we have

$$R = R_s \frac{L}{W} \tag{10.52}$$

Using material having $R_s = 100\ \Omega$ per square, we can make a 2000 Ω resistor of 0.025-mm width as

$$2000 = 100 \frac{L}{0.025}$$

$$L = \frac{0.025 \times 2000}{100} = 0.5\,\text{mm}$$

Resistances in the range from 40 to 400 Ω per square are possible.

10.21 Comments

The operational amplifier is probably the most widely employed electronic circuit device. Available in several hundred integrated circuit forms, it has many hundreds of applications of which we have indicated a very few. While we have discussed some of the basic circuits from which an integrated package is assembled, we are not normally much concerned with, and cannot even see, the tiny internal circuitry. It is the gain and phase characteristics between input and output terminals that are important to us; as long as the internal gain is very high, we are not greatly concerned with that figure.

But with gain controlled and stabilized with negative feedback, with high input resistance so that it takes negligible current from the driving circuits, with almost zero output resistance so that it can supply any reasonable output current, the operational amplifier in an integrated package is very nearly an ideal electronic device.

That the basic idea of the integration of devices on a silicon chip need not be restricted to complex circuits has been shown by the Darlington and the cascode circuits. Simple differential amplifiers are also available as single units or with as many as four units on a single chip.

REVIEW QUESTIONS

10.1 Why is the differential amplifier so well suited to integrated circuit production methods?

10.2 What conditions of equality make the differential amplifier operate as a balanced bridge?

10.3 What is meant by a differential signal?

10.4 What is meant by a common-mode signal?

10.5 What is the basic problem of amplification at dc or zero frequency?

10.6 What is the purpose of a single-ended differential amplifier?

10.7 Why is the differential amplifier circuit so well suited for dc amplification?

10.8 What is the common-mode rejection ratio?

10.9 Why is a large CMRR important in a differential amplifier?

10.10 Can we have a circuit with single-ended input and differential output?

10.11 What is the advantage of constant-current stabilization in a differential amplifier?

10.12 What is the action of the constant-current circuit when there is differential input to the amplifier?

10.13 What is the reason for the use of emitter follower circuits in integrated circuits?

10.14 What is an operational amplifier?

10.15 Why do we use negative feedback in an operational amplifier?

10.16 What is the ideal gain of an operational amplifier?

10.17 What is an inverting amplifier?

10.18 What is the gain in a noninverting amplifier?

10.19 What is the ideal output resistance of an operational amplifier?

10.20 Why do we use voltage reference circuits in an op amp?

10.21 What is the function of a dc level shifter in an operational amplifier?

10.22 Does V_i' at the input of an operational amplifier actually equal 0 V? Why not?

10.23 We sometimes say the input of a differential operational amplifier is at "virtual ground." What do we mean?

10.24 What is the value of the input current to an operational amplifier, in an ideal situation?

10.25 What is the basic requirement in selecting an integrated amplifier for operational amplifier use?

10.26 Can you explain why the CMRR is infinite if a truly constant-current source is used in place of R_E?

10.27 Sketch the transfer curve of a differential amplifier.

10.28 Draw an operational amplifier in block diagram form. Explain each functional block that you include.

10.29 Define input offset voltage.

10.30 At what rate of gain fall versus frequency do we expect to find instability in the operational amplifier?

10.31 When we compensate an operational amplifier, what happens to the bandwidth?

10.32 What components do we use to frequency compensate an operational amplifier?

10.33 We wish to compensate an amplifier with an RC circuit having a corner frequency of 8500 Hz. We have a resistor of 25,000 Ω; what C is needed?

10.34 What is a unity-gain isolator?

10.35 What is the purpose of an integrator?

10.36 Draw a circuit that would subtract two voltages, using an operational amplifier.

10.37 What is the slew rate?

10.38 Why are the frequency and peak voltage related in the slew rate?

10.39 In gain the Darlington connection is equivalent to what transistor circuit? What can you say about the relative input resistances?

10.40 Why do we use a Darlington connection at the input of a differential amplifier?

10.41 What are the advantages to be gained by use of a cascode circuit?

10.42 With what circuit is the voltage gain of a cascode equivalent?

10.43 Why is the bandwidth of the first stage of a cascode circuit so large?

10.44 What is the expected bandwidth of the second stage of a cascode?

10.45 Name two methods of forming integrated circuit capacitors.

10.46 Name two methods of forming integrated circuit resistances.

10.47 What is the function of the SiO_2 layer in an integrated gain block?

PROBLEMS

10.1 The circuit in Fig. 10.2 has $R_L = 10,000\ \Omega$, $R_E = 3000\ \Omega$, $R_s = 1000\ \Omega$, $h_{ie} = 1500\ \Omega$, and $h_{fe} = 60$. Find the differential input–differential output gain.

10.2 With $h_{fe} = 80$, $h_{ie} = 1200\ \Omega$ in the circuit in Fig. 10.25(a), find the differential voltage gain.

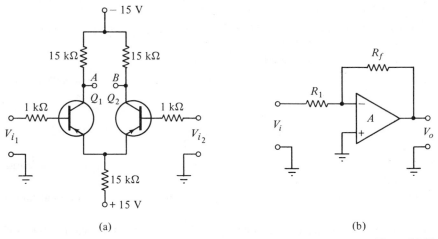

(a)   (b)

Figure 10.25

10.3 A differential amplifier has a differential gain of 150. To find the common-mode gain we determine that $V_i = 2$ V, $V_o = 20$ mV when the inputs are in parallel. What is the CMRR value in decibels?

10.4 For a differential amplifier with $R_L = 5000\ \Omega$, $R_E = 10,000\ \Omega$, $R_s = 2000\ \Omega$, $h_{fe} = 100$, and $g_m = 0.010$ mho, find A_{vd}, A_{vc}, and the CMRR value in decibels.

10.5 Find V_o for a differential amplifier with single-ended output and $V_{i_1} = 0.55$ mV, $V_{i_2} = 0.40$ mV, $A_{vd} = 6000$, and CMRR $= 5 \times 10^4$.

10.6 Find the output voltage of a noninverting operational amplifier for $V_i = 3.5$ mV, $R_1 = 40,000\ \Omega$, and $R_f = 240,000\ \Omega$; $A = 50,000$.

10.7 Find the output voltage for the circuit in Fig. 10.25(b), with $V_i = -4$ V to ground, if $R_1 = 0.1$ MΩ and $R_f = 0.5$ MΩ, if $A_v = 10^5$.

10.8 Find the output voltage for the circuit in Fig. 10.26(a), $A = 10^4$.

10.9 Find the circuit gain in Fig. 10.25(b), with $A = 10^4$, $R_1 = 100\ \Omega$, and $R_f = 10,000\ \Omega$.

10.10 Find the noninverting gain in Fig. 10.10(b), with $A = 10^4$, $R_1 = 100\ \Omega$, and $R_f = 10,000\ \Omega$. Compare your result to that of Problem 10.10.

10.11 With $R_f = 500,000\ \Omega$, choose R_1 values and set up the circuit to obtain an output

$$V_o = 3V_a - 6.5V_b - 4V_c$$

from three inputs $V_a = 1$ V, $V_b = 1.7$ V, $V_c = 2$ V. *Hint*: A unity-gain inverter might help.

10.12 Find the output voltage for the circuit in Fig. 10.26(b), with A large.

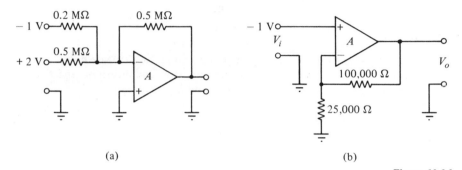

(a)   (b)

Figure 10.26

10.13 Find V_o if $V_i = 1$ mV in Fig. 10.25(b), gain $A = 50,000$, $R_1 = 100,000\ \Omega$, and $R_f = 50,000\ \Omega$.

10.14 Find the output voltage for a three-input summing circuit with $R_f = 0.2\ \mathrm{M}\Omega$, $R_a = 0.2\ \mathrm{M}\Omega$, $R_b = 0.35\ \mathrm{M}\Omega$, $R_c = 0.45\ \mathrm{M}\Omega$, $V_a = -2$ V, $V_b = 1$ V, and $V_c = 3.2$ V.

10.15 Determine the gain $A_{vd} = V_o/V_i$ for the circuit in Fig. 10.27(a), using A_1

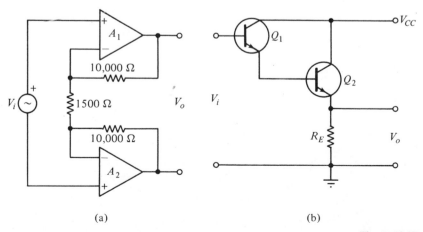

(a)   (b)

Figure 10.27

$= A_2 = 1500$. Remember that we can bisect the circuit into two circuits along the common line.

10.16 An operational amplifier has internal voltage gain of 80 dB, and at high frequencies the gain rolls off at a rate of -20 dB per decade. Unity gain (0 dB) is reached at 2 MHz; feedback $\beta = 0.03$. What is the gain A' in decibels, and what is the corner frequency?

10.17 The f_2 value of an operational amplifier occurs at 100 Hz and the no-feedback gain is 100 dB. What amount of feedback in decibels will be needed to reduce the gain to 40 dB, and what will the bandwidth become?

10.18 Assuming the transistors are not identical in the Darlington compound circuit in Fig. 10.27(b), show that

$$R_{i_1} \cong A_i R_E$$
$$A_i \cong h_{f e_1} h_{f e_2}$$

Find A_v.

10.19 Assuming the transistors of a Darlington pair, Fig. 10.27(b), are not identical, draw the circuit of an equivalent single transistor, and determine the h_i and h_f parameters in terms of $h_{f e_1}$, $h_{f e_2}$, $h_{i e_1}$, and $h_{i e_2}$.

11

Tuned and Video Amplifiers

To select and amplify desired signals at radio frequencies we combine transistors or FETs with resonant circuits. Ideally we would like to adjust those circuits to receive a narrow band of frequencies and reject all other frequencies but this is impossible at reasonable cost. So we define the *selectivity* of a radio receiver as a measure of its ability to separate a desired signal from other unwanted signals. In *radio-frequency amplifiers* we use a number of resonant circuits in cascade to raise the response to the desired band of signals and to reduce the response to unwanted signals near, but outside, the desired band. The transistor or FET isolates each resonant circuit and prevents the interaction that would occur if these circuits were placed directly in parallel. The transistors also provide signal gain.

Video amplifiers are designed with a frequency response extending from zero or near-zero frequency to 5 MHz or more for amplification of pulse signals. When designed with a response to 4.5 MHz, video amplifiers are used to amplify picture signals in television receivers. Transistors have gain falloff in these frequency regions due to their inherent capacitance. We compensate the transistor circuits with inductance to extend the frequency range, again employing the principles of resonant circuits.

11.1 Bandpass Amplifiers

In the *radio-frequency amplifier*, the usual input signal consists of a center frequency f_o around which are grouped a band of frequencies. The center frequency may range from one to many megahertz and the side frequencies may extend from a few to several hundred kilohertz on each side.

Radio-frequency amplifiers are expected to provide selectivity for a desired frequency band and rejection of other frequencies. An ideal response curve is indicated by the dashed rectangle in Fig. 11.1(a). Ideally all frequencies in the desired band would be amplified equally and all frequencies outside the desired band, such as the interfering signal at f_i, would produce zero response.

We also require *tunability* to shift the response rectangle over a wide range of frequencies at the will of the operator. Circuits must be simple since cascaded stages are simultaneously tuned and this requirement dictates that we depend primarily on the parallel-resonant LC circuit, with the capacitance element made variable for tuning.

Such LC circuits provide only an approximation to the desired response rectangle and the discrimination against adjacent frequencies provided by the "skirt" portions of the curve is often inadequate. A measure of the relative selectivity is the *shape factor* of the response curve, as the ratio of the frequency width of the curve at 60 dB down from the center frequency response to the bandwidth at which a signal is only 6 dB down.

For a single resonant circuit this shape factor may be 600 and, with 10-kHz bandwidth at 6 dB down from the resonant peak, the skirts cover 6 MHz for the 60-dB suppression. Several parallel-resonant circuits can be cascaded with their associated amplifiers to reduce the bandwidth at 60 dB suppression.

Other amplifiers operate at *intermediate frequencies* (I.F.) between the radio signal frequencies and the audio frequencies. For example, the usual I.F. in a broadcast receiver is 455 kHz. Such I.F. amplifiers are tuned to fixed frequencies and greater circuit complexity can be tolerated since the adjustments are usually made only by the manufacturer. These complex circuits provide wider bands and steeper skirts on the selectivity curve, as shown in Fig. 11.1(b).

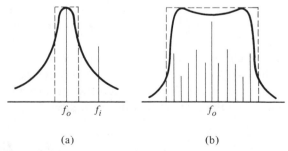

| (a) | (b) |

Figure 11.1 (a) Narrow-band R.F. response; (b) response of an I.F. amplifier.

11.2 The Parallel-Resonant Circuit

The parallel-resonant circuit in Fig. 11.2 is the common frequency-selective circuit in tuned amplifiers. The resistance R_p is usually that of the source, as shown in Fig. 11.2(b), where a transistor drives a resonant circuit. We consider the series resistance of the inductor and the capacitor as small.

At resonance the voltage V is in phase with the current I and the circuit at A,A' appears as a resistance; this is the *definition of the resonance condition*. To bring this about, the reactance of the capacitance must equal the reactance of the inductance:

$$X_C = X_L$$

and

$$\frac{1}{2\pi f_o C} = 2\pi f_o L \tag{11.1}$$

at the resonant frequency f_o, derived from Eq. 11.1 as

$$f_o^2 = \frac{1}{(2\pi)^2 LC}$$

$$f_o = \frac{1}{2\pi\sqrt{LC}} \tag{11.2}$$

A parameter Q is defined as the ratio of the parallel resistance to either reactance at resonance so

$$Q = \frac{R_p}{X} = \frac{R_p}{2\pi f_o L} = 2\pi f_o C R_p \tag{11.3}$$

$$= R_p \sqrt{\frac{C}{L}} \tag{11.4}$$

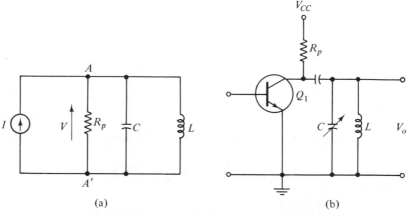

(a)   (b)

Figure 11.2   (a) Parallel-resonant LC circuit; (b) with a transistor.

A resonant circuit should be efficient in transmitting power to the next transistor or load. The power lost in the circuit is V^2/R_p since C and L are assumed to provide no loss. With R_p inversely proportional to the power lost, then Q is also inverse to the circuit losses. That is, a large Q indicates a low-loss circuit and we consider Q as a resonant circuit *figure of merit*;[1] Q values ranging from about 5 to 500 are found in such circuits.

The impedance of the parallel circuit at A,A' is

$$Z = \frac{R_p}{\sqrt{1 + R_p^2\left(\frac{1}{X_C} - \frac{1}{X_L}\right)^2}} \tag{11.5}$$

At resonance, $X_C = X_L$ and we have

$$Z_o = R_p \tag{11.6}$$

Using Eq. 11.3 this can also be written

$$Z_o = 2\pi f_o L Q = \frac{Q}{2\pi f_o C} \tag{11.7}$$

Upon change of the frequency from the resonant frequency, either X_C or X_L increases and the difference term in the denominator of Eq. 11.5 becomes larger and Z decreases. Since $V_{AA'} = IZ$, the voltage falls as frequency departs from the peak value at R_p, in Fig. 11.3. Driven by a transistor, the circuit provides a large voltage output at resonance and smaller voltages at all other frequencies.

The phase angle of Z changes from inductive at frequencies below resonance to capacitive at frequencies above resonance.

Equation 11.5 may be simplified for calculation purposes by rearrange-

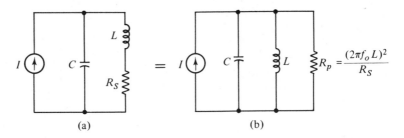

(a)   (b)

Figure 11.3 Equivalence of series and parallel forms for resistance connections.

[1]When the circuit resistance is placed in series with the inductance, the value of Q becomes

$$Q = \frac{2\pi f_o L}{R_s}$$

The two equations for Q are equivalent for the same circuit elements.

ment. Consider the denominator difference term

$$\frac{1}{X_C} - \frac{1}{X_L} = 2\pi f C - \frac{1}{2\pi f L}$$

$$= \sqrt{\frac{C}{L}}\left(2\pi f \sqrt{LC} - \frac{1}{2\pi f \sqrt{LC}}\right) = \sqrt{\frac{C}{L}}\left(\frac{f}{f_o} - \frac{f_o}{f}\right) \tag{11.8}$$

after noting that $2\pi \sqrt{LC} = 1/f_o$. Inserting this result in Eq. 11.5, we have

$$Z = \frac{R_p}{\sqrt{1 + R_p^2 \dfrac{C}{L}\left(\dfrac{f}{f_o} - \dfrac{f_o}{f}\right)^2}}$$

We recognize that $R_p^2(C/L) = Q^2$ from Eq. 11.4 and so have a simplified expression for the impedance

$$Z = \frac{R_p}{\sqrt{1 + Q^2\left(\dfrac{f}{f_o} - \dfrac{f_o}{f}\right)^2}} \tag{11.9}$$

In some cases, the resistance of the circuit is concentrated in series with the inductive branch, as in Fig. 11.3(a). For $Q > 7$, the series resistance R_s can be transformed to a parallel R_p, as

$$R_p = \frac{(2\pi f_o L)^2}{R_s} \tag{11.10}$$

or

$$R_s = \frac{(2\pi f_o L)^2}{R_p} \tag{11.11}$$

If a parallel resistance already is present, then the effective R_p is the parallel value of the two.

The various equations for the parallel RLC circuit, such as Q and bandwidth and resonant resistance, can be applied after R_s is transformed to an equivalent R_p.

11.3 Bandwidth of the Resonant Circuit

Figure 11.4 shows that the bandwidth of the resonant circuit, measured at the half-power points at which $V = 0.707V_o$, depends on the value of R_p. The exact nature of this dependence can be found by use of Eq. 11.9. We previously showed that the frequency of a bandwidth limit was found when the two terms of the denominator of an expression such as Eq. 11.9 were equal. Thus at the upper frequency limit f_2, identified in Fig. 11.4, we have

$$1 = Q^2\left(\frac{f_2}{f_o} - \frac{f_o}{f_2}\right)^2$$

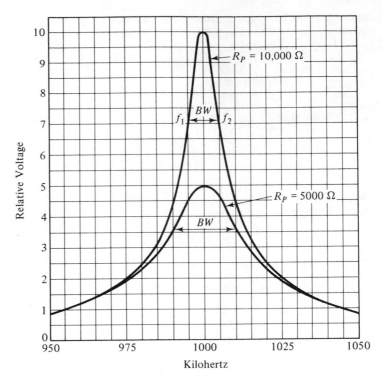

Figure 11.4 Resonance curves for two R_p, L, C parallel circuits.

Taking the square root,

$$1 = Q\left(\frac{f_2}{f_o} - \frac{f_o}{f_2}\right)$$

The lower limit at f_1 is below f_o and the term in parentheses must be reversed to remain positive. After taking the square root,

$$1 = Q\left(\frac{f_o}{f_1} - \frac{f_1}{f_o}\right)$$

If f_2 and f_1 are small departures from f_o, then we can say that

$$\frac{f_2}{f_o} - \frac{f_o}{f_2} \simeq \frac{2(f_2 - f_o)}{f_o} \tag{11.12}$$

Suppose that $f_2 = 1.1 f_o$; then

$$\frac{1.1}{1.0} - \frac{1.0}{1.1} = 1.1 - 0.91 \simeq 0.2 = \frac{2(1.1 - 1.0)}{1.0}$$

and the approximation is reasonable. Therefore, we have

$$1 = 2Q\left(\frac{f_2 - f_o}{f_o}\right) \tag{11.13}$$

$$1 = 2Q\left(\frac{f_o - f_1}{f_o}\right) \tag{11.14}$$

Adding these expressions, we have

$$2 = 2Q\left(\frac{f_2}{f_o} - \frac{f_o}{f_o} + \frac{f_o}{f_o} - \frac{f_1}{f_o}\right)$$

$$\frac{f_2 - f_1}{f_o} = \frac{1}{Q}$$

The bandwidth (BW) as $f_2 - f_1$ is

$$BW = f_2 - f_1 = \frac{f_o}{Q} \quad \text{(Hz)} \tag{11.15}$$

This expression again defines Q and while instruments for measurement of Q are available, the value of circuit Q is often calculated from a measurement of the bandwidth of a circuit while it is mounted in the equipment. That is, f_1 and f_2 are measured where the voltage response is $V_o/\sqrt{2}$, and Eq. 11.15 is used to find Q. High Q and low losses are indicative of narrow bandwidth and large R_p. By variation of the resistance R_p connected across the circuit, we can adjust Q and the bandwidth.

Equation 11.4,

$$Q = R_p\sqrt{\frac{C}{L}}$$

shows that the C/L ratio of a circuit also controls Q, that is, a large ratio of C to L gives a high Q circuit.

At the bandwidth limit frequencies the phase angle of Z is 45° below resonance and -45° above resonance at f_2.

Example: Choose L and C values for resonance at 1 MHz, with a resonant impedance of 100,000 Ω and bandwidth of 10 kHz.

We then have $Z_o = R_p = 100,000 \ \Omega$.

Also

$$Q = \frac{f_o}{BW} = \frac{10^6}{10^4} = 100$$

$$Z_o = 2\pi f_o L Q = R_p$$

$$L = \frac{R_p}{2\pi f_o Q} = \frac{100,000}{6.28 \times 10^6 \times 100} = 0.000159 \text{ H}$$

$$= 159 \ \mu\text{H}$$

We also have

$$Q = 2\pi f_o C R_p$$

$$C = \frac{Q}{2\pi f_o R_p} = \frac{100}{6.28 \times 10^6 \times 10^5} = 1.59 \times 10^{-10} \text{ F}$$

$$= 159 \text{ pF}$$

11.4 The Single-Tuned Transistor Amplifier

We frequently use a resonant circuit for both coupling and selectivity between two transistors, as in Fig. 11.5. The load on the resonant circuit at 2,2 is the input resistance of transistor Q_2, usually increased over h_{ie} by use of an unbypassed emitter resistor.

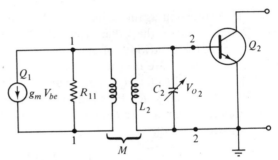

Figure 11.5 Single-tuned, inductively coupled circuit.

The secondary voltage V_{o_2} will be maximum when the circuit is tuned to resonance and

$$\frac{1}{2\pi f_o C_2} = 2\pi f_o L_2 \qquad\qquad (11.16)$$

With M as the mutual inductance between the coils L_1 and L_2, we choose

$$2\pi f_o M = \sqrt{R_{11} R_{i_2}} \qquad\qquad (11.17)$$

We find that the secondary voltage at resonance is

$$V_{o_2} = \frac{-2\pi f_o g_m M Q_e}{R_{11}} V_{be} \qquad\qquad (11.18)$$

where our effective circuit Q_e value relates to the circuit Q_2 of the secondary tuned circuit as

$$Q_e = \frac{Q_2}{1 + \dfrac{(2\pi f_o M)^2}{R_{11} R_{i_2}}} \qquad\qquad (11.19)$$

Using the condition of Eq. 11.17,

$$Q_e = \frac{Q_2}{2}$$

and the bandwidth is

$$BW = \frac{f_o}{Q_e} = \frac{2f_o}{Q_2} \tag{11.20}$$

which, with optimum M from Eq. 11.17, is twice the bandwidth of the secondary resonant circuit alone. By reducing M from the optimum value, we can control the bandwidth from twice the Q_2 value to that of Q_2.

The transformer should have the approximate relation

$$a = \sqrt{\frac{R_{11}}{R_{i2}}} \simeq \frac{n_1}{n_2} \tag{11.21}$$

where n_1/n_2 is the ratio of primary to secondary turns. This is only an approximate relation for such radio-frequency transformers, dependent on the completeness of the iron core surrounding the coils.

11.5 The Double-Tuned Transformer

For use in intermediate frequency amplifiers, where tuning is fixed by the manufacturer, we can widen and square up the response curve still further by tuning both primary and secondary windings of a transformer, as in Fig. 11.6. Usually we make $L_1 = L_2$ and $Q_1 = Q_2 = Q$; both sides of the transformer will also be independently resonant at f_o.

The magnetic coupling between L_1 and L_2 is measured by the *coefficient of coupling k*. With the coils far apart, $k = 0$ and the secondary voltage is low. As the coils are moved closer together, the secondary voltage rises to a maximum at the value of *critical coupling* at $k = k_c$ in Fig. 11.6(b). As the coils are moved still closer, the secondary voltage falls again. The critical coupling value of k is

$$k_c = \frac{1}{\sqrt{Q_1 Q_2}} \tag{11.22}$$

and for $Q_1 = Q_2 = Q$ we have

$$k_c = \frac{1}{Q} \tag{11.23}$$

The circuits are said to be *undercoupled* if $k < k_c$, *critically coupled* if $k = k_c$, and *overcoupled* if $k > k_c$.

When we plot secondary voltage against frequency, and the value of $k < k_c$, we have response curves with one peak, as shown for $k/k_c = 0.5$ in Fig. 11.6(c). The secondary voltage V_{o2} reaches a maximum peak value for $k/k_c = 1$. This is equivalent to $k = 1/Q$. Because of interactions between

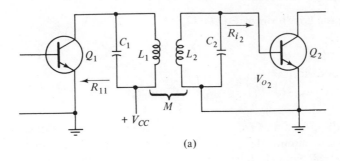

(a)

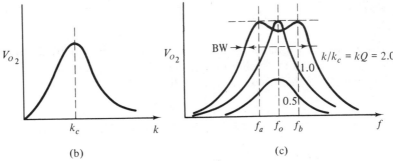

(b)   (c)

Figure 11.6   The double-tuned circuit.

the currents in the two resonant circuits when $k/k_c > 1$, $k > 1/Q$, the response curve splits and shows two frequencies of maximum response and the bandwidth widens.

The frequencies of peak response are

$$f_a = f_o\left(1 - \frac{1}{2Q}\sqrt{k^2Q^2 - 1}\right) \qquad \textbf{(11.24)}$$

$$f_b = f_o\left(1 + \frac{1}{2Q}\sqrt{k^2Q^2 - 1}\right) \qquad \textbf{(11.25)}$$

At critical coupling, with $k = 1/Q$, the expressions indicate that $f_a = f_b = f_o$ for the single peak, as they should.

If the dip in the center is allowed to fall to $0.707V_{\text{peak}}$, the bandwidth is $\sqrt{2}(f_b - f_a)$.

The gain at either frequency of peak response is

$$|A_{a,b}| = \frac{2\pi f_o g_m Q L_2}{2} \qquad \textbf{(11.26)}$$

with $L_1 = L_2$.

Such transformers cannot simply be aligned by tuning for a peak secondary voltage with overcoupling. One procedure is to reduce the Q value, thereby changing the coupling from overcoupled to undercoupled. This is done by shunting fixed resistors on both primary and secondary. If $k = 0.05$ and $Q_1 = Q_2 = Q = 50$, then $k_c = 0.02$ and the transformer is overcoupled. If we reduce Q to 15, then $k_c = 0.067$ and the transformer becomes under-coupled and gives only one response peak. Tuning for this peak will correctly establish both primary and secondary resonance. The loading resistors can then be removed.

The use of two tuned circuits produces a response over a greater band-width and with steeper skirts on the response curve so that the response more nearly approaches the ideal rectangle. More elaborate circuits are available with greater numbers of tuned circuits. Each circuit adds a small ripple of gain to the passband response but steepens the falloff at skirt frequencies. These are known as *stagger-tuned amplifiers*. The analysis is beyond the scope of this text.

11.6 The Pulse Waveform

In digital computation, data transmission, and radar we use the pulse wave-form, shown in ideal form in Fig. 11.7(a). After transmission through various circuits and amplifiers, the waveform may appear more as in Fig. 11.7(b) or (c).

The pulse starts at $t = 0$ but the received pulse is delayed by the transmission path as in Fig. 11.7(b). The exact starting time involves guessing as to when the received wave leaves the zero axis; to avoid this uncertainty the *rise time* of the pulse is arbitrarily defined as t_r, from 10 to 90 per cent amplitude. The *pulse width* is also difficult to determine so it is defined as t_w, measured between the 50 per cent rising level and the 50 per cent falling level. The top of the pulse may *sag* some percentage d or may *overshoot* by some percentage of the amplitude as shown in Fig. 11.7(c).

One of the major needs for the video amplifier arises in the amplification of pulses because of the large bandwidth needed for good waveform repro-duction. A general pulse train, as used in data transmission, is shown in Fig.

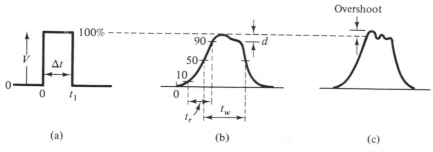

(a) (b) (c)

Figure 11.7 Pulse waveforms.

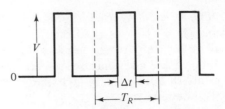

Figure 11.8 A pulse train.

11.8. Each pulse is Δt seconds long and the pulses are repeated at a *repetition rate* of $f_R = 1/T_R$ per second. Each pulse is made up of an infinite number of harmonic frequencies, starting with the fundamental or first harmonic at f_R, a second harmonic at $2f_R$, and more harmonics at nf_R frequencies to infinity. Of course, the very high-order harmonics are small in amplitude and are cut off by the amplifiers since no amplifier can have a passband to infinite frequency.

But the fidelity of the pulse waveform is dependent on the video amplifier's providing a uniform gain passband for these harmonics. Excellent reproduction is possible if the passband is wide enough to amplify frequencies to

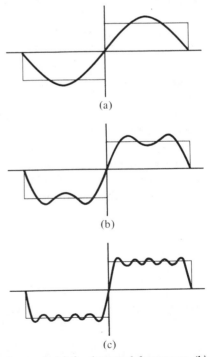

Figure 11.9 Approximation to a square wave: (a) fundamental frequency; (b) fundamental plus third harmonic; (c) fundamental plus third and fifth harmonics.

$4/\Delta t$, and reasonable reproduction is obtained with a passband reaching to $1/\Delta t$ hertz. Thus for a 1-μs pulse equal to Δt, repeated 400 times per second, the bandwidth for excellent reproduction would be 4 MHz and for reasonable reproduction we would need a 1-MHz bandwidth. In that 4-MHz passband there are harmonic signals every 400 Hz, or 10,000 signals must be amplified equally.

The process of building a wave through addition of the harmonic frequencies is shown in Fig. 11.9. Very high-order harmonics are needed to build the square corners. The harmonic amplitudes that are needed for the first few frequencies of a square wave are given in Table 11.1.

TABLE 11.1   **Relative Amplitudes of Harmonics in a Square Wave***

Harmonic	Amplitude	Harmonic	Amplitude	Harmonic	Amplitude
1	0.636	11	−0.058	45	0.014
3	−0.212	13	0.049	55	−0.011
5	0.127	15	−0.042	65	0.0098
7	−0.091	25	0.025	75	−0.0085
9	0.070	35	−0.018	85	0.0075

*Wave amplitude = 1. Negative signs indicate reversed phase.

11.7 The Shunt-Peaked Video Amplifier

The *video amplifier* has been defined as having a frequency range extending from near zero to a few megahertz. This takes the operating range into the frequencies at which the Miller-effect capacitance of the transistor or FET normally produces a fall in gain.

One way to widen the frequency response band of an amplifier is to reduce the load resistance. This reduces the gain and the Miller-effect capacitance and raises the f_2 limit of the amplifier. The f_2 limit may be further raised by use of an inductance in the load. The resulting resonant effect raises the load impedance at those frequencies where C_{ie}, in shunt, reduces the load. The gain is maintained uniform to a higher f_2 value.

In Fig. 11.10 we show the compensating inductance L, usually a fraction of a millihenry, connected in series with the FET load resistor R_s. In Fig. 11.10(b) we have the equivalent circuit of the transistor, including the Miller-effect capacitance of Q_2 and the wiring capacitance, which together form C_T. We have neglected and dropped the series blocking capacitor and the bias network from the circuit.

With $L = 0$ we have an uncompensated amplifier with load R_s and the

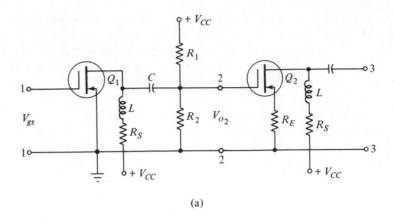

(a)

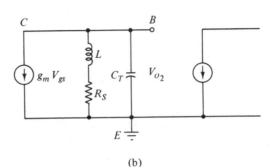

(b)

Figure 11.10   (a) Shunt compensation by L; (b) equivalent circuit.

upper limit frequency is

$$f_2 = \frac{1}{2\pi C_T R_s} \tag{11.27}$$

obtained by the methods of Chapter 8. We use this frequency as a norm, against which the higher f_2' of the compensated amplifier will be compared. The circuit is said to be *shunt-compensated* by the inductance.

We define a parameter as

$$q = \frac{2\pi f_2 L}{R_s} = \frac{L}{R_s^2 C_T} \tag{11.28}$$

after using Eq. 11.27. This equation expresses the ratio of inductive reactance at the f_2 frequency to the series resistance.

The inductance L is then varied and frequency-response curves plotted as in Fig. 11.11. The curve for $q = 0$ is for the uncompensated amplifier with $L = 0$. The condition of $q = 0.41$ is found to give the widest frequency band without a rise of gain above the mid-range value and for that value of q the bandwidth is increased to $f_2' = 1.72f$.

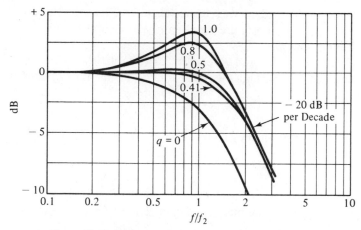

Figure 11.11 Shunt-peaked circuit response versus parameter *q*.

Values of $q > 0.41$ widen the response band further but introduce a peak at some frequency less than f_2. It should be noted that while one stage may show a peak of *G* decibel, *n* stages will increase that peak to *nG* decibels and such a high peak may not be acceptable.

We can find the time delay in transmitting a frequency through the amplifier from the phase angle θ. Using $\theta/360$ as the fraction of a cycle by which the signal is delayed and $1/f$ as the time of a cycle, we can write

$$\text{time delay} = \frac{\theta}{360f}$$

Figure 11.12 is plotted in terms of time delay. If all frequencies are given an equal time delay, the curve will be flat and the whole wave will be delayed

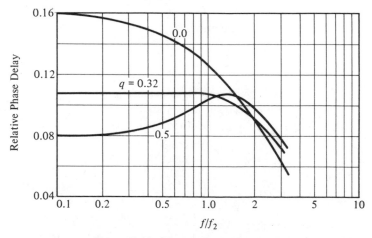

Figure 11.12 Time delay for the shunt-peaked amplifier.

but is unchanged in waveform. Therefore, $q = 0.32$ offers the best phase response. A compromise between $q = 0.41$ and $q = 0.32$ is often made at about $q = 0.35$.

With $q = 0.35$ and R_s chosen for suitable low-frequency gain and bandwidth f_2, we can find the value for the compensating inductance from Eq. 11.28:

$$L = qR_s^2 C_T \tag{11.29}$$

The high-frequency gain of the compensated amplifier is given by a complicated expression:

$$A_{v(\text{hi})} = -g_m R_s \left[\frac{1 + q^2(f/f_2)^2}{1 + (1 - 2q)(f/f_2)^2 + q^2(f/f_2)^4} \right]^{1/2} \tag{11.30}$$

where $-g_m R_s$ is the mid-range gain. The phase angle expression is also complicated:

$$\theta = -\tan^{-1} \frac{f}{f_2} \left[1 - q + q^2 \left(\frac{f}{f_2} \right)^2 \right] \tag{11.31}$$

The rise in gain at the high frequencies is due to resonance between L and C_T, the lower peaks being produced as a result of a low value of circuit q.

Example: An FET has $g_m = 0.003$ mho and is to be used in an amplifier to provide a mid-frequency gain of 10, with the gain extended to the highest possible frequency. The FET has $C_{gs} = 1$ pF and $C_{gd} = 2$ pF.

We find R_s from the mid-frequency gain:

$$A_{v(\text{mid})} = -g_m R_s = 10$$

$$R_s = \frac{10}{0.003} = 3300 \ \Omega$$

The Miller-effect capacitance is

$$\begin{aligned}
C_{ie} &= C_{gs} + (1 + g_m R_s)C_{gd} \\
&= 1 + (1 + 0.003 \times 3.3 \times 10^3) \times 2 \\
&= 1 + 22 = 23 \text{ pF}
\end{aligned}$$

We then add 2 pF for wiring capacitance so that $C_T = 25$ pF.

The limit frequency of the uncompensated amplifier is

$$\begin{aligned}
f_2 &= \frac{1}{2\pi C_T R_s} = \frac{1}{2\pi \times 25 \times 10^{-12} \times 3.3 \times 10^3} \\
&= 1.93 \text{ MHz}
\end{aligned}$$

By use of

$$\begin{aligned}
L = qR_s^2 C_T &= 0.41(3.3 \times 10^3)^2 \times 25 \times 10^{-12} \\
&= 1.12 \times 10^{-4} \text{ H} = 0.112 \text{ mH}
\end{aligned}$$

we can extend the frequency range to

$$f_2' = 1.72f_2 = 1.72 \times 1.93 = 3.32\,\text{MHz}$$

11.8 The Series-Peaked Video Amplifier

A more elaborate circuit that also raises the high-frequency response limit is the *series-peaked circuit* in Fig. 11.13. We add a small inductance in series with C_T, with which it resonates to raise the gain at the high frequencies.

For the uncompensated amplifier with $L = 0$, the mid-frequency gain is

$$A_{v(\text{mid})} = -g_m R_s \tag{11.32}$$

and the f_2 frequency limit is

$$f_2 = \frac{1}{2\pi(C_1 + C_T)R_s} \tag{11.33}$$

where C_1 is the output and wiring capacitances of the first transistor Q_1. The optimally flat response is obtained when

$$q = \frac{2\pi f_2 L}{R_s} = 0.67 \tag{11.34}$$

and

$$C_1 = \frac{C_T}{3} \tag{11.35}$$

giving a bandwidth of $f_2' = 2f_2$.

Using Eq. 11.33 in 11.34, we have

$$L = \frac{qR_s}{2\pi f_2} = qR_s^2(C_1 + C_T) \tag{11.36}$$

Often the output capacitance of Q_1 is not exactly one-third of C_T, as called for in Eq. 11.35. To increase either C_1 or C_T to obtain this ratio will reduce f_2 as calculated by Eq. 11.33. But a reduction in f_2 is not the result desired. Accordingly, the shunt-peaked circuit, with its lessened complexity, is generally used in video amplifiers.

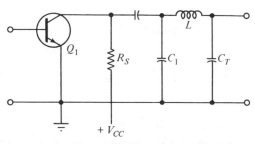

Figure 11.13 The series-peaked circuit.

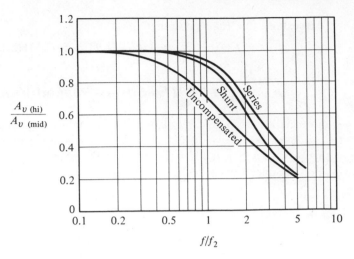

Figure 11.14 Comparison of shunt and series compensation.

A comparison of the uncompensated and compensated responses is made in Fig. 11.14.

11.9 Review

Q is a most important and useful parameter of the parallel-resonant circuit. It is defined as

$$Q = 2\pi f_o C R_p = \frac{R_p}{2\pi f_o L} = R_p \sqrt{\frac{C}{L}}$$

where

$$f_o = \frac{1}{2\pi\sqrt{LC}}$$

for the circuit-resonant frequency.

We then know that the bandwidth is

$$BW = f_2 - f_1 = \frac{f_o}{Q}$$

and that at resonance

$$Z_o = R_p = 2\pi f_o L Q$$

For frequencies near resonance,

$$Z = \frac{R_p}{\sqrt{1 + Q^2 \left(\frac{f}{f_o} - \frac{f_o}{f}\right)^2}}$$

For the tuned coupled circuit the bandwidth can be increased to $2f_o/Q$ with critical coupling k_c for the coils and ωM equal to $\sqrt{R_{11}R_{i2}}$. Further widening and steepening of the skirts is possible by tuning of both primary and secondary and making $k > k_c$. These methods lead to a response that is closer to the ideal rectangular shape for the response curve.

The shunt-compensated circuit is most usually employed in video amplifiers.

REVIEW QUESTIONS

11.1 Define resonance.

11.2 Why do we desire a rectangular frequency-response curve?

11.3 What is the condition of the circuit reactances at resonance?

11.4 What is the resonant frequency?

11.5 Define selectivity.

11.6 What is the shape factor of a resonant circuit?

11.7 What range of bandwidths do we expect to find in radio-frequency amplifiers?

11.8 How is Q defined in terms of circuit elements?

11.9 How is Q defined in terms of bandwidth?

11.10 How do we often measure Q?

11.11 What is the resonant resistance of a parallel circuit?

11.12 Why is the resonant resistance a maximum at resonance? What happens to the currents in L and C?

11.13 What is the bandwidth of a resonant circuit?

11.14 A circuit has a resonant frequency of 100 MHz and a bandwidth of 5 MHz. What is the Q?

11.15 A circuit has a shape factor of 100 : 1 and is 6 kHz wide at 6 dB down from the peak. What is the width of the curve at 60 dB down?

11.16 What are the phase angles of the resonant impedance at the bandwidth limit frequencies?

11.17 How does the C/L ratio affect the bandwidth?

11.18 What is the tuning result of
(a) Critical coupling?
(b) Overcoupling?
(c) Insufficient coupling?

11.19 In a tuned coupled circuit, on what factors does V_{o_2} depend?

11.20 On what factors does the bandwidth depend in a tuned coupled circuit?

11.21 What is the maximum value of the coupled Q_e in terms of the circuit Q?

11.22 Define critical coupling in terms of Q.

11.23 What is a video amplifier?

11.24 How do we shunt compensate a video amplifier?

11.25 How do we series compensate a video amplifier?

11.26 Why do we avoid a gain peak in a cascaded shunt-compensated amplifier?

11.27 What is the parameter q in a shunt-compensated amplifier?

11.28 What is a good compromise value for q in a shunt-peaked amplifier?

11.29 What is the requirement on the two capacitances in a series-peaked amplifier?

11.30 How would you tune a double tuned overcoupled transformer?

11.31 In a parallel-resonant circuit, in order to make the circuit inductive, should the generator frequency be raised or lowered from resonance?

11.32 To raise the resonant frequency of a parallel circuit, which is it necessary to do?
  (a) Increase the capacitance
  (b) Increase the resistance
  (c) Decrease the inductance
  (d) Increase the inductance

11.33 What is the repetition rate?

11.34 How is pulse rise time measured?

11.35 How is received pulse width measured?

11.36 What is overshoot of a pulse?

11.37 How many harmonics are in a pulse wave?

11.38 How many harmonic frequencies are needed for excellent reproduction of a pulse wave with repetition rate of 1000 per second, at 10-MHz frequency?

PROBLEMS

11.1 A circuit is resonant at 2000 Hz. If the coil is 0.120 H, what capacitance is being used?

11.2 A parallel circuit is resonant at 5.35 MHz and uses a 40-pF capacitor. What is the value of the inductance?

11.3 In a parallel-resonant circuit the reactance of the capacitance is 1450 Ω and $R_p = 20,000$ Ω. What is the Q of the circuit?

11.4 In Problem 11.3, the inductance is 250 μH. What is the frequency of resonance?

11.5 At resonance in the circuit in Fig. 11.15(a), $I = 10$ mA and $V = 7.5$ V. What is the resonant resistance of the circuit?

11.6 In Problem 11.5 the inductive reactance at resonance is 455 $\Omega = 2\pi f_o L$; what is the value of Q?

11.7 Capacitor C in Fig. 11.15(a) is 350 pF. Using data from Problems 11.5 and 11.6, what is the resonant frequency?

11.8 The half-power points of the frequency response of a resonant circuit are 647.5 and 662.5 kHz.

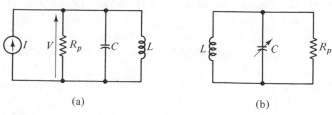

Figure 11.15

(a) What is the resonant frequency?
(b) What is the circuit Q?

11.9 A parallel-resonant circuit as in Fig. 11.15(a) has $Q = 150$ and $I = 0.6$ mA. The voltage V is 100 V.
(a) Find the reactance of L and C at resonance.
(b) What power is being supplied to the circuit?

11.10 Using the circuit in Fig. 11.15(a), the voltage V is 55 V; $R_p = 4500\ \Omega$, $L = 5\ \mu$H, and $C = 0.001\ \mu$F.
(a) What is the resonant frequency?
(b) What is the circuit Q?
(c) What is the current I at resonance?

11.11 A parallel RLC circuit is resonant at 27 kHz. The circuit contains a 0.015-H inductance, a 0.002316-μF capacitor, and a parallel resistance of 40,000 Ω.
(a) What is the circuit impedance at resonance?
(b) What is the circuit Q?
(c) What is the bandwidth?
(d) What is the circuit impedance at f_2?

11.12 The circuit in Fig. 11.15(b) is to tune the broadcast band from 550 to 1600 kHz. The capacitor is varied and has a maximum value of 375 pF. What is the value of inductance used, and what should be the minimum value of C?

11.13 A circuit is resonant at 455 kHz and has a 10-kHz bandwidth. The reactance of the inductance is 1250 Ω; what is R_p of the circuit?

11.14 A parallel-resonant circuit is resonant at 20 MHz, the Q is 150, and the reactances are each 750 Ω. What is the value of R_p and of an R_s in series with L that might replace R_p?

11.15 In a parallel-tuned circuit, the resistance in series with the inductor is 12 Ω and the inductive reactance is 1450 Ω at resonance. Find the Q of the circuit.

11.16 The resonant frequency of a parallel RLC circuit is 7.3 MHz. If the Q is 80, find the half-power frequencies.

11.17 A parallel RLC circuit is resonant at 20 MHz. The Q is 200 and the reactances are 750 Ω.
(a) What is R_p?
(b) What resistor should be paralleled with R_p to bring the Q down to 50?
(c) What are the bandwidths, with and without this resistor?

11.18 In a resonant circuit, $C = 60$ pF and L is 130 μH. The Q of the circuit is 150 and $R_p = 100,000$ Ω. What is the bandwidth in hertz?

11.19 A parallel circuit is resonant at 2 MHz. In order to have the circuit resonate at 10 MHz, what must be the ratio of the new capacitance to the original capacitance?

11.20 The circuit in Fig. 11.3(a) has $L = 0.01$ H, $C = 1$ μF, and $R_s = 3$ Ω. Determine f_o, BW, and Q.

11.21 Two circuits of the type of Problem 11.18 are critically coupled. Determine k_c.

11.22 In a double-tuned circuit, $k = 2k_c$ and $Q = 80$. If $f_o = 10.7$ MHz,
(a) What are the frequencies of peak response?
(b) What is the circuit bandwidth if the gain is down 3 dB at the dip at f_o?

11.23 A double-tuned transformer has $L_1 = L_2 = 100$ μH, $Q_1 = Q_2 = 100$, with coupling k at 150 per cent of critical. The source has $g_m = 4500$ μmhos.
(a) Find the values for C_1 and C_2 for resonance at 1.59 MHz.
(b) Compute the frequency separation of f_a and f_b.
(c) What is the voltage gain at resonance?

11.24 An FET amplifier is to have a gain of 20, extending to the highest possible frequency. Find that frequency when shunt-compensated, with $q = 0.41$. Also find L needed if $g_m = 0.0025$ mho, $C_{gs} = 1$ pF, $C_{gd} = 2.5$ pF, wiring capacitance $= 3$ pF.

11.25 A shunt-peaked video amplifier uses an FET with $g_m = 0.0025$ mho, $C_{gd} = 2$ pF, $C_{gs} = 1.5$ pF, $C_w = 2$ pF, and $R_s = 4000$ Ω. When designed for $q = 0.41$ find the f_2' frequency and the needed value of L. What is the low-frequency gain?

11.26 A pulse chain of 0.5-μs pulses is sent at a rate of 4000 per second. What amplifier bandwidth will be needed for reasonable pulse wave form at the output? For excellent wave form?

12
Power Amplifiers

We now want to put our signals to work, and large power outputs are required for such purposes as driving loudspeakers or servomechanisms or transmitting radio signals through space. Amplifiers providing large power outputs require large input signals. The small-signal equivalent circuit cannot be used and we must go back to the graphical method of transistor or tube analysis. Since the output curves are not linear for large-signal excursions, we have to determine the output wave distortion and devise circuits for reducing that distortion.

For maximum power output the loads should match the output resistances of the amplifiers. Direct matching of output resistance and load does not always happen. More usually we have to use an impedance-transforming circuit or transformer to supply the matched load conditions necessary for maximum power transfer.

Therefore, in our analysis of power amplifiers we shall be interested in such items as distortion levels, power efficiency of the transistor, elimination of the heat losses, and the impedance-matching circuits.

12.1 Defined Operating Conditions

Three broad areas of operating conditions are defined for power amplifiers, dependent on the chosen bias and the input voltage amplitude. We use the letters A, B, and C to designate these conditions, as demonstrated for a transistor in Fig. 12.1.

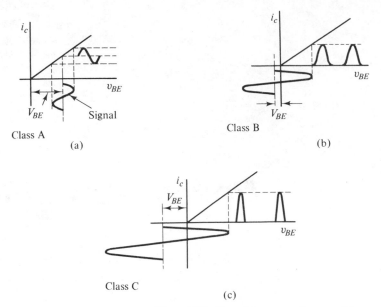

Figure 12.1 Transistor operating conditions.

For *Class A operation*, the bias is selected to place the Q point near the center of the transfer curve relating v_{BE} and i_C. Collector current is present for all values of signal input, as shown. This is the operating condition of the small-signal amplifiers in our preceding discussions. Distortion is low but the power output is also low because of the small input signal. The efficiency of conversion of dc power to ac power is limited to 50 per cent. Practical transistor amplifiers can reach 45 per cent.

With *Class B conditions*, the bias is selected to place the Q point on the cutoff line and collector current is present only on the forward half cycle of the input voltage. During the reverse half cycle, collector current is not present in the output of the transistor. Distortion is therefore high but the efficiency of power conversion can reach a theoretical maximum of 78.5 per cent. *Push-pull circuits* are employed to supply the missing half cycle and to reduce the distortion to usable levels. Practical efficiencies reach 65 per cent.

For operation *in Class C*, the bias is set to two or more times the cutoff level and current occurs only in short pulses near the forward peak of the input wave. Theoretical efficiency reaches 100 per cent but the distortion is so high as to limit the circuit's application to radio frequencies where the distortion harmonics can be filtered out by resonant circuits. Practical efficiencies reach 80 per cent.

Transistor power amplifiers employ the C-E circuit because of its high power gain; similarly, tube power amplifiers use the grounded-cathode circuit. Class A and B conditions yield amplifiers suited to audio-frequency power amplification and Class B and C amplifiers are used at radio frequencies. Because of the special methods required for the analysis of Class C amplifiers with current pulses of varying length and because of their limited application, the Class C amplifier will not be discussed here.

12.2 The Ideal Transformer

The *transformer* in Fig. 12.2, with a laminated iron magnetic core, is often used as an impedance-transforming device in amplifiers operating at audio frequencies from 50 to 15,000 Hz.

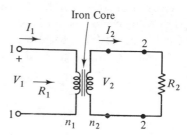

Figure 12.2 The ideal transformer.

The transformer is efficient in handling power. The voltages and currents present in the primary and secondary are related by the turns ratio

$$a = \frac{N_1}{N_2} \tag{12.1}$$

giving the voltage and current ratios as

$$\frac{V_1}{V_2} = \frac{I_2}{I_1} = a \tag{12.2}$$

The secondary load is

$$R_2 = \frac{V_2}{I_2} \tag{12.3}$$

On the primary side of the transformer we have

$$R_1 = \frac{V_1}{I_1} \tag{12.4}$$

But from Eq. 12.2 we can write

$$V_1 = aV_2$$

$$I_1 = \frac{I_2}{a}$$

and inserting these values into Eq. 12.4 we have

$$R_1 = \frac{aV_2}{I_2/a} = a^2 \frac{V_2}{I_2}$$

But $V_2/I_2 = R_2$ by Eq. 12.3 so

$$R_1 = a^2 R_2 \qquad\qquad (12.5)$$

A load in the secondary *appears* in the primary as a resistance $a^2 R_2$, with ac voltage applied. We say "appears" because R_1 is present only when alternating voltages are present; a direct current does not affect a transformer.

Thus a loudspeaker of $4\ \Omega$ can be made to appear as $400\ \Omega$ on the primary side if we use a transformer with the turns ratio

$$a = \sqrt{\frac{400}{4}} = \sqrt{100} = 10$$

Such an impedance-transforming device is called an *ideal transformer*. Actual transformers closely approach the ideal transformer performance. The transformer also isolates the dc component of collector current from the load. For dc the transformer primary appears as the low dc resistance of the primary winding, often considered to be zero.

12.3 Power Relations in the Class A Amplifier

For the large-signal amplifier with an output-matching transformer and under Class A conditions, the circuit is that in Fig. 12.3. Calculation of performance must be derived from the graphical output characteristics because of the nonlinearity of the transistor curves for large-signal excursions.

To obtain large power output we operate the transistor with large power input and the internally developed heat is the limiting factor. The input power is obtained from the collector supply and partially converted to signal

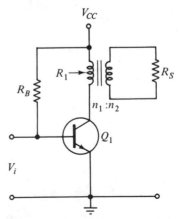

Figure 12.3   The C-E power amplifier.

output power and delivered to R_s, through the transformer. We have

$$\text{power input} = \text{ac output} + \text{losses}$$
$$V_{CC}I_C = I_c^2 R_1 + (V_{CE}I_C + I_C^2 R_{dc}) \tag{12.6}$$

where R_{dc} is the dc resistance in the collector circuit, outside the transistor, $V_{CE}I_C$ being the steady transistor loss at the Q point. Equation 12.6 shows that the transistor loss or dissipation is

$$P_d = V_{CE}I_C = V_{CC}I_C - I_C^2 R_{dc} - I_c^2 R_1 \tag{12.7}$$

The signal output term is $I_c^2 R_1$ and as it increases, the transistor loss $V_{CE}I_C$ must decrease and the transistor operates cooler in Class A.

We must design for the worst possible case of no signal, however, so we have the Q-point loss as

$$\max P_d = V_{CC}I_C - I_C^2 R_{dc} \tag{12.8}$$

The primary winding of the transformer usually has negligible dc resistance and in most amplifiers

$$R_{dc} = R_E \tag{12.9}$$

as the value of the emitter bias resistor. Resistance R_E is also kept small, however, since its resistance reduces the overall power efficiency and we shall neglect it in this somewhat idealized case. Therefore,

$$\max P_d = V_{CC}I_C - I_C^2 R_E \cong V_{CC}I_C \tag{12.10}$$

and this establishes the expected transistor loss in the circuit.

The limiting hyperbola in Fig. 12.4 is drawn for an allowable transistor loss at a specified operating temperature; it may be less than the indicated

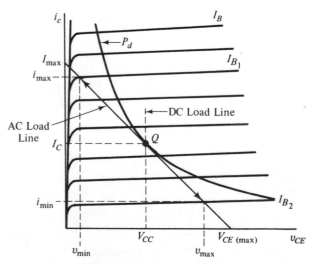

Figure 12.4 Construction of the ac load line.

loss from Eq. 12.10 and selection of that allowable loss is a thermal problem that will be discussed shortly. The Q point will be located on or below the limiting dissipation curve.

The presence of the transformer introduces a new step in the construction of the load line since with a transformer the dc and ac loads are not the same in resistance. We first select a V_{CC}, usually one-half the maximum rated collector-emitter voltage $V_{CE(max)}$. We then draw the dc load line, which always starts at V_{CC}. Since the dc resistance of the collector circuit in Fig. 12.3 is assumed zero, the slope will be $-1/R_{dc} = -1/0$, and the dc load line will be vertical as shown dashed in Fig. 12.4. The intersection with the locus will establish the Q point.

The presence of R_E would have given the dc load a slight slope to the left but would not alter the general situation. This is discussed in the example that follows.

Draw the *ac load line* from $2V_{CC}$ on the x axis and through the Q point. The y intercept of the ac load line establishes I_{max} and the slope of the line is that of R_1, which is the primary load to be supplied by the transformer. That is, from the slope

$$R_1 = \frac{2V_{CC}}{I_{max}} = \frac{V_{CC}}{I_C} \tag{12.11}$$

since $I_{max} = 2I_C$ by the geometry of the figure. The power loss is $P_d = V_{CC}I_C$ and we can also write

$$R_1 = \frac{V_{CC}^2}{P_d} \tag{12.12}$$

The base current will be driven symmetrically to some values at I_{B_1} and I_{B_2} as shown, and the transistor will reach peak collector currents as i_{max} and i_{min}. The power output is found from these peak values of a sine wave as

$$P_o = \left(\frac{i_{max} - i_{min}}{2\sqrt{2}}\right)^2 R_1 = \frac{(i_{max} - i_{min})^2 R_1}{8} \tag{12.13}$$

From the figure $I_C = (i_{max} - i_{min})/2$ and the efficiency of conversion of the dc power to signal power is

$$\text{eff.} = \frac{P_o}{P_d} \times 100\% = \frac{(i_{max} - i_{min})R_1}{4V_{CC}} \times 100\% \tag{12.14}$$

The efficiency depends on the amplitude of the output signal. For the *greatest possible output signal* that swings over the complete length of the load line, we have $i_{min} = 0$ and $2V_{CC}/R_1 = i_{max}$. Then

$$\text{eff.} = \frac{i_{max} - 0}{4V_{CC}/R_1} = \frac{i_{max}}{2i_{max}} = 50\%$$

and this is the *maximum theoretical power conversion efficiency* for the transformer-coupled Class A amplifier.

The saturation voltage limits i_{max} to values less than the theoretical but because of the low saturation voltage of most transistors a practical amplifier can approach the theoretical figure. Vacuum tubes, with much higher saturation voltages, rarely exceed 30 per cent in conversion efficiency.

Example: Select a load and determine the power output and conversion efficiency for a signal $i_B = \pm 60$ mA, maximum possible, applied to the transistor in Fig. 12.5. It is rated $V_{CE(max)} = 65$ V, $P_d = 30$ W under the expected operating conditions and with $R_E = 1\ \Omega$.

We first draw the maximum-power dissipation curve for 30 W as shown and select

$$V_{CC} = \frac{V_{CE}}{2} = \frac{65}{2} = 32.5 \text{ V}$$

A bias resistor $R_E = 1\ \Omega$ and so the dc load line is drawn with a $-1/1$ slope to 1 A and $32.5 - 1 = 31.5$ V. The Q point is then found at $I_C = 0.94$ A.

The ac load line is drawn from $2V_{CC} = 65$ V through the Q point and R_{ac} is found from the slope as

$$R_{ac} = \frac{V_{CE}}{I_C} = \frac{31.5}{0.94} = 34\ \Omega$$

but $R_1 = R_{ac} - R_E = 34 - 1 = 33\ \Omega$. We need a transformer to transform

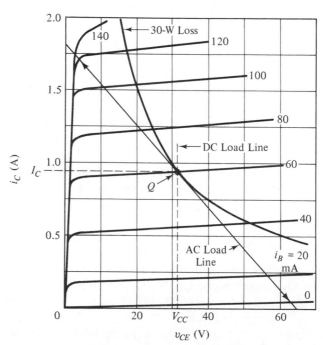

Figure 12.5 AC load line.

$R_s = 4\ \Omega$ of a loudspeaker to $33\ \Omega$. The turns ratio must be

$$a = \sqrt{\frac{33}{4}} = 2.87$$

With the designated input base signal we swing along the ac load line up to $i_B = 120$ mA and down to $i_B = 0$, giving $i_{max} = 1.70$ A and $i_{min} = 0.05$ A. The power output is

$$P_o = \frac{(1.70 - 0.05)^2 \times 33}{8} = 11.2 \text{ W}$$

The power input to the circuit is

$$P_{in} = 32.5 \times 0.94 = 30 \text{ W}$$

as expected and the power conversion efficiency is

$$\text{eff.} = \frac{11.2}{30} \times 100 = 37\%$$

12.4 Voltage Limitations

The manufacturer specifies a maximum safe value for V_{CE} but the transistor has some physical limitations on the voltage that may be applied.

The collector-base depletion layer widens as the collector voltage is increased and the depletion layer may extend completely through the thin base region at some high voltage. This is known as *punch-through*. The transistor under this condition appears to have a short circuit between emitter and collector. The base loses control until the collector voltage is reduced.

Avalanching of charges may occur in the collector region at a high voltage; however, no physical damage results. This is known as *first breakdown*. If the avalanching current channels into small areas, a hole may be melted through the base. This is known as *second breakdown* and the transistor is destroyed.

The maximum voltage for vacuum tubes is determined by the insulation limits and is specified by the manufacturer.

12.5 Effect of the Thermal Environment

In order to keep the collector-base junction leakage current small in comparison to the signal current in the collector, it is necessary to keep the temperature of the junction below 200°C (392°F) for silicon transistors and 100°C (212°F) for germanium transistors.

The rate of heat removal from the collector junction is proportional to the difference in temperature between the collector and the ambient surroundings of the transistor. With T_J as the junction temperature and T_A as the ambient temperature in °C,

$$T_J - T_A = \theta_{JA}P_d \qquad (12.15)$$

where θ_{JA} is the *thermal resistance* of the transistor case and mounting in units of °C/W. The thermal resistance really states the temperature differential needed per watt of heat removed from the transistor.

The power level for which the dissipation curve of a transistor is drawn is dependent on the thermal resistance of the transistor case and its mounting and on the ambient temperature surrounding the equipment. In small-signal amplifiers, P_d is usually at milliwatt level and air convection and conduction by the transistor leads is sufficient to remove the generated heat. For higher-power amplifiers the transistor mounting is designed to dissipate the heat by metallic conduction and convection by the air or by forced-air cooling.

The collector junction is usually in good thermal contact with the case. Electrical insulation is obtained by mounting on a base plate with a thin mica insulator with the air pockets filled with silicone grease. The base plate may be fitted with convection fins and is then known as a *heat sink*. This improves the heat transfer to the air. But even the heat sink is not able to reduce the case temperature to that of the surrounding air and the junction temperature will be above that of the case. The air may be considerably above 25°C as well. The characteristics of a small heat sink are given in Fig. 12.6, with $\theta_{CA} = 3$°C/W.

Suppose that we have a transistor in which 6-W dissipation produces the junction limit temperature when the case is held at 25°C in a water bath. But in air at 25°C the case temperature will be much above 25°C because of the

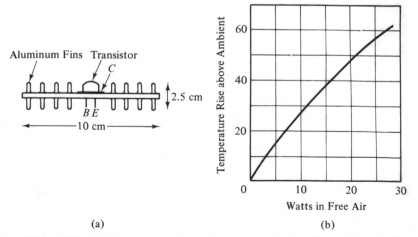

(a)   (b)

Figure 12.6 (a) Finned heat sink and transistor, MS-10; (b) temperature rise, thermal resistance = 3°C/W.

poor heat transfer from case to air. The junction temperature must rise by
the amount the case temperature exceeds 25°C in order to continue to conduct
6 W from the junction. Therefore, with a higher temperature for its surroundings, we must *derate* the allowable transistor power level. This is confirmed if we write Eq. 12.15 as

$$T_J = \theta_{JA}P_d + T_A \tag{12.16}$$

which indicates that the junction seems to float at some constant differential
above the ambient. For constant power loss, a 5°C rise in ambient will cause
a 5°C rise in junction temperature. In fact, the increase may be greater because
as T_J rises, I_{CBO} and h_{FE} will rise, giving a further increase in I_C and junction
temperature.

The heat removal situation is demonstrated by the electrical analog in
Fig. 12.7(a), with two resistances in series. The thermal resistance circuit is

$$\theta_{JA} = \theta_{JC} + \theta_{CA} \quad (°C/W) \tag{12.17}$$

where

θ_{JA} = total thermal resistance, junction to ambient
θ_{JC} = junction to case thermal resistance (supplied by the manufacturer)
θ_{CA} = case to ambient resistance of mounting

The manufacturer supplies a value for θ_{JC} as well as a *thermal derating curve*,
as shown in Fig. 12.7(b) for a transistor with 30-W dissipation at 25°C case
temperature. From the slope of the curve above 25°C we can find that

$$\theta_{JC} = \frac{T_{J(max)} - T_C}{P_d} \tag{12.18}$$

For this transistor we have $T_J = 200°C$, $T_C = 25°C$, and $P_d = 30$ W so that
$\theta_{JC} \cong 6°C/W$.

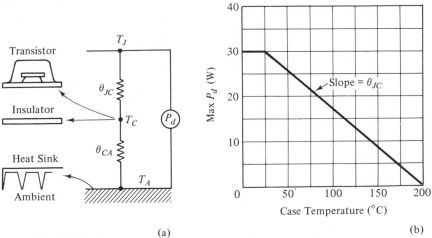

(a)   (b)

Figure 12.7 (a) Analog of the thermal circuit; (b) derating curve for a transistor,
30 W at 25°C (77°F) or below.

Substitution of Eq. 12.17 in 12.15 gives us

$$\theta_{CA} = \frac{T_{J(\text{max})} - T_A}{P_d} - \theta_{JC} \qquad (12.19)$$

which determines the allowable thermal resistance for the transistor mounting and heat sink.

The use of the derating curve can be shown in the following examples.

Example 1: What power rating can we assign to the transistor in Fig. 12.7(b) when used in a mounting and heat sink having $\theta_{CA} = 3°C/W$ and with $T_A = 45°C$?

We have

$$\theta_{JA} = \theta_{JC} + \theta_{CA}$$
$$= 6 + 3 = 9°C/W$$

From Eq. 12.15,

$$P_d = \frac{T_J - T_A}{\theta_{JA}} = \frac{200 - 45}{9} = \frac{155}{9}$$
$$= 17.2 \text{ W}$$

and this establishes a value for the limit hyperbola for this transistor under the specified temperature conditions.

The case temperature at 17.2 W can be found from the derating curve as 100°C.

Further data indicate that this transistor (RCA 40316) has a thermal resistance when operating in free air of 30°C/W and a maximum dissipation in that condition of 6 W. Thus the value of $\theta_{JA} = 9°C/W$ for the mounting system provides a considerable improvement in power rating over the free-air condition.

Example 2: Consider the transistor (in the previous example) operating at an ambient temperature of 35°C with $P_d = 20$ W. What is the maximum allowable value for θ_{CA}?

By Eq. 12.19,

$$\theta_{CA} = \frac{200 - 35}{20} = 8.3°C/W$$

From the curve the case temperature will be 82°C.

12.6 Determination of Output Distortion

Distortion of the output waveform occurs by reason of the nonlinear nature of the characteristics of the transistors employed. A current transfer curve is plotted in Fig. 12.8, resulting from points chosen along the load line for the amplifier in Fig. 12.5. For large-signal excursions we would anticipate

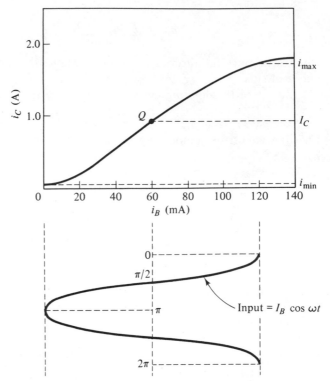

Figure 12.8 Current transfer curve for the load line of Fig. 12.5.

some distortion of the i_C waveform, or flattening of positive and negative peaks with the Q point shown.

The distorted waveform can be described by a fundamental frequency sinusoidal wave added to various amplitudes of harmonically related sine waves. The harmonics are at integer multiples of the fundamental frequency. The wave may be studied by use of a wave analyzer, which provides an amplitude reading for each harmonic frequency present.

We can also analyze amplifier outputs by reading amplitudes from the transfer curve, however, using i_C values resulting from an assumed sinusoidal base current signal, as shown in Fig. 12.8. We write the collector current wave as a Fourier series of cosines:

$$i_C = A_o + A_1 \cos \omega t + A_2 \cos 2\omega t + \cdots \qquad (12.20)$$

The amplitudes of the several frequencies present can be determined by evaluation of the $A_o, A_1, A_2, \ldots$ coefficients. The transfer curve can be used to find i_C at three points in time, spaced over the positive and negative halves of the input base current cosine wave. Substitution of these values into Eq.

12.20 results in three equations:

At $\omega t = 0$, $i_C = i_{\max}$

$$i_{\max} = A_o + A_1 \cos(0) + A_2 \cos 2(0)$$
$$i_{\max} = A_o + A_1 + A_2 \qquad (12.21)$$

since $\cos(0) = 1$.

At $\omega t = \pi/2$, $i_C = I_C$:

$$I_C = A_o + A_1 \cos\left(\frac{\pi}{2}\right) + A_2 \cos 2\left(\frac{\pi}{2}\right)$$
$$I_C = A_o - A_2 \qquad (12.22)$$

since $\cos(\pi/2) = 0$, $\cos \pi = -1$.

At $\omega t = \pi$, $i_C = i_{\min}$:

$$i_{\min} = A_o + A_1 \cos(\pi) + A_2 \cos 2(\pi)$$
$$i_{\min} = A_o - A_1 + A_2 \qquad (12.23)$$

since $\cos \pi = -1$, $\cos 2\pi = 1$.

Summarizing, we have

$$i_{\max} = A_o + A_1 + A_2$$
$$I_C = A_o - A_2$$
$$i_{\min} = A_o - A_1 + A_2$$

and solving simultaneously

$$A_2 = A_o = \frac{i_{\max} + i_{\min} - 2I_C}{4} \qquad (12.24)$$

$$A_1 = \frac{i_{\max} - i_{\min}}{2} \qquad (12.25)$$

The second-harmonic distortion present is the ratio

$$\% \, D_2 = \frac{A_2}{A_1} \times 100 \qquad (12.26)$$

Since A_1 is the peak value of the fundamental frequency, the fundamental output power is

$$P_1 = \left(\frac{A_1}{\sqrt{2}}\right)^2 R_1 = \frac{A_1^2}{2} R_1 \qquad (12.27)$$

Had we used five points along the input current wave, adding two currents as i_x and i_y at half amplitude at $\omega t = \pi/3$ and $\omega t = 2\pi/3$, we could calculate additional harmonics. The equations are

$$A_o = \tfrac{1}{6}(i_{\max} + i_{\min}) + \tfrac{1}{3}(i_x + i_y) - I_C \qquad (12.28)$$
$$A_1 = \tfrac{1}{3}(i_{\max} - i_{\min}) + \tfrac{1}{3}(i_x - i_y) \qquad (12.29)$$
$$A_2 = \tfrac{1}{4}(i_{\max} + i_{\min}) - \tfrac{1}{2}I_C \qquad (12.30)$$
$$A_3 = \tfrac{1}{6}(i_{\max} - i_{\min}) - \tfrac{1}{3}(i_x - i_y) \qquad (12.31)$$
$$A_4 = \tfrac{1}{12}(i_{\max} + i_{\min}) - \tfrac{1}{3}(i_x + i_y) + \tfrac{1}{2}I_C \qquad (12.32)$$

With

$$D_2 = \frac{A_2}{A_1} \times 100\%, \qquad D_3 = \frac{A_3}{A_1} \times 100\%, \qquad D_4 = \frac{A_4}{A_1} \times 100\%,$$

the total-harmonic distortion is

$$D = \sqrt{D_2^2 + D_3^2 + D_4^2 + \cdots} \tag{12.33}$$

It is convenient to be able to recognize harmonic content from the wave-form and so we present Fig. 12.9. The waveform in Fig. 12.9(a) in which the positive and negative waves are not similar contains predominantly even-order harmonics, whereas that in Fig. 12.9(b) with the two halves as mirror images contains odd-order harmonics.

Levels of allowable distortion are subject to individual judgment but a total distortion of 5 per cent is usually tolerated. The allowable level is reduced to less than 1 per cent when high-fidelity equipment is involved. In most power amplifiers this low level of distortion can only be achieved through use of negative feedback.

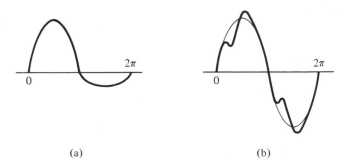

(a)   (b)

Figure 12.9 (a) Even harmonics present; (b) odd harmonics present.

Example: The values for waveform analysis can be read from Fig. 12.8 as

$$i_{max} = 1.70 \qquad i_x = 1.40$$
$$i_{min} = 0.05 \qquad i_y = 0.55$$
$$I_c = 0.94$$

We have

$$i_{max} - i_{min} = 1.65$$
$$i_{max} + i_{min} = 1.75$$
$$i_x - i_y = 0.85$$
$$i_x + i_y = 1.95$$

Using Eq. 12.28 to 12.32 we have

$$A_o = \frac{1.75}{6} + \frac{1.95}{3} - 0.94 = 0.002$$

$$A_1 = \frac{1.65}{3} + \frac{0.85}{3} = 0.833$$

$$A_2 = \frac{1.75}{4} - 0.47 = -0.033$$

$$A_3 = \frac{1.65}{6} - \frac{0.85}{3} = -0.0083$$

$$A_4 = \frac{1.75}{12} - \frac{1.95}{3} + \frac{0.94}{2} = -0.034$$

The distortion is predominantly even order, due to the second and fourth harmonics. We find

$$D_2 = \frac{0.033}{0.833} \times 100 = 4\%$$

$$D_3 = \frac{0.0083}{0.833} \times 100 = 1\%$$

$$D_4 = \frac{0.034}{0.833} \times 100 = 4.1\%$$

Total distortion:

$$D = \sqrt{4^2 + 1^2 + 4.1^2} = 5.7\%$$

12.7 The Push-Pull Circuit and Class B Operation

If we move the Q point down the transfer curve toward the origin, our forward swing can still drive the transistor up to i_{max}. Therefore we have a greater positive i_C current swing and greater power output. The other half of the input wave drives the transistor toward cutoff, however, and we have an unsymmetrical positive and negative wave, as in Fig. 12.10. While achieving greater power output from a given transistor, we have generated greater even-harmonic distortion.

The bias conditions as shown in Fig. 12.10 place the transistors at cutoff for part of a half cycle, the operating condition is intermediate between Class A and Class B, and it is known as Class AB.

We can cancel most of the distortion if we add a second transistor Q_2, oppositely connected as in Fig. 12.11(a), and achieve the power output of two transistors. The input transformer T_1 supplies opposite polarity signal voltages to the two bases, with respect to the midpoint of the winding. On the first half cycle, transistor Q_1 is driven in the forward direction by the input signal and transistor Q_2 is driven toward cutoff by the opposite polarity input voltage from the other half of the winding. On the next half cycle, the input voltages reverse and Q_1 is driven toward cutoff and Q_2 gives a full forward output. These output currents combine in the output transformer T_2; as the transistors appear to operate oppositely on each half cycle, the circuit is called a *push-pull amplifier*.

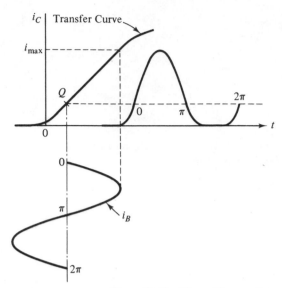

Figure 12.10   Class *AB* operation.

The circuit action is similar to that of a differential input–differential output amplifier; had in-phase signals been applied to the two inputs, the net output signal would be zero. The opposite-phase input signals provide an output after being subtracted in the output transformer, however; that is, $+A - (-A) = 2A$. The even-harmonic components are generated in phase and cancel in the output. Figure 12.12 illustrates how the fundamental waves add and the second harmonics cancel upon subtraction in the transformer.

Accuracy of harmonic cancellation is assured only with balanced transistor parameters and equal input voltages of 180° phase relation.

If we move the Q point to cutoff on the transfer curve, we have complete cutoff of one-half of the input wave and can show the action of the second transistor by an opposite transfer curve drawn to the common origin, as in Fig. 12.13. With cutoff bias, the operating condition is Class *B*. Each transistor supplies an independent half wave and these are combined in the output. Use of the push-pull connection cancels the inherent even-order distortion of the Class *B* operation and allows us to utilize the higher efficiency and power output of Class *B*.

The current components i_{C_1} and i_{C_2} in the output transformer are

$$i_{C_1} = I_a \sin \omega t + I_{C_1} \tag{12.34}$$

$$i_{C_2} = -I_a \sin \omega t + I_{C_2} \tag{12.35}$$

where $I_a \sin \omega t$ and $-I_a \sin \omega t$ are the oppositely phased ac signals and I_{C_1} and I_{C_2} are the steady Q-point currents. Passing in the windings of transfor-

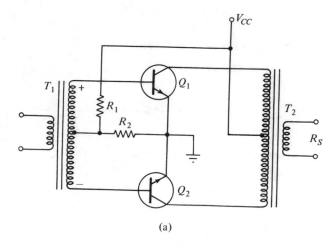

(a)

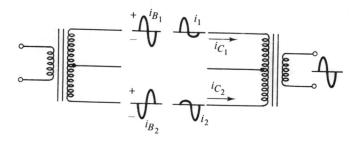

(b)

Figure 12.11 The push-pull amplifier.

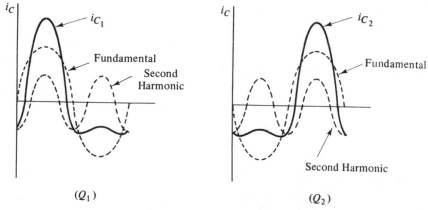

(Q_1)

(Q_2)

Figure 12.12 Waveforms in the push-pull output.

299

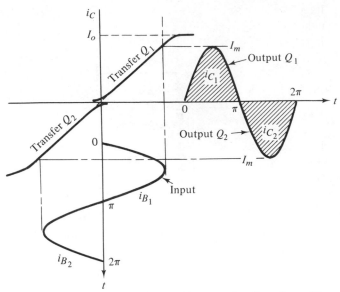

Figure 12.13 The action of two transistors under Class *B* conditions.

mer T_2 in opposite directions, these currents are subtractive in their effect on the secondary load so that if $I_{C_1} = (-)I_{C_2}$, then

$$i_2 = K(i_{C_1} - i_{C_2}) = 2KI_a \sin \omega t \qquad (12.36)$$

The signal currents appear added in the secondary and any dc components are canceled. Thus the steady magnetomotive force that might saturate the iron core is removed and the transformer core is more effectively used.

Since the Q-point current is essentially zero in a Class *B* amplifier, the power supply is called upon for sudden current surges as the signal input varies. Power supplies for Class *B* circuits should maintain constant voltage through these surges or have good *voltage regulation*. The use of shunt-C filters is advisable for such applications.

Since the dc input power is low for zero and small signals, Class *B* amplifiers are also preferred for battery-operated equipment.

12.8 Performance of a Class B Push-Pull Amplifier

For study of a Class *B* push-pull amplifier we have the output characteristics for Q_1 as drawn in the upper half of Fig. 12.14, and with Q_2 in a subtractive relation to Q_1 in that its characteristics are drawn upside down. The factor common to both transistors is V_{CC}, and the v_{CE} axes are aligned at that voltage. Since the transistors operate with zero i_C at cutoff, the V_{CC} point at which the curves are aligned is also the Q point of the amplifier. Signal swings occur up and down the load line from that quiescent point. The ac load line shown represents the largest output swing, from the upper knee of Q_1 to the same knee position for Q_2. The slope of the load line represents the primary transformer load R_1, presented to each transistor, as

$$R_1 = \frac{V_{CC}}{I_o} \tag{12.37}$$

We have assumed the saturation voltage to be small, with a current I_o.

On the first half cycle, the input signal voltage is forward to Q_1 and drives the operating point up the load line toward I_o; this represents operation below cutoff for Q_2. On the next half cycle the input voltage is forward to Q_2 and its operating point is driven along the load line toward its I_o value; this is below cutoff for Q_1.

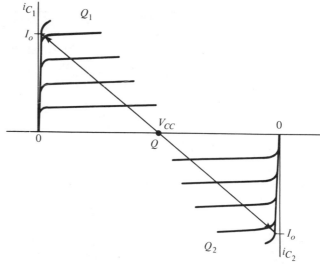

Figure 12.14 The Class B composite load line.

The output current consists of two separate half sine waves, combined into a full sine wave by the output transformer, and gives the output waveform shown in Fig. 12.12. Using what we have learned about such waveforms in the study of diode rectifiers, the dc value of current for one transistor is

$$I_{dc_1} = \frac{I_m}{\pi}$$

where I_m is the peak of the sine wave. For both transistors

$$I_{dc} = 2I_{dc_1} = \frac{2I_m}{\pi} \tag{12.38}$$

and the dc power input to the amplifier is

$$P_{dc} = \frac{2I_m V_{CC}}{\pi} \tag{12.39}$$

The ac current represented by the composite output wave of Fig. 12.12 is

$$I_{rms} = \frac{I_m}{\sqrt{2}} \tag{12.40}$$

and the ac power output to the transformer is

$$P_{ac} = \frac{I_m^2}{2} R_1 \tag{12.41}$$

The power conversion efficiency of a Class B amplifier is

$$\text{eff.} = \frac{P_{ac}}{P_{dc}} \times 100\% = \frac{I_m^2 R_1/2}{2I_m V_{CC}/\pi} \times 100\%$$

Because of Eq. 12.37, we can write

$$\text{eff.} = \frac{(\pi I_m^2/4)(V_{CC}/I_o)}{I_m V_{CC}} \times 100\% = \frac{\pi I_m}{4 I_o} \times 100\% \tag{12.42}$$

It should be possible to drive the peak of the signal output wave to where $I_m = I_o$ on the transistor curves. Under the maximum signal producing this large peak current we have

$$\text{max eff.} = \frac{\pi}{4} \times 100 = 78.5\%$$

as the maximum theoretical conversion efficiency. This is a considerable improvement over Class A conditions and is one of the reasons for the widespread use of Class B amplifiers.

By use of Eq. 12.38 we can write the signal power output as

$$P_{ac} = \frac{\pi^2 I_{dc}^2 R_1}{8} \tag{12.43}$$

which shows that the dc power input increases with increasing input signal. The peak transistor dissipation does not occur at maximum output but instead occurs at a signal level that is 40 per cent of that maximum. The peak power loss for use in transistor selection is

$$\text{max } P_d = \frac{2}{\pi^2} \frac{V_{CC}^2}{R_1} = 0.20 \frac{V_{CC}^2}{R_1} \tag{12.44}$$

at which point the conversion efficiency is 50 per cent.

12.9 Output Circuits without Transformers

The output load of most power amplifiers is a loudspeaker and these usually have a resistance of 3 to 16 Ω. At this low level of resistance there is no need for impedance matching and loudspeakers can be directly employed as loads for power transistors. Thus the output transformer, with its problems of weight, size, cost, and frequency distortion, can be eliminated in appropriate push-pull circuits which cancel the dc currents which might otherwise appear in the speaker circuits.

The usual circuits are derived from the bridge circuit shown in Fig. 12.15(a), in which two push-pull circuits drive a common load R_1 between y and z. These circuits are, respectively, Q_1 and Q_3 as the upper amplifier and Q_2 and Q_4 as the lower amplifier. The four input windings have a common primary, not shown, and the dots indicate the simultaneously positive terminals. Transistors Q_1 and Q_4 are simultaneously driven upward in current and Q_2 and Q_3 are driven downward at the same time. Thus point y is raised in potential and z is lowered; on the next half cycle y goes down and z goes up in potential so that we have an ac voltage across the load. Points y and z

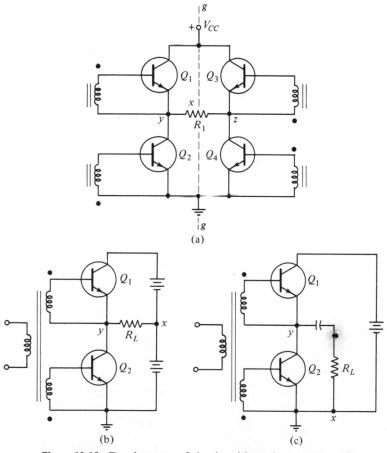

(a)

(b)   (c)

Figure 12.15   Development of circuits without the output transformer.

appear to teeter-totter in voltage around point x as a fulcrum. With no signal, points y and z are in dc balance and no dc current passes through the load.

The circuit in Fig. 12.15(b) is formed by splitting the first circuit along the common or ground line at g, g; transistors Q_3 and Q_4 are eliminated. Point y moves up and down as Q_1 and Q_2 are driven forward and reversed and an ac signal is present. Point x is maintained at its previous dc potential by splitting the power supply into two sections. The value for R_L is one-half of load resistance R_1.

The circuit in Fig. 12.15(c) evolved to avoid the expense of two power supplies. Since point x was at zero signal potential to ground in Fig. 12.15(b), it can be connected to actual ground if the large blocking capacitance C is used to avoid short-circuiting the dc supply. Capacitance C is usually of several thousand microfarads so that its reactance will be small compared to the resistance R_L of 3 to 16 Ω. This $R_L C$ combination establishes a low-frequency limit for the amplifier as

$$f_1 = \frac{1}{2\pi R_L C}$$

12.10 Phase Inverters for Push-Pull Input

Two equal voltages at 180° in phase are needed for push-pull amplifier input. The size and cost of input transformers can be avoided by use of circuits known as *phase inverters;* examples are shown in Fig. 12.16.

The circuit in Fig. 12.16(a) is called a phase splitter; it consists of an in-

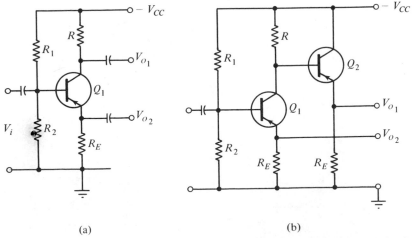

(a)   (b)

Figure 12.16 Phase inverter circuits.

phase output across R_E at V_{o_2} and a reversed-phase output across R at V_{o_1}. In reality the circuit is an emitter follower with a collector load added. Choosing R and R_E to provide a gain of unity to V_{o_1}, we have balanced voltages since the emitter follower gain is also near unity.

A second phase inverter is that in Fig. 12.16(b), which consists of the preceding circuit with Q_2 added. Both outputs are now taken from low-output resistance emitter followers and this is better for providing balanced and good waveform signals for Class B amplifiers, where there may be demands for large currents to drive the power transistors.

A differential amplifier may also be used as a phase inverter to drive a push-pull output stage.

12.11 Complementary Symmetry Circuits

The requirement for equal and oppositely phased input voltages for push-pull amplifiers is eliminated by use of a matched pair, *npn* and *pnp*, *complementary transistors*. As shown in Fig. 12.18(a), a signal of positive polarity to ground will simultaneously drive the *npn* unit Q_1 into forward conduction and the *pnp* unit Q_2 into cutoff. Thus a common input voltage will give Class B operation of the push-pull circuit. Figure 12.18(a) also employs a Darlington compound connection for higher input resistance. The Darlington transistors are also of complementary form. The push-pull circuit is basically that shown in Fig. 12.15(b), with R_1 as the load.

Figure 12.18(b) shows a driver stage using Q_1 and Q_2 as complementary symmetry transistors with a common input voltage. Their outputs, as emitter followers, provide low-resistance and high-current sources for normal *pnp* transistors Q_3 and Q_4 in a Class B push-pull circuit whose basic circuit form is that in Fig. 12.15(c).

Crossover distortion, as in Fig. 12.17, can appear in the output currents of such amplifiers because of the lack of symmetry in the *pnp* and *npn* characteristics near cutoff. One unit will not turn off at exactly the same currents and voltages as the other turns on. A similar phenomenon may be encountered in Class B amplifiers using the same transistor types and is corrected by bias adjustment and use of ample negative feedback.

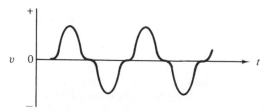

Figure 12.17 Crossover distortion in a waveform.

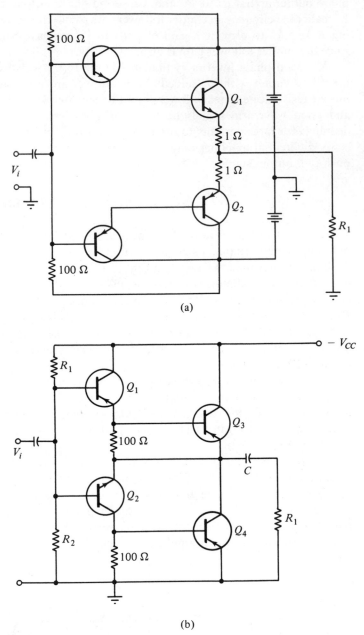

(a)

(b)

Figure 12.18 (a) Complementary symmetry power amplifier; (b) complementary symmetry driver for a Class *B* push-pull amplifier.

12.12 The Class B Linear
Radio-Frequency Amplifier

A Class *B* amplifier, with its accompanying high-power efficiency, may be operated single-ended or push-pull when used at radio frequencies with a resonant load. Because of the near linearity of the transfer curve between input i_B and output i_C, the output voltage is proportional to the driving voltage. The title of *Class B linear amplifier* is intended to emphasize this point as the amplifier is used to develop radio-frequency power when driven by varying amplitude or modulated radio-frequency voltages.

Since the resonant frequency of the tuned load circuit is high at f_o, in Fig. 12.19, the distortion components generated will have even higher frequencies at $2f_o$, $3f_o$, ... Having frequencies of two or more times the tuned frequency of the circuit, the harmonics can be well filtered out by a resonant load circuit. To discriminate against these harmonics the load circuit Q is usually maintained in the range from 10 to 15.

To retain its property of linearity, the transistor should not be overdriven

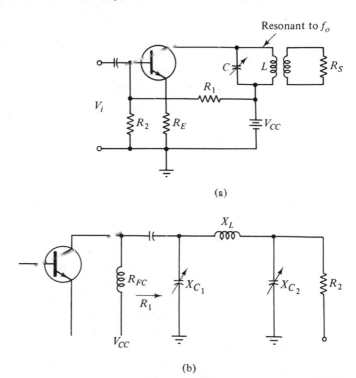

(a)

(b)

Figure 12.19 (a) A Class *B* linear radio-frequency amplifier with tuned load; (b) the π-matching network.

since that will take it into the flattened saturation portion of the transfer curve above i_{max} in Fig. 12.8.

A common modification of the resonant circuit, the π-*matching network* in Fig. 12.19(b), provides increased filtering action for the undesirable harmonics. The output power is normally supplied to an antenna of 50 to 75 Ω as R_2 and the load facing the transistor is R_1 of several hundred ohms. With Q greater than 10, the circuit elements of the π network should be

$$X_{C_1} = \frac{R_1}{Q} \qquad X_{C_2} = \sqrt{\frac{R_1 R_2}{Q^2 - R_1/R_2}} \qquad (12.45)$$

$$X_L = \frac{R_1}{Q^2}\left(\frac{R_1}{X_{C_1}} + \frac{R_2}{X_{C_2}}\right) \qquad (12.46)$$

Since the reactances of the shunt capacitances decrease with higher frequencies, they provide low impedance paths which short-circuit the higher-order harmonics to ground. This is why the π circuit is preferred to the T circuit.

12.13 Summary

The power output from transistors and tubes is limited by the internal losses that must be removed as heat. Only cooling by conduction or convection is possible at the low operating temperatures of transistors, although heat may be removed by radiation at the temperatures encountered with large-power vacuum tubes. The internal collector loss of the transistor must be removed through the transistor mounting and this is a point needing careful thermal design. Heat sinks are available to improve the heat transfer at temperatures only 100°C (212°F) to 200°C (400°F) above the ambient air temperature.

Large-signal excursions for the transistor create *nonlinear distortion* and harmonic frequencies in the output waveform. Suitable loads are chosen for maximum output conditions on the transistor output characteristics and the resulting harmonic distortion can be calculated directly from the transfer curve, relating i_B and i_C for a given load.

Class A operation has relatively poor power efficiency and low power output from a given transistor or tube but it has small waveform distortion. Class B conditions lead to higher efficiency and greater power output but the even-order distortion is very large. Push-pull circuits can reduce the even-order distortion and negative feedback can be added to reduce the remaining odd-order harmonics. These several circuit modifications reduce the distortion of the Class B condition to allowable levels and the Class B amplifier is especially valuable in the generation of large-power outputs from small transistors or tubes or when the dc power supply is limited in capability, as in battery-operated equipment.

12.1 What is meant by I_{CBO}?

12.2 Why must we limit the junction temperature of a transistor?

12.3 Define Class A operation; where would you establish the Q point?

12.4 Define Class B operation; where would you establish the Q point?

12.5 Define Class C operation; at what value would you establish the base bias?

12.6 Compare the relative amounts of power output and distortion from a given transistor in Class A, Class B, and Class C operation.

12.7 What is the theoretical maximum conversion efficiency for Class A, Class B, and Class C amplifiers?

12.8 How close to the theoretical conversion efficiency can you expect to operate a transistor Class A amplifier? A vacuum-tube Class A amplifier?

12.9 What is meant by the variable a as related to a transformer?

12.10 A certain transformer is listed as "400 to 3.2 Ω." What is meant by this, and what turns ratio would you find?

12.11 What is an ideal transformer?

12.12 What is meant by the locus of collector loss?

12.13 How does I_{CBO} vary with temperature?

12.14 With fixed base-bias current, what happens to the Q point on a load line when h_{FE} increases?

12.15 Define *thermal resistance*.

12.16 How do we use a transistor loss derating curve?

12.17 How can we find the thermal resistance of the transistor case from the derating curve?

12.18 How would θ_{SA} change if we add an air blower on the heat sink?

12.19 Why are fins put on a heat sink?

12.20 Define heat sink.

12.21 What is second breakdown of a transistor? Is it damaging?

12.22 What limits the value of anode voltage applied to a triode?

12.23 What is punch-through in a transistor?

12.24 Explain the purpose of an output transformer in a Class A amplifier. How would you determine the turns ratio?

12.25 Define total-harmonic distortion percentage.

12.26 Define second-harmonic distortion.

12.27 What is the difference in form of a wave having even-order harmonics and a wave having odd-order harmonics?

12.28 How many current points would be needed on a transfer curve to find the value of the sixth harmonic?

12.29 What are the advantages of Class B push-pull operation over Class A push-pull operation?

12.30 What is the advantage of Class *AB* operation over Class *A* operation?

12.31 What feature of Class *B* push-pull operation makes it especially valuable for battery-operated equipment?

12.32 What is complementary symmetry in transistor amplifiers?

12.33 What is crossover distortion?

12.34 What is the purpose of a phase inverter?

12.35 Why is the phase inverter in Fig. 12.16(b) superior to that in Fig. 12.16(a)?

12.36 Why does a Class *B* linear radio-frequency amplifier use a tuned circuit for a load?

12.37 What kind of signals can a Class *B* linear radio-frequency amplifier handle?

12.38 At what value of output power does maximum collector dissipation occur:
(a) In a Class *A* amplifier?
(b) In a Class *B* amplifier?

PROBLEMS

12.1 What turns ratio is needed for a transformer to couple an 8-Ω load to a transistor requiring 450 Ω?

12.2 An amplifier with an output resistance of 425 Ω is assembled with a transformer having a turns ratio $a = 2.57$. What should the secondary load be?

12.3 A transformer is found mounted on an 8-Ω loudspeaker. The turns ratio is measured as 10.5: 1. What ac primary resistance will be present?

12.4 A given transistor type has a dissipation rating of 20 W in a certain heat sink. What is the greatest possible power output when one of these transistors is used in Class *A* service? In Class *B* service?

12.5 A transformer provides a load of 25 Ω to a transistor with the output curves

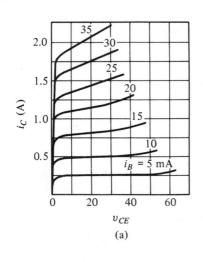

(a)

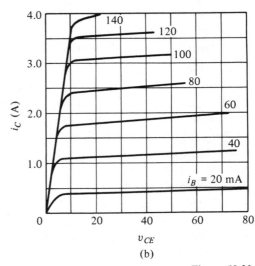

(b)

Figure 12.20

shown in Fig. 12.20(a). The primary has a dc resistance of 5 Ω. Find the true Q point, power output, and efficiency for $V_{CC} = 30$ V, collector loss of 25 W, and driving signal of ± 15 mA peak at the base.

12.6 Determine the second-harmonic distortion for the amplifier of Problem 12.5.

12.7 Draw the transfer i_B, i_C curve for the load line of Problem 12.5.

12.8 A transistor is derated according to the curve in Fig. 12.21. What is the thermal resistance θ_{JC}?

12.9 The heat sink and mounting for the transistor shown in Fig. 12.21 establishes the case temperature at 75°C. What is the power loss in the transistor?

12.10 A silicon transistor is derated according to the curve in Fig. 12.21. For a case temperature of 60°C, what is the allowable power dissipation?

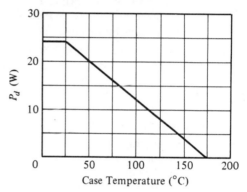

Figure 12.21

12.11 If the ambient air temperature is 35°C, what is the allowable value of θ_{CA} for the transistor of Problem 12.10?

12.12 The junction temperature of a transistor is 130°C. The dissipation at a case temperature of 25°C is 5 W; at a 25°C ambient air temperature it is 2 W. What is the value of θ_{CA}?

12.13 A silicon transistor is rated at a thermal resistance $\theta_{JC} = 0.9°C/W$ with $T_{J(\max)} = 160°C$.
 (a) Find the allowable power dissipation if the case is maintained at 50°C.
 (b) Find the power that could be dissipated if $\theta_{CA} = 2°C/W$ and the ambient air temperature is 35°C.

12.14 A silicon transistor has $T_{J(\max)} = 180°C$ and $\theta_{JC} = 0.7°C/W$. If mounted so that $\theta_{CA} = 0.9°C/W$, find the power dissipation allowable if the ambient temperature is 30°C.

12.15 A transformer-coupled Class A amplifier drives an 8-Ω loudspeaker through a transformer having $a = 4.3:1$. With a power supply of $V_{CC} = 36$ V, the amplifier delivers 3 W to the loudspeaker.
 (a) Find the ac voltage across the transformer primary.
 (b) Find the rms value of loudspeaker voltage.
 (c) Find the rms value of the loudspeaker current.

12.16 For the following current measurements from a waveform, find the second-harmonic distortion percentage: $i_{max} = 0.9$ A, $i_{min} = 0.47$ A, and $I_C = 0.65$ A. The load resistance through which this current passes is 95 Ω. What is the fundamental power output?

12.17 If the Q-point dc current is 0.22 A in Problem 12.15, find the conversion efficiency.

12.18 Find the collector dissipation for a Class A operated transistor with $V_{CC} = 30$ V, $I_C = 2.0$ A, and ac current of 0.7 A. The load is supplied by a transformer with $a = 2.2 : 1$; the secondary load is 10 Ω. What is the signal power output?

12.19 An amplifier has only second-harmonic distortion.
  (a) If $i_{max} = 250$ mA, $i_{min} = 5$ mA, and $I_C = 100$ mA, find the value of the second-harmonic distortion D_2.
  (b) If a transformer of $a = 4 : 1$ couples a 10-Ω load to this amplifier, what is the fundamental frequency power output?

12.20 Using the transistor in Fig. 12.20(b) with $V_{CC} = 40$ V, $I_B = 60$ mA at the Q point, with 20 Ω given by the transformer load, find
  (a) The fundamental power output with an ac sinusoidal input of ± 40 mA peak.
  (b) The second- and third-harmonic percentages.
  (c) The conversion efficiency.

12.21 A Class B push-pull amplifier is supplied by $V_{CC} = 50$ V and the signal swings the collector voltage to $v_{min} = 10$ V. The dc loss in both transistors is 40 W.
  (a) Find the power being delivered to the load.
  (b) Find the conversion efficiency.

12.22 With a transformer load of 30 $\Omega = R_1$ for each transistor, a Class B push-pull amplifier takes 0.75 A from the dc supply with a particular input signal.
  (a) What is the ac power output?
  (b) If the dc supply is at 40 V, what is the transistor power loss and the conversion efficiency?

13

Oscillator Principles

Oscillator circuits are the generators of our radio frequencies. A primary requirement is that the frequency be stable but, under conditions of varying supply voltages and varying temperature, sufficient stability is difficult to achieve.

There is a variety of circuits available but fundamentally those to be studied here all depend on positive feedback for their operation.

13.1 Oscillator Feedback Principles

In the circuit in Fig. 13.1 the voltage fed back from the amplifier output supplies the total input

$$V_f = V_i = \beta V_o$$

and since $V_o = -AV_i$, then

$$V_i = -A\beta V_i \qquad\qquad (13.1)$$

from which

$$(1 + A\beta)V_i = 0$$

If an output is present however, then $V_i \neq 0$ and therefore

$$1 + A\beta = 0$$
$$A\beta = -1 \qquad\qquad (13.2)$$

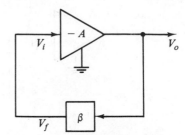

Figure 13.1 The basic feedback oscillator.

which was the condition we found in Chapter 9 that would lead to oscillation in a feedback amplifier.

Here we *want* the circuit to oscillate and so the expression above states two requirements for oscillation to occur:

1. That $A\beta = -1$.
2. That the net phase shift around the feedback loop be 0° or 360° or $2n \times 180°$, where n is 0, 1, 2, 3, ...

In Fig. 13.1 the amplifier provides its own input and initially the gain must be such that $|A\beta| > 1$. An initial turn-on surge or noise voltage provides V_i at the input and this is amplified to the output. This amplified output is fed back to the input as a larger signal. The process is repeated at successively greater amplitudes until the output becomes limited at cutoff and saturation. By operation to those limits the gain is reduced to an average level called for by Eq. 13.2 and a steady level is maintained.

The frequency of oscillation adjusts itself so that the phase shift requirement is satisfied.

Limitation of amplitude by cutoff and saturation implies distortion. The resultant harmonic frequencies can best be filtered out by use of a resonant circuit for the β network. High Q (high C) there provides better discrimination against the harmonic frequencies and also causes the circuit to oscillate more precisely at the resonant frequency of the tuned circuit.

The reactance network in Fig. 13.2(a) provides feedback and 180° of phase shift between the output and the input; the FET provides 180° of additional phase shift to meet the phase requirement of 360° for oscillation.

Reactances X_1, X_2, and X_3 must be resonant and the frequency of oscillation adjusts to make this occur. We know that the resonant frequency will make

$$X_1 + X_2 + X_3 = 0 \qquad (13.3)$$

For this equation to be true, one or two but not all three of the reactances must be negative and capacitive. This shows us the possibility of two basic

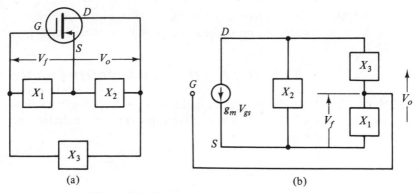

Figure 13.2 (a) Tuned-circuit feedback; (b) the equivalent circuit.

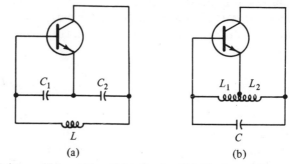

Figure 13.3 (a) Colpitts oscillator without bias circuits; (b) Hartley oscillator circuit.

circuits as in Fig. 13.3, the *Hartley oscillator* with two inductances (a tapped coil) and the *Colpitts oscillator* with two capacitances.

For the FET circuit in Fig. 13.2(b) we have

$$A = -g_m R$$

where R is the resonant resistance of X_2 in parallel with $X_1 + X_3$. The circuit as drawn also shows us that

$$\beta = -\frac{X_1}{X_1 + X_3}$$

from our previous work on feedback amplifiers. From these equations it is possible to show that the gain requirement for oscillation leads to

$$g_m \geq \frac{X_2}{X_1} \tag{13.4}$$

which can be satisfied. The frequency requirement of Eq. 13.3 satisfies the phase requirement and the circuit will oscillate.

In order that g_m be positive, Eq. 13.4 shows that X_1 and X_2 must be the same type of reactance; X_3 must therefore be of the opposite type to satisfy Eq. 13.3.

13.2 The Hartley and Colpitts Oscillators

The Hartley and Colpitts circuits in Fig. 13.3 are the basic feedback oscillator circuits. Analysis of the Colpitts oscillator provides an equation for the frequency of oscillation:

$$f_0 = \frac{1}{2\pi \sqrt{L\dfrac{C_1 C_2}{C_1 + C_2} + C_1 C_2 \dfrac{h_{oe}}{h_{ie}}}} \tag{13.5}$$

The first term under the radical is the resonant frequency of the L and equivalent C values of the tuned circuit. The second term under the radical shows that variation of the transistor parameters can have an effect on the frequency of oscillation. The effect is small because $h_{oe} \ll h_{ie}$. For design, we use

$$f_o = \frac{1}{2\pi \sqrt{L\dfrac{C_1 C_2}{C_1 + C_2}}} = \frac{1}{2\pi \sqrt{LC}} \tag{13.6}$$

We are normally concerned with oscillator frequencies of 1 MHz or over and a variation of 0.01 per cent at 1 MHz represents a shift of only 100 Hz; later we shall discuss oscillators in which such a shift is not negligible.

The gain requirement for oscillation for the Colpitts circuit gives

$$h_{fe} \geq \frac{C_2}{C_1} + h_{ie} h_{oe} \frac{C_1}{C_2} \tag{13.7}$$

and, with $C_2 = C_1$, we have no difficulty in meeting this requirement.

The Hartley oscillator can be similarly analyzed for frequency of oscillation. Again a small additive term is dependent on h_{oe}/h_{ie} and the frequency can be slightly affected by transistor parameters. Basically the frequency of oscillation is

$$f_o = \frac{1}{2\pi \sqrt{LC}} \tag{13.8}$$

where L is the total inductance of the tapped coil.

The necessary gain requirement makes h_{fe} a function of L_2/L_1; again we have no problem in meeting this requirement since we may make $L_2 = L_1$.

13.3 Practical Transistor Oscillators

The basic circuits in Fig. 13.3 do not include the bias circuits and associated blocking and bypass capacitors that are required for normal operation. These oscillator circuits are shown complete in Fig. 13.4.

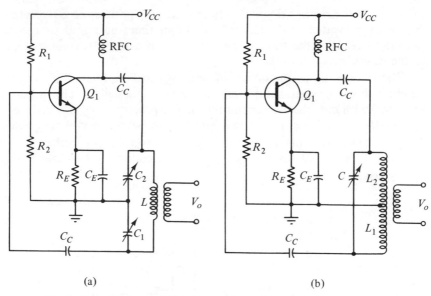

Figure 13.4 (a) Practical Colpitts oscillator circuit; (b) Hartley oscillator circuit.

Resistors R_1, R_2, and R_E provide the bias with initial design values setting the Q point a little more toward cutoff than for normal Class A operation. Full Class B or Class C biasing cannot be used because we would then have zero initial current. This would mean that there would be no way for the oscillations to start. Our design values for these bias resistors are determined by the methods of Chapter 5. Capacitor C_E is selected to have a reactance less than $R_E/10$ at the operating frequency.

The two blocking capacitors C_C are placed to isolate the bias voltages to the transistor and should have reactances of only a few hundred ohms.

The *radio-frequency choke*, RFC, passes the dc collector current but by reason of its inductive reactance it prevents current at oscillator frequency from reaching the power supply. It should have an inductive reactance that is much larger than the reactance of a blocking capacitor, or

$$X_L \geq 10X_C \qquad (13.9)$$

The frequency of oscillation in each circuit will be very close to the resonant frequency of the tuned circuit when stray capacitances of the wiring are included. The tuned circuit is sometimes called a *tank circuit* because it serves as a reservoir of radio-frequency energy, part of which is inductively coupled to the output as V_o.

While values of g_m and h_{f_e} have been indicated as minimums to assure that oscillations will start, the value of C_2 is usually equal to C_1 in the Colpitts circuit because equal capacitance split-stator tuning capacitors are generally available. In the Hartley circuit the choice of the tap position on L is arbitrary. It is usually located so that L_2 is two to four times greater than L_1; this does not overdrive the transistor and results in a lower harmonic content in the waveform.

The Armstrong oscillator circuit is shown in Fig. 13.5(a). If examined carefully, it can be seen that it is basically a Hartley oscillator circuit (tapped inductor) with only the base circuit tuned. The power supply is introduced in *series* with the collector inductance, and the radio-frequency choke prevents high-frequency currents from entering the power supply. Instead, these currents are provided with an easy path to the emitter and ground through the bypass capacitor C_C.

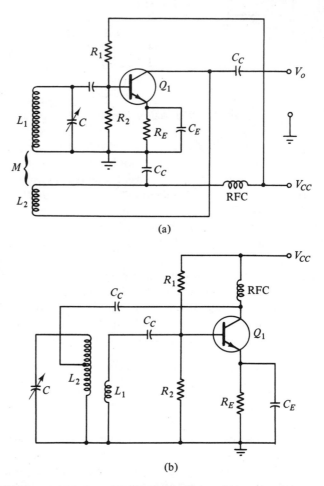

(a)

(b)

Figure 13.5 (a) Armstrong variation of the Hartley oscillator; (b) tuned-collector oscillator, tapped-collector coil.

The circuit in Fig. 13.5(b), is a tuned-collector oscillator, also basically a Hartley circuit. In order to raise the Q of the tuned circuit, the collector is connected to a tap on the circuit inductance.

13.4 *Crystal Control of Frequency*

Piezoelectric quartz crystals are of hexagonal form, with atomic plus and minus charges arranged in the unit crystal as in Fig. 13.6(a). Being symmetrical, the charges balance and the crystal is electrically neutral. Horizontal pressure from the left and right reduces angle ϕ and the negative charge (1) moves up and the positive charge (2) moves down. This movement destroys the symmetry and the crystal shows a negative charge at the top surface and a positive charge on the bottom surface. Horizontal tension enlarges angle ϕ and the charges move in the opposite directions. Therefore the charge

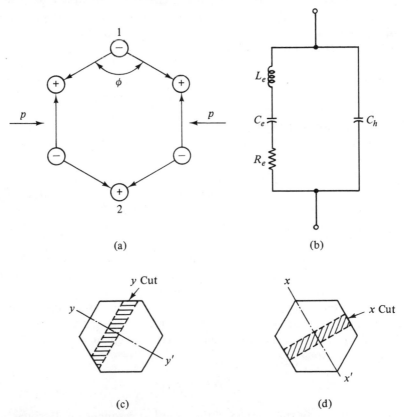

(a) (b)

(c) (d)

Figure 13.6 (a) Piezoelectric charge orientation; (b) crystal equivalent circuit; (c) crystal in the holder; (d) axes of the basic crystal cuts.

appearing on the upper and lower crystal surfaces alternately changes with variation of horizontal pressure. Conversely, if electrical charges are placed on the upper and lower surfaces by applying a voltage, a mechanical deformation of the crystal occurs in the horizontal direction.

An alternating voltage applied to the top and bottom crystal surfaces will cause the crystal to vibrate. If the time of travel of the mechanical vibration through the crystal is equal to a half cycle of the alternating voltage, then mechanical resonance occurs. The amplitude of the vibrations and the electrical voltage can be large.

Rochelle salt is a very active piezoelectric material. It is easily damaged by moisture, however, and its use is confined to microphones. Quartz gives smaller output voltages but is mechanically very stable. The oscillation is electromechanical in nature and the frequency of resonance is dependent on a thickness dimension of the crystal.

Electrical circuits employing piezoelectric crystals can be analyzed by replacing the crystal with its equivalent electrical network, Fig. 13.6(b). The magnitudes of L_e, C_e, and R_e of the network depend on the way the crystal slice is cut and its thickness, as in Fig. 13.6(c). Capacitance C_h is that of the mounting electrodes, usually electroplated onto the faces of the thin quartz slab. The valuable property of the quartz crystal is its sharp resonant response, which gives it a high equivalent Q. The Q is typically 30,000 but can reach 500,000 in units mounted in evacuated cans to eliminate air damping of the vibrations. With a Q of 30,000, a crystal resonant at 4 MHz will have a bandwidth of only 133 Hz.

The x and y axes of the crystal are shown in Fig. 13.6(c), with x- and y-cut crystals sliced perpendicular to the designated axes. These crystals have frequencies that vary with temperature but by an appropriate orientation of the angle of cut a zero temperature coefficient of frequency can be obtained. Resonant frequencies from about 10 kHz to above 20 MHz are possible with usable crystal thicknesses and higher frequencies can be generated as harmonic frequencies.

Several crystal oscillator circuits are shown in Fig. 13.7. That in Fig. 13.7(a) uses the crystal as a feedback element, with maximum positive feedback and oscillation at the crystal series or low-resistance resonance. The tuned circuit is adjusted near the crystal frequency and serves as an output waveform filter, the exact oscillation frequency being set by the crystal.

Series resonance occurs when the left branch of the equivalent circuit appears resistive, at

$$f_o = \frac{1}{2\pi\sqrt{L_e C_e}} \tag{13.10}$$

The resistance of this branch of the circuit at resonance is R_e since the reac-

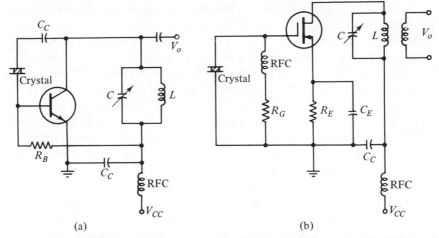

Figure 13.7 (a) Transistor crystal oscillator, collector tuned; (b) FET-tuned output oscillator.

tances of L_e and C_e are equal, opposite in sign, and cancel. This is the frequency of maximum positive feedback.

The crystal electrodes form the capacitance C_h with quartz as a dielectric. Capacitance C_h causes a parallel resonance, with a resonant frequency f_p determined when the reactances of the two branches are equal, as

$$X_{L_e} - X_{C_e} = X_{C_h}$$

$$2\pi f_p L_e - \frac{1}{2\pi f_p C_e} = \frac{1}{2\pi f_p C_h} \tag{13.11}$$

But this can also be written as

$$2\pi f_p L_e = \frac{1}{2\pi f_p C_h} + \frac{1}{2\pi f_p C_e}$$

$$= \frac{1}{2\pi f_p}\left(\frac{1}{C_h} + \frac{1}{C_e}\right)$$

$$f_p = \frac{1}{2\pi\sqrt{L_e C}} \tag{13.12}$$

where C is the equivalent series value of C_e and C_h,

$$C = \frac{C_e C_h}{C_e + C_h} \tag{13.13}$$

The effective circuit resistance is

$$R_p = \frac{Q}{2\pi f_p C} \tag{13.14}$$

and is very high. The parallel-resonant frequency is very slightly higher than f_o.

The circuit in Fig. 13.7(b) employs the crystal in its parallel-resonant high-resistance mode as a resonant circuit in the gate, in a *drain-tuned FET oscillator*. Again, the tuned circuit supplies a frequency-selective filter but the crystal parallel resonance fixes the oscillating frequency.

Changes in the load on the oscillator will alter the gain and thereby cause a change in the Miller-effect capacitance. This C_{ie} is in parallel with the tuned circuit of an *LC* oscillator and shifts the frequency. The shift can be reduced by the use of large tank capacitance and small *L* in those circuits. Greater stability is given by use of a crystal since the Miller capacitance appears in parallel with C_h. With an alteration of C_h, which is in series with C_e, the effective tuning capacitance *C* is not appreciably altered. In one example, a change of 10 per cent in C_h changed the crystal frequency only 0.003 per cent.

To isolate an oscillator from such changes in load that might affect the frequency, the oscillator is often followed by a *buffer amplifier*. This is designed to present a constant and high impedance load to the oscillator. An emitter follower is suitable for this purpose.

13.5 Resistance-Capacitance Feedback Oscillator

Variable *LC* oscillators tune over a frequency range that is proportional to $1/\sqrt{C}$. Available variable capacitors rarely have a maximum-to-minimum capacitance ratio greater than 10:1 so that *LC* oscillators usually tune over ranges of $\sqrt{10} \cong 3.1$. The function of the *LC* circuit is to provide the needed phase shift, however, and *RC* circuits can do this with a frequency range proportional to *C*, or over a 10:1 frequency range. This decade frequency range is of advantage in laboratory oscillators, used in a circuit like that in Fig. 13.8.

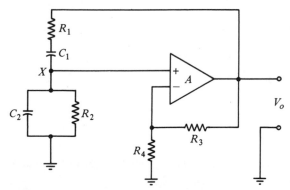

Figure 13.8 Basic Hewlett-Packard resistance-capacitance laboratory oscillator.

An operational amplifier provides $+A$ gain internally, with $360°$ of phase shift; the β circuit of the series RC and parallel RC elements need provide no additional shift at the feedback frequency. With

$$\beta = \frac{Z_2 \angle \theta_2}{Z_1 \angle \theta_1 + Z_2 \angle \theta_2} = \frac{Z_2}{Z_1 + Z_2} \qquad (13.15)$$

and the zero phase shift requirement can be met if $\theta_1 = \theta_2$. With

$$\theta_1 = \tan^{-1}\left(-\frac{1}{2\pi f_o R_1 C_1}\right); \qquad \theta_2 = \tan^{-1}\left(-\frac{1}{2\pi f_o R_2 C_2}\right)$$

the zero phase shift occurs with $R_1 = R_2 = R$ and $C_1 = C_2 = C$ at a frequency

$$f_o = \frac{1}{2\pi RC} \qquad (13.16)$$

The needed gain, with negative feedback provided by R_3 and R_4 in the configuration of the operational amplifier, is

$$A = \frac{R_3}{R_4} \geq 3 \qquad (13.17)$$

The negative feedback should provide a gain slightly greater than 3 so that oscillations will start. For instance, we might choose $R_3 = 350,000\ \Omega$ and $R_4 = 100,000\ \Omega$. With negative feedback present to limit the gain close to the critical value, the oscillator does not operate far into cutoff and saturation; as a result, excellent sine waves can be obtained at the output.

The R_3, R_4 resistors are often changed to a series resistor R_3 that supplies current to a low-power lamp as R_4. Feedback is taken across the lamp and is small when the lamp is cold at low output. As the output increases, the lamp resistance increases much more rapidly than the current; the gain is reduced and the output regulated to a constant level. The gain is maintained just over the critical value of 3, cutoff and saturation are avoided, and excellent output waveform is obtained.

13.6 Comments

The oscillator serves as the source for high-frequency voltages. The feedback circuits discussed here are not the only forms but they are most generally used.

The stability of oscillator frequency is the most important criterion of performance. The problem of stability of frequency would be a simple one if the oscillator were operated in isolation, at constant temperature, with unchanging components and constant voltages, and with no power taken from the circuit. Difficulties arise when we supply power from oscillators built with practical components and operating in normal environments.

At one time a crystal oscillator was almost the only means of obtaining stable operation; advanced materials and isolation of the oscillator function has now led to drift rates of only a few hertz per hour at megacycle frequencies, when using the tuned-circuit forms.

REVIEW QUESTIONS

13.1 Is the feedback negative or positive in an oscillator?

13.2 What two requirements must be satisfied to make an oscillator from a feedback amplifier?

13.3 In a feedback amplifier, what is the minimum value of β needed for oscillation?

13.4 What is the value of $A\beta$ for sustained oscillation?

13.5 An amplifier has a gain of 5; how is this gain reduced to the limiting value when the amplifier oscillates?

13.6 Explain the process of buildup of oscillations in an LC oscillator.

13.7 Explain the process of buildup of oscillations in an Armstrong oscillator.

13.8 What circuit determines the frequency of oscillation in a Hartley oscillator? The Colpitts oscillator? The Armstrong oscillator?

13.9 How is feedback provided in an Armstrong oscillator?

13.10 In what way is a crystal-controlled oscillator better than an LC oscillator?

13.11 Why is quartz a suitable material for oscillating crystals?

13.12 What is the effect of variation of output load on the oscillating frequency of an oscillator?

13.13 How does output load affect the frequency of an oscillator?

13.14 What is meant by the temperature coefficient of a quartz crystal?

13.15 What is the phase shift requirement for the β circuit of an RC oscillator?

13.16 What gain must an amplifier have to be used in an RC oscillator?

13.17 What is the purpose of using a tungsten incandescent bulb in the emitter or ground lead of an RC oscillator?

PROBLEMS

13.1 In Fig. 13.4(a), the Colpitts oscillator has $C_1 = C_2 = C$ and $L = 150\ \mu$H. Find $C_1 = C_2$ for oscillation at 1.5 MHz.

13.2 In a Colpitts oscillator in Fig. 13.4(a), $C_1 = 250$ pF, $C_2 = 100$ pF, and $L = 300\ \mu$H. What is the frequency of oscillation?

13.3 A transistor Colpitts oscillator has $L = 37\ \mu$H, $C_1 = C_2 = 310$ pF, $Q = 15$, $h_{oe} = 5 \times 10^{-6}$ mho, and $h_{ie} = 800\ \Omega$. Find the frequency of oscillation; show the effect in hertz due to the correction term involving h_{oe}/h_{ie}.

13.4 In the circuit in Fig. 13.4(b), oscillating at 3.7 MHz, determine if $C_C = 0.05$ μF, $L_{RFC} = 1.5$ mH, and $C_E = 0.1$ μF are suitable values.

13.5 A quartz crystal has equivalent $L_e = 3.66$ H, $C_e = 0.032$ pF, $C_h = 6$ pF, and $R_e = 4500$ Ω. What is the Q at the series resonant frequency?

13.6 A quartz crystal has $L_e = 0.6$ H, $C_e = 0.022$ pF, $C_h = 5.42$ pF, and $Q = 20{,}000$. Find f_o and f_p.

13.7 The crystal of Problem 13.6 has a trimmer capacitor C_m across C_h to vary the oscillator frequency. If the trimmer is variable from 2.8 to 9.8 pF, find the possible range of variation of crystal frequency. (Use a calculator).

13.8 A quartz crystal has $L_e = 250$ H, $C_e = 0.04$ pF, $R = 1800$ Ω, and $C_h = 7$ pF. When used to tune the base circuit of a transistor with $C_{ie} = 30$ pF, find the frequency of oscillation. The equivalent circuit is that shown in Fig. 13.9(b).

13.9 In Fig. 13.9(a) the circuit is to oscillate and $C = 0.1$ μF, $L = 0.15$ H, and $R_p = 20{,}000$ Ω.
(a) Find the gain A.
(b) Is the phase requirement satisfied at resonance?
(c) What is the value of β?
(d) What is the output frequency?

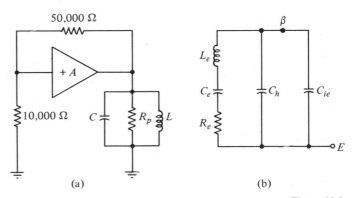

(a) (b)

Figure 13.9

13.10 Choose the RC elements for an oscillator as in Fig. 13.8, for operation at $f_o = 2$ kHz. Minimum gain is to be 3.2.

13.11 If $R_1 = R_2 = 10{,}000$ Ω, $C_1 = C_2 = 0.05$ μF, $R_3 = 5000$ Ω, and $R_4 = 1500$ Ω, what is the frequency of oscillation of the RC oscillator of Fig. 13.8? What is the amplifier gain with feedback?

13.12 In Problem 13.11, if $+A = 350$, choose values for R_3 and R_4 to assure that oscillations will just start.

13.13 In Problem 13.12, C_1 and C_2 are changed to a split-stator or double variable capacitor with a range from 30 to 330 pF (each section). With $R_1 = R_2 = 10^5$ Ω, what is the tuning range of the RC oscillator?

14

Modulation
and Detection

Electromagnetic radio waves propagate well through space only if the frequency is high, at least above 200 kHz. Speech and music frequencies lie in the band below 15 kHz; television picture signals utilize a video band ranging from 30 Hz to 4.5 MHz. To radiate such signals through space adequately requires that the base band frequencies be *translated* to appropriate channels in the radio-frequency spectrum that will carry the signals. These channels presently cover the frequency range from 200 kHz to many gigahertz. Such translation or conversion of frequency is accomplished by the process of *modulation*.

Detection, or *demodulation*, is the name for the reverse process by which the desired signals are recovered from the radio-frequency carriers and made audible or visible as in television.

14.1 Fundamentals of Modulation

An alternating voltage can be expressed as a function of time as

$$v = A \cos(2\pi ft + \theta) \qquad (14.1)$$

where
  A = peak amplitude of the wave
  f = frequency in hertz
  θ = phase angle
  t = time

In telephone and radio transmission we want to convey information to the receiver. The wave of Eq. 14.1 can carry information only through its presence or absence. The telegraph is an example of such an on-off signal but the telegraph has been limited in its *rate* of transmission of information. To send information as fast as it is generated, or in *real time*, requires that a characteristic of the radio wave be varied at the real time rate.

Equation 14.1 has two characteristics capable of being varied with time in accordance with the information we want to transmit. Thus we have two basic methods of *modulation* of an ac wave, as

1. *Amplitude modulation* (AM), in which the wave amplitude A is caused to vary in accordance with the amplitude of the modulating signal.

2. *Frequency modulation* (FM), in which the frequency f of the wave is changed in accordance with the amplitude of the modulating signal.

Actually, FM is a subprocess of a more general form known as *angle modulation*, as is *phase modulation* (PM) in which θ is caused to vary with the modulating signal; however, FM is most generally used.

Another class of systems does employ the method of the telegraph but turns the signal on and off at a very high rate and generates very short pulses for transmission. The information signal is sampled at a rate of several thousand per second and a characteristic of the pulse is varied to represent the amplitude of each sample. We have several possibilities:

1. PAM, *pulse-amplitude modulation*, in which the amplitude of the pulse is varied by the sample amplitude.

2. PDM, *pulse-duration modulation*, in which the duration of the pulse represents the sample amplitude.

3. PCM, *pulse-code modulation*, in which a coded train of pulses represents the sample amplitude.

The PCM system is becoming widely used because of its freedom from noise and distortion in the transmitting path, wire, cable, or space. In the reception of PCM we do not need to receive an accurate pulse waveform. By transmitting the sample amplitude in a code of pulses, we need only to determine that a pulse was sent or not sent. The binary code of Chapter 16 is generally employed in generating the pulse trains, which are decoded back to sample amplitudes at the receiver.

14.2 The Frequency Spectrum in AM

The frequency of the radio-frequency wave will be designated f_c, as the *carrier frequency* on which the informational signal is to be modulated. The informational signal from speech, music, or the dots of a TV picture will be assigned a frequency f_s. The frequency f_s will be only one of a large band of frequencies making up the complete music or TV picture spectrum; we use this one frequency as an example of what happens to every such frequency. The signal frequency f_s is smaller than the carrier frequency f_c on which it is to be modulated.

The signal frequency can be written as

$$v_s = V_s \cos 2\pi f_s t \tag{14.2}$$

and the carrier frequency on which we wish to *modulate* the information signal is

$$v_c = A \cos 2\pi f_c t \tag{14.3}$$

from Eq. 14.1, after dropping the constant angle as having no meaning here.

In *amplitude modulation* we vary the coefficient A with the informational signal so that

$$A = V_c + V_s \cos 2\pi f_s t$$
$$= V_c\left(1 + \frac{V_s}{V_c} \cos 2\pi f_s t\right) \tag{14.4}$$

The *modulation factor* is

$$m_a = \frac{V_s}{V_c} \tag{14.5}$$

and

$$A = V_c(1 + m_a \cos 2\pi f_s t) \tag{14.6}$$

In AM systems it is not desirable for m_a to exceed 1.0 or 100 per cent because of excessive distortion that is generated.

Substitution of Eq. 14.6 for A into Eq. 14.3 gives us an expression for the amplitude-modulated wave:

$$v = V_c(1 + m_a \cos 2\pi f_s t) \cos 2\pi f_c t \tag{14.7}$$
$$= V_c \cos 2\pi f_c t + m_a V_c \cos 2\pi f_s t \cos 2\pi f_c t \tag{14.8}$$

The first term is the carrier and the second term can be simplified if we use the trigonometric identity

$$\cos a \cos b = \tfrac{1}{2} \cos (a + b) + \tfrac{1}{2} \cos (a - b)$$

showing that the second term actually consists of two waves. We can then state the amplitude-modulated wave as

$$v = V_c \cos 2\pi f_c t + \frac{m_a}{2} V_c \cos 2\pi (f_c + f_s)t + \frac{m_a}{2} V_c \cos 2\pi (f_c - f_s)t \tag{14.9}$$

The amplitude-modulated wave consists of three frequencies, the original carrier at f_c and two side frequencies. The *upper side frequency* appears as the sum of the carrier and modulation frequencies, $f_c + f_s$, and the *lower side frequency* appears as the difference of the carrier and the modulation frequencies, $f_c - f_s$. If f_s is small, then the three frequencies are closely grouped and centered on f_c.

The resultant waveform is shown in Fig. 14.1, with the constant amplitude carrier in Fig. 14.1(a) and the modulated wave in Fig. 14.1(b). The modulated wave is *not* the result of simple addition of two frequencies but is the sum of *three* frequencies of Eq. 14.9. Amplitude modulation occurs because of the *product* of two frequencies as found in Eq. 14.8. Such a product is a necessary condition for any method of amplitude modulation.

If the modulating signal is one of many frequency components in speech or music, as examples, then many side-frequency pairs exist, and the groups of side frequencies are called the *upper* and *lower sidebands*. Figure 14.2(a) shows the spectrum of an AM waveform in which a carrier at f_c has been modulated by a frequency f_s, generating two side frequencies. In Fig. 14.2(b) we have an AM spectrum in which a carrier at f_c is simultaneously modulated by three signals at 1000, 2000, and 4000 Hz.

The information-carrying signal has been *translated* by amplitude modula-

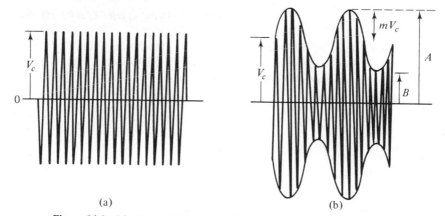

(a) (b)

Figure 14.1 (a) Unmodulated waveform; (b) amplitude-modulated wave, $m_a = 0.5$.

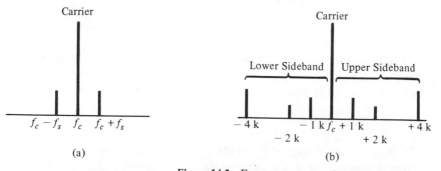

(a) (b)

Figure 14.2 Frequency spectra for AM modulation.

tion to a different frequency f_c, ideally without distortion. The carrier f_c can be placed anywhere in the radio spectrum for ease in transmission over wires or through space or for better conditions of amplification. The bandwidth occupied by the signal has been doubled since we now have two sidebands. That is, the modulated wave represented by the spectrum in Fig. 14.2(b) with $f_c = 1$ MHz would have sidebands extending from 0.996 to 1.004 MHz, a total bandwidth of 8 kHz and twice the frequency of the highest-frequency modulating signal. With frequencies scarce in our crowded radio bands, this doubling of bandwidth is one of the deficiencies of the AM system of modulation.

The waveform of an AM wave would appear on a cathode-ray oscilloscope as in Fig. 14.1(b). Measurements A and B may be made to find the value of m_a from the pattern; that is

$$m_a = \frac{A - B}{A + B} \tag{14.10}$$

14.3 The Power Spectrum in AM

Equation 14.9 tells us that the carrier has an amplitude V_c and when the modulated wave is applied to a resistive load R, the power due to the three components in double sideband AM (DSB) is
Carrier power:

$$P_c = \frac{V_c^2}{R} \tag{14.11}$$

Upper sideband:

$$P_s = \frac{m_a^2}{4} \frac{V_c^2}{R} = \frac{m_a^2}{4} P_c \tag{14.12}$$

Lower sideband:

$$P_s = \frac{m_a^2}{4} \frac{V_c^2}{R} = \frac{m_a^2}{4} P_c \tag{14.13}$$

The total power in an AM wave is therefore

$$P = P_c \left(1 + \frac{m_a^2}{2} \right) \tag{14.14}$$

The total power in the sidebands is $m_a^2/2$ times the power in the carrier and is divided among the many frequency components, each frequency having its own value of m_a such that

$$\frac{m_{a_1}^2}{2} + \frac{m_{a_2}^2}{2} + \frac{m_{a_3}^2}{2} + \cdots = \frac{m_a^2}{2}$$

At $m_a = 1.0$ or 100 per cent modulation, the sideband average power with sinusoidal modulation is 50 per cent of the carrier power and the total average power is 150 per cent of the unmodulated carrier power. The signal sidebands employ only one-third of the total power of an AM wave at 100 per cent modulation, the other two-thirds being present in the carrier.

In Fig. 14.1(b) the peak above V_c is mV_c and at $m = 1.0$, the peak voltage is $2V_c$ and double the carrier level. The equipment must be designed to withstand such a voltage at peak modulation.

With $2V_c$ at the peak on full modulation, the power is $4V_c^2/R = 4P_c$ and power peaks of four times the normal unmodulated carrier must be supplied.

A radio station is rated at a carrier power of 1000 W. To reach 100 per cent modulation we must supply 500 W of average modulating power and there will then be 250 W average in each sideband. At 10 per cent modulation ($m_a = 0.10$) the carrier power is still 1000 W but by Eq. 14.14 the average sideband power is only 2.5 W. Since only the sidebands are usable power at the receiver, this low modulation percentage gives a very poor power efficiency for information transmittal. Accordingly, the modulation factor m_a is normally maintained near 1.0.

14.4 The Diode Modulator for AM

Amplitude modulation is often generated by use of the nonlinear voltage-current curve of a diode or that of the transistor emitter-base junction. Let us approximate the diode curve in Fig. 14.3(a) by use of a linear curve added to a parabolic curve, as

$$i = a_1 v + a_2 v^2 \qquad (14.15)$$

In Fig. 14.3(b) two voltages are applied to the diode and load circuit, resonant at frequency f_c:

$$v = V_c \cos 2\pi f_c t + V_s \cos 2\pi f_s t \qquad (14.16)$$

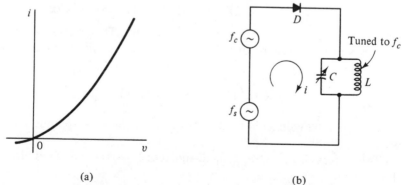

(a) (b)

Figure 14.3 (a) A diode characteristic; (b) simplified diode modulator.

where $V_s < V_c$ and $f_s \ll f_c$. Substituting this voltage expression into Eq. 14.15, we can develop an equation for the circuit current as

$$i = a_1 V_c \cos 2\pi f_c t + a_1 V_s \cos 2\pi f_s t + a_2 V_c^2 \cos^2 2\pi f_c t$$
$$+ a_2 V_s^2 \cos^2 2\pi f_s t + 2a_2 V_c V_s \cos 2\pi f_s t \cos 2\pi f_c t \qquad \textbf{(14.17)}$$

By trigonometric identity, we have

$$\cos^2 a = \tfrac{1}{2} \cos 2a + \tfrac{1}{2}$$

and the third and fourth terms of Eq. 14.17 can be modified to

$$a_2 V_c^2 \cos^2 2\pi f_c t = \frac{a_2 V_c^2}{2}\left[\cos 2\pi (2f_c)t + \frac{1}{2}\right]$$

$$a_2 V_s^2 \cos^2 2\pi f_s t = \frac{a_2 V_s^2}{2}\left[\cos 2\pi (2f_s)t + \frac{1}{2}\right]$$

But $2f_c$ and $2f_s$ are double frequencies or second harmonics.

Rearranging Eq. 14.17 and using these equations involving the second-harmonic frequencies, we have

$$i = a_1 V_c \cos 2\pi f_c t + \frac{a_2 V_c^2}{2}\left[\cos 2\pi (2f_c)t + \frac{1}{2}\right]$$

$$+ a_1 V_s \cos 2\pi f_s t + \frac{a_2 V_s^2}{2}\left[\cos 2\pi (2f_s)t + \frac{1}{2}\right] \qquad \textbf{(14.18)}$$

$$+ 2a_2 V_c V_s \cos 2\pi f_s t \cos 2\pi f_c t$$

This current passes through the tuned circuit, resonant and having a high resistance only near frequency f_c. The frequency $2f_c$ is far removed from f_c, being a second harmonic, and so are frequencies f_s and $2f_s$ as modulating frequencies, and the dc terms $a_2 V_c^2/2$ and $a_2 V_s^2/2$. Only the first and last terms of Eq. 14.18, involving frequency f_c, will produce appreciable voltages across the resonant circuit. Thus that circuit acts as a filter to remove the frequencies we do not want in the modulator output.

Therefore, the effective output voltage is

$$v = a_1 R_p V_c \cos 2\pi f_c t + 2a_2 R_p V_s V_c \cos 2\pi f_s t \cos 2\pi f_c t \qquad \textbf{(14.19)}$$

where R_p is the resonant impedance of the circuit. This expression shows the frequency product term predicted as necessary for amplitude modulation. We can reduce Eq. 14.19 to

$$v = a_1 R_p V_c \left(1 + \frac{2a_2}{a_1} V_s \cos 2\pi f_s t\right)\cos 2\pi f_c t \qquad \textbf{(14.20)}$$

which shows the voltage in the form of Eq. 14.7 as an amplitude-modulated wave.

This process is known as a small-signal or low-power method of modulation.

14.5 High-Power-Level AM Modulation

A method of amplitude modulation better suited to high-power use is that of power conversion in the Class *C modulated amplifier* shown in Fig. 14.4. With the input at f_c, the high-frequency current output is linearly related to the supply voltage to the amplifier. If that supply voltage is varied by a Class *B*

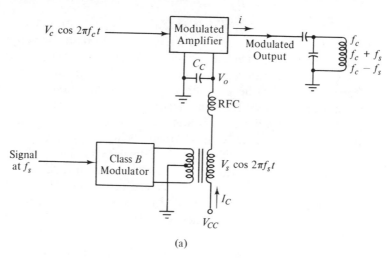

(a)

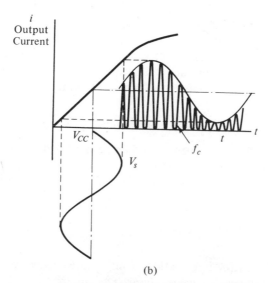

(b)

Figure 14.4 (a) High-level AM modulator circuit; (b) resultant output current.

amplifier as *modulator*, we have

$$v_o = V_{CC} + V_s \cos 2\pi f_s t$$
$$= V_{CC}(1 + m_a \cos 2\pi f_s t) \qquad (14.21)$$

The current to the output tuned circuit varies in a similar manner, giving

$$i = \frac{V_{CC}}{R_p}(1 + m_a \cos 2\pi f_s t) \cos 2\pi f_c t \qquad (14.22)$$

and this represents an amplitude-modulated wave, with carrier at f_c and sidebands $f_c + f_s$ and $f_c - f_s$, Eq. 14.7 and 14.9. With the supply voltage varying at the f_s rate, then the carrier frequency current in the tank circuit will have an envelope shape corresponding to the modulating signal.

The current from the dc supply is I_C since the modulation-frequency variation is up and down from the steady level and averages out. The dc supply furnishes the steady carrier power $V_{CC}I_C$ and the modulator must furnish the power for the sidebands, at maximum being $V_{CC}I_C/2$. The output transformer of the Class B amplifier must be adequate for this level of power.

The secondary load resistance for the Class B modulator must be known in order to specify the turns ratio of the transformer. The resistance into which this transformer delivers the sideband power is that of the modulated amplifier at the modulation frequency. This is

$$R_b = \frac{\text{modulation component of secondary voltage}}{\text{modulation component of secondary current}}$$
$$= \frac{m_a V_{CC} \cos 2\pi f_s t}{m_a I_C \cos 2\pi f_s t} = \frac{V_{CC}}{I_C} \qquad (14.23)$$

which is simply the dc resistance represented by the modulated amplifier.

Another form of amplitude modulation forces a current of one frequency through a resistance or impedance whose magnitude is varied at a second frequency. Thus

$$v = I_s \cos 2\pi f_s t \times R \cos 2\pi f_c t \qquad (14.24)$$

This shows directly the frequency product term needed for amplitude modulation to take place. This method is used in low-frequency servo systems by variation of an inductance at the carrier rate.

14.6 Linear Detection for AM Signals

To recover the useful information from an AM wave we normally use a *diode detector* or *envelope demodulator*. The circuit in Fig. 14.5(a) acts much as a half-wave diode rectifier with a shunt-capacitor filter.

Capacitor C is chosen so that the parameter $f_c RC$ is in the range of 30 to

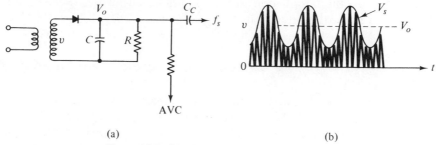

(a)   (b)

Figure 14.5 (a) Linear diode detector circuit; (b) output waveform.

200, as done for the rectifier. The capacitor charges to the peak of each carrier cycle and holds that voltage until the next positive half cycle. With an amplitude-modulated wave as an input signal,

$$v = V_c(1 + m_a \cos 2\pi f_s t) \cos 2\pi f_c t$$

The voltage of the capacitor at the carrier peaks is

$$V_o = V_c(1 + m_a \cos 2\pi f_s t) \tag{14.25}$$

as shown in Fig. 14.5(b). Since $m_a = V_s/V_c$, we have

$$V_o = V_c + V_s \cos 2\pi f_s t \tag{14.26}$$

This shows that in the voltage across the load R we have recovered the original modulation signal and also have a dc term equal to the carrier amplitude of the received signal.

The voltage V_o appears across R and C_C blocks the dc component but passes the modulation frequency f_s to the output of the detector. Practical circuit values are shown in Fig. 14.6 for $f_c \approx 2$ MHz.

While R and C are chosen so that $f_c RC$ is large, they should be chosen so that $f_s RC$ is approximately given by

$$\max f_s RC < \frac{1}{2\pi m_a} \tag{14.27}$$

Such a choice ensures that RC is small enough that at frequency f_s the capacitor C can discharge between cycles of f_s and thus its voltage is able to follow

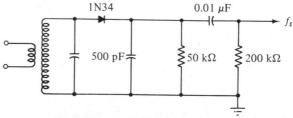

Figure 14.6 A practical detector circuit for AM.

changes in V_s, the information signal amplitude. Distortion results if Eq. 14.27 is not approximately satisfied.

14.7 Automatic Volume Control (AVC)

In radio reception it is desired that weak signals be amplified more and strong signals be amplified less so that the output level of all signals is about the same and remains so even though signals *fade* and weaken in transmission. The dc term in Eq. 14.26, equal to the received carrier level, provides a

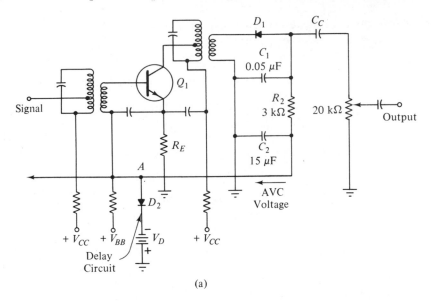

(a)

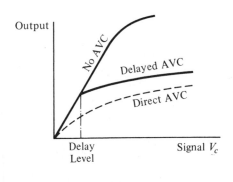

(b)

Figure 14.7 (a) An AVC circuit; (b) automatic volume control action.

measure of received signal strength that is used to achieve *automatic volume control* (AVC).

Shown in the circuit in Fig. 14.7(a) is an $R_2 C_2$ filter at the diode detector output. The $R_2 C_2$ product is made large, a fraction of a second, to remove any modulation at low f_s frequencies from the dc voltage. Connected as shown, the output of D_1 makes this dc voltage negative to ground. After filtering, this dc voltage is applied to the base of the *npn* transistor Q_1 and other amplifier transistors, to reduce I_E. Both h_{fe} and h_{ie} are functions of I_E and the gain of the amplifiers can be reduced as the strength of the received signal increases, as measured by V_c.

Without AVC the output of the detector increases in proportion to the input signal as shown in Fig. 14.7(b). The AVC action decreases the gain with increased signal strength, giving an output change as shown by the dashed curve and holding all signals more nearly constant in output.

All signal levels produce an AVC voltage, however, and the gain is reduced even for weak signals, where the full gain is needed. Accordingly we use *delayed AVC*, applied by the *diode clamp* circuit at point A on the AVC line of the amplifier. Until the AVC line reaches $-V_D$, the delay bias diode D_2 is closed and transmits $-V_D$ volts on the AVC line; this is the desired bias level for all amplifiers. But when the AVC voltage from D_1 becomes more negative than $-V_D$, diode D_2 opens and the varying AVC bias is used to control the amplifier gain. The result is a high gain for weak signals and a more ideal AVC characteristic for large signals, as shown in Fig 14.7.

14.8 The Single-Sideband System of Modulation

At 100 per cent modulation in an AM signal, the carrier requires two-thirds of the power but conveys no useful information. The carrier can be viewed as a power waste and the second sideband as needless duplication, except for its rare aid when selective fading distorts one sideband in radio transmission. If we use only one sideband and suppress the carrier, we have the system generally known as *single-sideband* (SSB) *transmission*. We are able to transmit the information with reduced power requirements, half of the bandwidth and lessened interference between signals, but at some expense in complexity of equipment.

A carrier must be introduced at the receiver, closely adjusted to the original carrier frequency since a Δf carrier difference produces a Δf shift in all signal frequencies and introduces distortion. The carrier must be accurate to within 10 or 20 Hz for intelligibility of voice signals, and stable oscillators are required to generate the local carrier.

The first step in generation of an SSB signal is to develop an AM signal

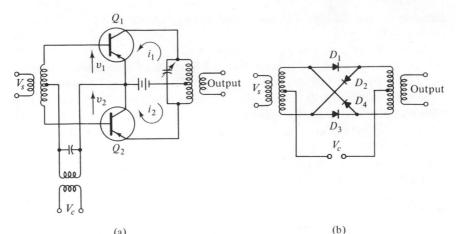

(a)   (b)

Figure 14.8 (a) A balanced modulator; (b) a diode balanced modulator.

without carrier. A *balanced modulator* is used, as shown in Fig. 14.8(a). The respective inputs to Q_1 and Q_2 have the f_c signal common and the f_s signal differentially connected, giving

$$v_1 = V_c \cos 2\pi f_c t + V_s \cos 2\pi f_s t \qquad (14.28)$$
$$v_2 = V_c \cos 2\pi f_c t - V_s \cos 2\pi f_s t \qquad (14.29)$$

The emitter-base junctions of the transistors can be assumed to have volt-age-current relations that can be represented by Eq. 14.15, as

$$i = a_1 v + a_2 v^2$$

Use of Eq. 14.28 and 14.29 as the applied voltages produces mixing of the two frequencies; with the push-pull connection giving a subtractive output, the output expression will contain only the terms

$$i_o = 2a_1 V_s \cos 2\pi f_s t + 4a_2 V_s V_c \cos 2\pi f_s t \cos 2\pi f_c t \qquad (14.30)$$

Now in Sec. 14.2 we used the trigonometric identity

$$\cos a \cos b = \tfrac{1}{2} \cos (a + b) + \tfrac{1}{2} \cos (a - b)$$

Expanding the product term of Eq. 14.30 according to this relation, we have

$$i_o = 2a_1 V_s \cos 2\pi f_s t + 2a_2 V_s V_c [\cos 2\pi (f_c + f_s)t + \cos 2\pi (f_c - f_s)t]$$

The output circuit is parallel resonant and of high resistance R_p only near frequency f_c. Frequency $f_s \ll f_c$ by assumption and the first term does not produce an appreciable voltage across the circuit because the circuit imped-ance is negligible at f_s. Thus, with R_p as the resonant impedance, the voltage across the output circuit is only

$$v_o = 2a_2 R_p V_s V_c [\cos 2\pi (f_c + f_s)t + \cos 2\pi (f_c - f_s)t] \qquad (14.31)$$

where $f_c + f_s$ and $f_c - f_s$ represent the sidebands of an AM wave and with the carrier absent. This result should not be surprising since the carrier represented a common input to the two transistors, in push-pull or differential connection.

Another form of balanced modulator is shown in Fig. 14.8(b), using diodes instead of transistors. The action of the circuit is similar to that of the transistor circuit and the output consists solely of the two sideband frequencies. The circuit is commonly used in telephone carrier-current transmission and requires accurate center-tap connections on the transformers for balancing out the carrier.

The sideband outputs are shown schematically in Fig. 14.9(a). A sharp cutoff filter must be used to select one of the sidebands, yielding the SSB signal in Fig. 14.9(b). The filter must separate signals in the two sidebands that differ only by twice the lowest modulation frequency. For voice frequencies the amplifiers are usually designed to cut off at about 200 Hz so that the lowest frequencies will differ by at least 400 Hz, as indicated by the separation of the bands in Fig. 14.9(a). Filters that can achieve this amount of skirt selectivity and place the rejected sideband at least 30 dB below the level of the accepted frequencies are usually designed with piezoelectric quartz elements of high Q and operate at f_c frequencies of 2 to 5 MHz.

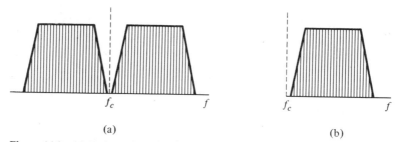

(a)   (b)

Figure 14.9 (a) Balanced modulator output spectra; (b) output after lower sideband is removed by filtering.

14.9 Frequency Translation; the Product Detector

By use of the principle of modulation it is possible to translate a band of frequencies centering at f_a to a band of similar frequencies centering at another frequency f_b. As for modulation, an input signal of frequency f_c and a locally generated signal at f_x are simultaneously applied to a diode, transistor, or tube having a parabolic characteristic to produce the frequency-product term necessary for modulation. An output voltage then appears in a circuit tuned to $f_c \pm f_x$.

The signal to be translated may be an AM wave, having a carrier f_c and side frequencies at $f_c + f_s$ and $f_c - f_s$. The translation process yields sums and differences of all input frequencies and if we mix a local signal at frequency f_x, we shall have the following output frequencies:

$$
\begin{array}{ll}
f_c & f_x \\
f_c + f_x = f_q & f_c - f_x = f_k \\
f_c + f_s + f_x = f_q + f_s & f_c + f_s - f_x = f_k + f_s \\
f_c - f_s + f_x = f_q - f_s & f_c - f_s - f_x = f_k - f_s
\end{array}
$$

The sum and difference terms will have amplitudes proportional to the products of the individual wave amplitudes.

The frequency groups centered at f_q and f_k constitute two translated AM waves, each containing the original side frequencies but moved up and down onto new carrier frequencies. This principle of *frequency translation* receives wide application in receivers and transmitters.

Additional frequencies of small amplitude are generated in the process. These spurious frequencies can create interference signals in other channels unless suppressed by high-Q resonant filters.

The translator in Fig. 14.10 employs one transistor for local generation of f_x and the mixing of the frequencies. The resonant circuit of L_x, C_x and the feedback coil L_M form an emitter-tuned oscillator operating at f_x. The internal interaction through the base-emitter junction with f_c produces an output in the L_k, C_k circuit tuned to f_k. The f_k output is predicted by a *conversion transconductance* g_c and the *conversion gain* is

$$A_c = g_c R \tag{14.32}$$

where R is the resonant resistance of the output circuit at frequency f_k. The circuit is useful as a translator even if A_c is less than unity.

The process of frequency translation is also employed in the *product detector* by which we reinsert the carrier and recover the SSB signals. Suppose that the input at f_c in Fig. 14.10(b) is modulated as an upper sideband $f_c + f_s$. Then suppose that the locally generated oscillation is adjusted to $f_x = f_c$. The mixed output will include frequencies at

$$
\begin{array}{l}
f_c \\
f_c + f_s \\
f_c + f_s + f_c \\
f_c + f_s - f_c = f_s
\end{array}
$$

The first three output frequencies are high and are bypassed by C but the last frequency at f_s is that of the original modulation; it is passed to the output through C_c.

This is the method for reintroduction of the carrier, and for good intelligibility the local oscillation must be within 10 or 20 Hz of the f_c used in generating the transmitted sideband. Severe requirements are seen to be placed on the stability of the local oscillator.

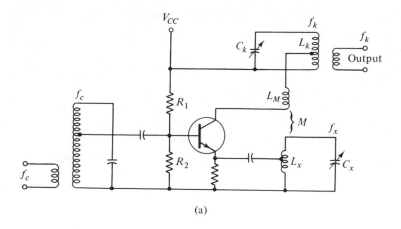

(a)

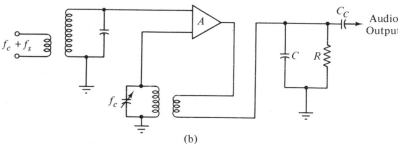

(b)

Figure 14.10 (a) Transistor frequency translator; (b) product detector.

14.10 The Frequency Spectrum of FM Signals

Most natural and man-made radio noise is in the form of amplitude variations of the signal and a system that eliminates amplitude variation in its received signals can also eliminate most radio noise. The system of *frequency modulation* operates with constant amplitude signals and is effective in reducing noise in radio reception.

If we are to design circuits for FM equipment, we must know the frequency spectrum required by the frequency components of a frequency-modulated (FM) wave. We shall again use as the information to be transmitted a frequency f_s and a signal voltage

$$v_s = V_s \cos 2\pi f_s t \qquad (14.33)$$

We vary the *frequency* of our FM wave in proportion to the *amplitude* of this signal.

We define the *frequency deviation* as

$$f_d = k_f V_s \quad \text{(kHz)} \qquad (14.34)$$

where k_f is the proportionality factor in kilohertz per volt. This relates the amplitude of the modulating signal to the frequency variation of the FM wave.

Suppose that a 1-V signal produces a 10-kHz deviation, then $k_f =$ 10 kHz/V. This is applied to a 50-MHz frequency and the frequency is shifted to 50.01 MHz. A 2-V signal produces a 20-kHz deviation to 50.02 MHz; a -2-V signal swings the frequency to 49.98 MHz. If the frequency f_s is 1000 Hz, then the swings between 50.02 and 49.98 MHz occur 1000 times per second for a 2-V peak cosine signal.

Maximum values of f_d are assigned for various radio services, as 75 kHz for sound broadcasting and 25 kHz for the sound channel in television. Maximum signal frequencies are $f_{s(max)} = 15$ kHz for the audio range. Then we define the *deviation ratio* as

$$m_f = \frac{f_{d(max)}}{f_{s(max)}} = \frac{k_f V_{s(max)}}{f_{s(max)}} \tag{14.35}$$

Thus $m_f = 75 \text{ kHz}/15 \text{ kHz} = 5$ for sound broadcasting.

With modulation by the signal of Eq. 14.33, our FM wave can be written as

$$v_o = V_o \cos\left(2\pi f_{oc} t + m_f \sin 2\pi f_s t\right) \tag{14.36}$$

The frequency f_o is called the *center frequency* because the FM signal deviates up and down from the f_o value, in accordance with the modulation term in the parentheses. The center frequency may go to zero for some values of m_f and this is the reason we do not call f_o a carrier frequency. There are side frequencies in pairs above and below the center frequency at every harmonic of f_s all the way to infinite frequency. The side frequencies are not restricted to $\pm f_s$ as for AM.

Practically, we do not have to contend with FM signals of infinite bandwidth. The amplitudes of the high-order side frequencies decrease quite rapidly and become negligible; however, we do use bandwidths of ± 75 kHz for FM sound broadcasting. The wider bandwidths give greater suppression of noise in FM transmission through space.

14.11 Bandwidth of FM Signals

The necessary FM bandwidth for good waveform reproduction can be assumed as

$$\text{BW} \cong 2f_{d(max)} \tag{14.37}$$

Spectrum frequencies and amplitudes for an unmodulated amplitude of unity are plotted in Fig. 14.11 for a variety of signal and m_f conditions; all components with amplitudes over 1 per cent are shown.

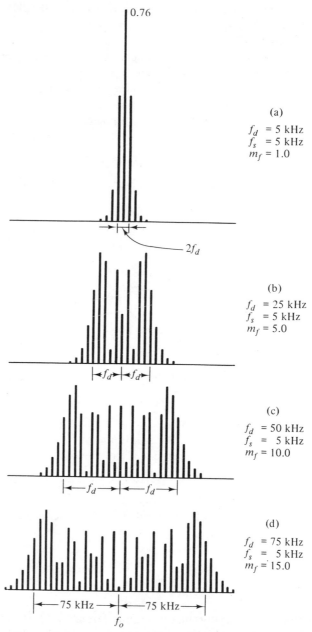

Figure 14.11 Spectra of FM waves for various m_f values.

In FM broadcasting, $f_{d(\max)}$ has been established at $\pm 75\,\text{kHz}$ and the usual design makes the receiver bandwidth $\pm 100\,\text{kHz}$, which accommodates all side frequencies that are greater than 1 per cent in amplitude, according to Fig. 14.11(d).

Since the amplitude V_o is unchanged by frequency modulation, the average powers in unmodulated and modulated waves are equal. Modulation simply spreads the available energy among the sidebands.

The spectrum for $m_f = 1.0$, shown in Fig. 14.11(a), closely approaches that of an AM wave and it is found that for $m_f = 0.6$ or less the spectrum reduces to

$$v_o = V_o[A_o \cos 2\pi f_o t + A_1 \cos 2\pi (f_o + f_s)t - A_1 \cos 2\pi (f_o - f_s)t] \qquad (14.38)$$

which is similar to double-sideband AM. The bandwidth is then

$$\text{BW} = 2f_{s(\max)}$$

and the system is said to be one of *narrow-band* FM. For $m_f > 0.6$, the system is considered *wide-band* in nature.

14.12 Generation of FM Signals

An FM signal may be generated by causing the capacitance of a tuned-circuit oscillator to vary with the modulating signal amplitude. A semiconductor diode may be connected across the tuned circuit, with the diode under reversed bias. The capacitance of the junction will vary if a bias voltage

$$v_B = -V_{BO} + V_s \cos 2\pi f_s t \qquad (14.39)$$

is applied. With the capacitance of the diode proportional to the square root of the applied voltage, the amplitude of V_s must be kept small for linearity. But with V_s small, the frequency of the oscillator will be

$$f = f_o\left(1 + \frac{V_s \cos 2\pi f_s t}{4V_B}\right) \qquad (14.40)$$

and this conforms to the needs for FM generation.

14.13 The Limiter and Discriminator for FM Detection

Noise and varying amplitude interference may arrive with the FM signals. Since FM signals have constant amplitude, this interference may be removed by *limiting* all signals to a common amplitude. With this done, the only varying property of the input is frequency and we can proceed to convert the frequency variations back to the amplitude variations of the original speech or music.

One form of amplitude limiter employs an amplifier with low collector voltage so that the load line of the transistor is short and all incoming signals are large enough to drive the amplifier to cutoff and to saturation. Since this distance on the load line is fixed, all signals above a threshold value will appear at a uniform output level; this is illustrated in Fig. 14.12(a).

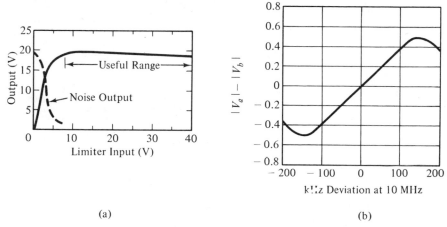

(a)   (b)

Figure 14.12 (a) Limiter performance; (b) discriminator action.

Another circuit that adopts operational amplifier construction is shown in Fig. 14.13(a). The transistors are normally biased into equal conduction. For any input V_i more negative than some value $-V_x$, Q_1 is cut off and the bias from R_E is reduced because of the smaller common current. This reduction in bias raises the current in Q_2 and its output voltage becomes V_{sat}. Then as V_i rises, Q_1 begins to conduct and the emitter voltage rises, reducing the current in Q_2. At some positive $V_i = +V_x$ on Q_1, the bias across R_E becomes large enough to cut off Q_2 and its output voltage goes to V_{CC}.

For any input signal with peak-to-peak amplitude exceeding $2V_x$, the output is limited at a peak-to-peak value of $V_{CC} - V_{sat}$. This action is demonstrated in Fig. 14.13(b).

The *Foster-Seeley discriminator* shown in Fig. 14.14 is one form of detector for FM signals. The input transformer serves as a load for a limiter circuit and the primary voltage is in series with the secondary voltage to ground, the reactance of the blocking capacitor C being neglected. The diodes D_a and D_b have applied voltages

$$V_a = V_1 + \frac{V_2}{2} \qquad V_b = V_1 - \frac{V_2}{2} \qquad \textbf{(14.41)}$$

The primary and secondary voltages of the overcoupled circuit, at a

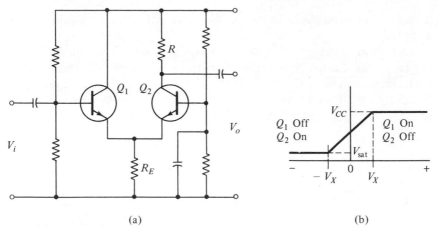

(a)   (b)

Figure 14.13 An emitter-coupled limiter.

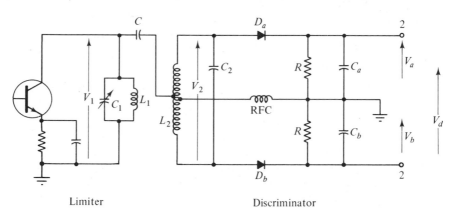

Limiter   Discriminator

Figure 14.14 A frequency discriminator.

resonant frequency equal to the FM center frequency, are

$$V_1 = 2\pi f_o g_m Q L_1 \left(\frac{1}{1 + k^2 Q^2}\right) V_o \qquad (14.42)$$

$$V_2 = jkQ\sqrt{\frac{L_2}{L_1}} V_1 \qquad (14.43)$$

The j factor in V_2 shows it to be in phase quadrature to V_1. Thus we draw the phasor diagram in Fig. 14.15(a) with V_1 and V_2 at right angles at resonance.

Diode voltages V_a and V_b are shown as the sum and difference of V_1 and $V_2/2$ according to Eq. 14.41. At frequencies below resonance the phase of V_2 changes toward the diagram in Fig. 14.15(b) and above resonance the phase of V_2 shifts oppositely toward the diagram in Fig. 14.15(c).

The diodes provide an output equal to $V_a - V_b$ and at resonance this is

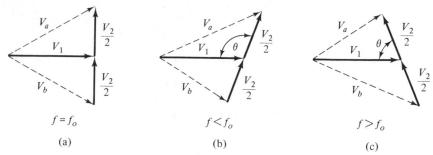

Figure 14.15 Phasor description of discriminator performance.

zero. At deviations above and below resonance the angle θ becomes smaller and larger, respectively, and the difference $V_a - V_b$ progressively changes. This voltage difference is translated into the diode output curve in Fig. 14.12(b). Linear output voltage versus frequency is available for a range of nearly ± 150 kHz at a center frequency of 10 MHz.

Equation 14.42 shows that the output is proportional to the input amplitude V_o; thus the output would vary with amplitude or noise if the circuit were not preceded by a limiter.

The transformer is overcoupled and the separation of the response peaks in Fig. 14.12(b) is dependent on the Q of the primary and secondary circuits. With $Q_1 = Q_2 = Q$, the needed value is

$$Q = \frac{f_o}{2f_{d(\text{max})}}$$

For speech and music $2f_{d(\text{max})} \cong 200$ kHz and at $f_o = 10$ MHz, the value of Q will be 50.

14.14 The Ratio Detector for FM

Another detector for FM is the *ratio detector* drawn in Fig. 14.16. The full secondary voltage charges C_c to the peak of signal V, through the two diodes in series. The values of R_2 and C_c are large so that V is constant for a given signal but varies for different signal amplitudes.

As in the discriminator, the primary voltage and one-half of the secondary voltage is applied to each diode. At the center frequency these voltages are equal and there is equal charge in C_1 and C_2 but there is no voltage across R_1. This is true regardless of the magnitudes of V_a and V_b. When frequency deviations occur, the unequal diode voltages charge C_1 and C_2 unequally but the total of V_a and V_b is constant at V. Voltage at A is dependent on the proportional change in V_a and V_b and so we have the ratio detector. A varying signal then appears at A to ground.

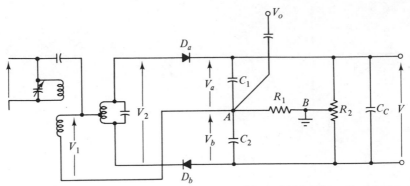

Figure 14.16 The ratio detector.

The performance of the circuit is somewhat affected by signal amplitude and the operation is improved if the circuit is preceded by a limiter.

14.15 Automatic Frequency Control (AFC)

At the high frequencies employed with FM a crystal cannot always be used and *automatic frequency control* (AFC) circuits are used to improve frequency stability.

A discriminator circuit is connected to the oscillator output as in Fig. 14.17. When the oscillator frequency corresponds to the center frequency for which the discriminator is tuned, the output from the discriminator is zero. When the oscillator is off frequency, a positive or negative voltage is developed at the discriminator output. This voltage is filtered to remove signal components and applied to a reverse-biased diode capacitor across the oscillator tuned circuit, where the voltage adds or subtracts to a fixed reverse bias. A bias change in one direction causes an increase in diode capacitance and a reduction of oscillator frequency; a bias change in the opposite direction causes a reduction in diode capacitance and an increase in oscillator frequency.

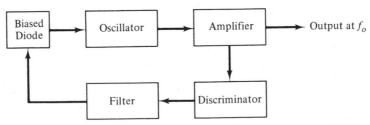

Figure 14.17 Automatic frequency control (AFC).

The directional bias from the discriminator is always so polarized as to shift the oscillator frequency toward its proper frequency. FM receivers are stabilized against frequency drift with temperature by such circuits.

14.16 Comments

The transmission of information by radio waves requires that those waves be controlled by the informational signal. We can vary any one of three characteristics of the radio wave, its amplitude in AM, its frequency in FM, and its phase in PM. Thus low-frequency audio and picture signals can be transferred to a higher frequency, which will radiate efficiently through space.

AM involves simple equipment but is susceptible to noise and interference. The SSB-AM system is most economical in its use of frequency space. FM requires more complex equipment and employs wide frequency bands but noise interference does not limit its usefulness. For equally intelligible signals, SSB-AM and FM use much less power than does the AM system. Thus there is no one system that is better than the others for all purposes but we can say that double-sideband AM is generally less useful than the other systems and its application is declining.

The process of frequency translation to different frequency bands is employed in receivers as well as in transmitters and requires the development of the product of two frequencies. The output always contains more frequencies that does the input, sometimes a great many more. Tuned circuit filters are ordinarily used to separate the wanted frequencies from the ones not wanted.

REVIEW QUESTIONS

14.1 What are two fundamental methods of modulating a wave of frequency f?

14.2 What is the process of amplitude modulation?

14.3 What is the process of FM?

14.4 What is PM?

14.5 What are the types of pulse modulation?

14.6 What is the modulation factor in AM?

14.7 Sketch an AM wave as modulated by a square wave.

14.8 Sketch an FM wave as modulated by a square wave.

14.9 Why is the bandwidth of an AM signal twice the highest frequency present in the modulating signals?

14.10 What are upper and lower sidebands of an AM wave? How are their frequencies related to the modulating signal?

14.11 What is the relation of carrier power to sideband power at 50 per cent modulation? At 100 per cent modulation?

14.12 Why do we filter the output of a diode modulator?

14.13 At $m_a = 1.0$, what carrier power must be supplied if the modulator output is 750 W? What is the power in the upper sideband?

14.14 A broadcast station is assigned a carrier frequency of 990 kHz and a bandwidth of 10 kHz. What range of frequencies can it transmit in the sidebands?

14.15 What is per cent modulation in AM?

14.16 What is meant by SSB? By DSB?

14.17 For the same power in the information signal, compare the total power in AM-DSB and in SSB.

14.18 How is an SSB signal generated?

14.19 Compare DSB-AM and SSB on the basis of bandwidth.

14.20 Explain the operation of a diode detector.

14.21 What is frequency translation?

14.22 Explain a product detector. When is it used?

14.23 What is the most serious problem in SSB reception?

14.24 What is meant by AVC? Describe AVC circuit performance.

14.25 What is delayed AVC and why is it used?

14.26 What is the purpose of a balanced modulator?

14.27 What is frequency deviation in FM?

14.28 Compare AM and FM on bandwidth needs.

14.29 Why does the power in an FM wave not vary?

14.30 What is the approximate bandwidth of an FM signal?

14.31 An FM signal has a deviation of 90 kHz for a $+10$-V signal. What is the value of k_f?

14.32 An FM station has a channel from 90.8 to 91 MHz.
   (a) What is its center frequency?
   (b) What is the maximum permissible deviation ratio for a max modulating frequency of 10 kHz?

14.33 What is a limiter? Why is it needed in FM reception?

14.34 Explain one form of limiter circuit.

14.35 Explain the action of a discriminator.

14.36 A diode modulator has input frequencies at 2.75, 2.80, and 2.95 MHz. The mixing frequency is 2.345 MHz. What output frequencies are obtained?

14.37 Explain how to translate a frequency from 30 MHz to 1.65 MHz; give frequencies involved in the output and explain their separation.

14.38 What is narrow-band FM?

14.39 What is wide-band FM? What is its advantage over AM in a noisy channel?

14.40 How can per cent AM be found from an oscilloscope pattern?

14.41 What is AFC?

PROBLEMS

14.1 The carrier of an AM wave is at 25 W. What is the average power in each sideband at 40 per cent modulation?

14.2 A carrier is generated at 150 W and the average modulator power output is 45 W. What per cent modulation is possible?

14.3 An AM wave is stated by

$$v = 100[1 + 0.20 \cos 2\pi(1000)t + 0.05 \cos 2\pi(3000)t] \cos 10^6 t$$

State all frequencies present in hertz, and give the per cent modulation for each.

14.4 A modulated amplifier operates at 2250 V and a carrier power input of 500 W.

(a) What modulator power output will be required for 80 per cent modulation?

(b) What is the ratio needed for the Class B output transformer for the modulator if the Class B stage needs a 7500-Ω output primary load?

14.5 An AM transmitter has a carrier of 1000 W. What average output must the modulator have to reach 80 per cent modulation? What is the total average power required by the wave?

14.6 In Problem 14.5, what would be the per cent saving in power if the same information was transmitted by SSB?

14.7 An AM carrier is at $f_c = 10$ kHz and is modulated 50 per cent by $f_a = 400$ Hz and 20 per cent by $f_b = 800$ Hz. Plot a frequency spectrum showing proper amplitudes if the carrier is shown at 2.5-cm length.

14.8 If the carrier in Problem 14.7 is at 100 W, what is the total power in the AM wave?

14.9 The current in a diode modulator is given by

$$\text{mA}, \ i = 10 + 2.5v + 1.0v^2$$

The voltage applied to the modulator is

$$v = [3.0 \cos 2\pi(60,000)t + 0.2 \cos 2\pi(1000)t]$$

Calculate the amplitude and frequency of all output current components.

14.10 A carrier of an AM wave has $V_c = 2.75$ V at the detector. What voltage is available for AVC action? If $m_a = 0.5$, find the voltage of the audio signal at the detector output.

14.11 A 100-kHz carrier is amplitude modulated by a 5-kHz signal and the upper sideband is transmitted. The receiver mixes a signal of 100.3 kHz. What is the frequency of the recovered audio signal? Has distortion occurred?

14.12 An SSB signal $f_c + f_s$ is applied to a frequency translator along with $V_q \sin 2\pi f_q t$. Show that the SSB signal may be translated to a higher frequency or a lower frequency without distortion of the side frequencies.

14.13 A 10,000-Hz modulating signal gives $m_f = 15$ in an FM transmitter. What is the minimum desirable bandwidth for the transmitter resonant circuits?

14.14 In an FM transmitter, when the audio frequency is 400 Hz and the audio voltage is 2.5 V, the deviation of frequency is 5.7 kHz. What audio voltage will cause a deviation of 10 kHz?

14.15 If the maximum audio frequency applied to an FM transmitter is 5.0 kHz and $m_f = 20$, what is the maximum frequency deviation from 10 MHz?

14.16 A steady carrier at 4.350 MHz is transmitted adjacent to one on 4.354 MHz, modulated at 2000 Hz. What output frequencies are found in a remote receiver? Do you see any problems arising?

14.17 An FM signal has a maximum frequency deviation of 50 kHz and is modulated by audio signals up to a maximum of 10 kHz. If a receiver has a bandwidth of 50 kHz, is this adequate?

14.18 An FM station with $f_{s(\text{max})} = 15$ kHz and $m_f = 7$ operates at 10-MHz center frequency. What should be the -3-dB bandwidth of a tuned circuit to pass the major spectrum components of this signal? What Q would be required of the resonant circuit?

14.19 An FM system uses 15 MHz as a center frequency. The modulating signal when at 10 kHz generates a maximum frequency deviation of 5 kHz. Find the bandwidth to pass the necessary components for good fidelity of the received signal.

15
Radio Systems

Radio systems are designed to transmit information and the noise appearing along the transmission path is the basic low-signal limit for reception. We now have sufficient knowledge of circuits and the basic processes of modulation and detection to understand the overall design of radio receivers and transmitters and can discuss some of the system considerations that include noise, bandwidth of signal, power, and method of modulation.

As in most engineering problems, there is rarely a unique answer to be found and trade-offs are necessary.

15.1 Noise

Random variations of current in electronic circuits are called *noise*. External noise caused by atmospheric discharges or *static* limits radio reception of weak signals below about 20 MHz. At higher frequencies the external noise decreases and random currents internal to the circuits produce a noise or hiss and this limits weak signal reception. Circuit noise is predominantly due to impacts of electrons and atoms in thermal agitation in the materials of the circuits. Each impact sends out a short energy pulse with a very broad frequency range. The thermal noise produced by a resistance R (an antenna, for example) appears as the voltage of an equivalent noise generator

$$V_{\text{noise}} = 7.4 \times 10^{-12}\sqrt{TR(\text{BW})} \qquad (15.1)$$

where

T = temperature, degrees absolute (°C + 273°)

R = resistance of the circuit

BW = frequency band (3-dB) included in the noise measurement

For a resistance of 100 Ω at room temperature (300° absolute), the noise voltage is 1.28 μV per MHz of bandwidth.

Since the noise is dependent on temperature, when receiving very weak space signals the input stages are sometimes immersed in a case at the temperature of liquid helium. This reduces T in Eq. 15.1 and therefore the circuit noise. A weaker signal can then be received.

The amount by which the signal power overrides the noise power in the circuit determines the understandability of the signal and we speak of *signal-to-noise ratio* in evaluating circuit performance. The noise contributed by an amplifier is measured by the *noise figure* (N.F.), stated as the ratio of the signal power to noise power at the input, S_i/N_i, to the same ratio at the output, S_o/N_o. That is,

$$\text{N.F.} = \frac{S_i/N_i}{S_o/N_o} \tag{15.2}$$

The measurement is usually expressed in decibels. The amplifier noise added in the circuit degrades the signal, making S_o/N_o less than S_i/N_i. Some transistors are less noisy than others and less noise is usually produced when the transistor is operated at low currents and voltages. Noise figure for input stages at high frequencies is usually in the range of 4 to 6 dB, although 2 to 3 dB can be obtained by careful transistor selection.

15.2 Information in Signals

We now need a measurement of the information contained in a signal.

In a pulsed data transmission system we need to recognize only two levels of output per pulse interval: on and off. The *quantity of information* is assumed to depend on the number of such levels or conditions that we must recognize in a unit signal; that is,

$$I_o = \log_2 L \quad \text{bits} \tag{15.3}$$

where I_o is the number of *bits* (basic units of signal), L is the number of recognizable levels or signal conditions, and the logarithmic base of 2 results from the fact that each signal level has an even probability of being recorded correctly or incorrectly. Admittedly this is an arbitrary definition but it leads to useful and comparative results.

The single pulse is our unit of information content since with $L = 2$, on and off, then

$$I_o = \log_2 2 = 1 \quad \text{bit}$$

If we transmit 500,000 pulses per second, each 1 μs long, the rate of information is high, as

$$R = 500,000 \text{ bits/s}$$

With a pulse length of 1 μs, the bandwidth must be 1 MHz for reasonable accuracy of waveform in this data system.

Speech is our most usual form of information transmittal. With speech we can encompass a range of 1000 : 1, or 30 dB, between the weakest intelligible sound and the loudest sound in the human voice. It takes an intensity change of about 2.5 dB before the ear can notice a change of intensity; with an intensity range of 30 dB, we are able to recognize 12 significant levels of speech intensity. Thus $L = 12$ for speech.

While speech contains frequencies up to about 6000 Hz, good intelligibility is possible with the channel narrowed to 3000 Hz, as in the telephone. The basic interval in speech is considered to be a half cycle of the highest frequency present, or 6000 intervals per second for a 3000-Hz speech transmission. With 12 recognizable intensity levels in speech, the maximum information rate in speech is

$$R = 6000 \log_2 12 = 2000 \times 3.58 = 20,500 \text{ bits/s}$$

Channel bandwidth is, of course, 3000 Hz.

A television signal can be similarly analyzed. The difference between white and black on a television screen is not very great and the eye can only distinguish about 10 levels of gray between the black and white limits. The picture is formed in dots on horizontal lines and the eye can resolve about 500 dots of white, black, or color per line. In the United States television system there are 525 lines per picture and 30 pictures or frames per second. Putting this data together, we can calculate the information rate in a television signal as

$$R = 500 \times 525 \times 30 \log_2 10 = 2.62 \times 10^7 \text{ bits/s}$$

We see that the speech rate of 20,500 bits/s is low but only requires a bandwidth of 3000 Hz. Data pulses transmit information at 500,000 bits/s or more and require a few megahertz of bandwidth. Television transmits information at very high rates and utilizes 4.5 MHz of bandwidth.

Thus we conclude that *the bandwidth occupied by a signal is proportional to the rate at which information is transmitted.*

15.3 Information Capacity of a Channel

We now need to measure the information capacity of a transmission channel. The more levels we try to separate, the smaller are the differences between signal levels and the lower is the accuracy of discrimination in the presence of noise. A quiet channel will have a greater information capacity than a noisy one and to measure the information capacity of a channel we use the *signal-to-noise ratio*.

Suppose we have a given bandwidth and a received signal power S and noise power N. The power input to the receiver is $S + N$ in the presence of N units of noise power. The number of distinguishable levels of signal is assumed to increase as the ratio of voltages or

$$L = \sqrt{\frac{S + N}{N}} = \sqrt{1 + \frac{S}{N}} \qquad (15.4)$$

since a voltage V is proportional to $\sqrt{Power}$.

The information received through the channel per basic signal interval is

$$I_o = \log_2 \sqrt{1 + \frac{S}{N}} = \frac{1}{2} \log_2 \left(1 + \frac{S}{N}\right) \qquad (15.5)$$

The number of basic intervals per second is the rate of sampling, or twice the bandwidth. The information rate is

$$R = 2B \times \frac{1}{2} \log_2 \left(1 + \frac{S}{N}\right)$$

The total information that can be transmitted over a noisy channel in total time T is

$$C = BT \log_2 \left(1 + \frac{S}{N}\right) \quad \text{bits} \qquad (15.6)$$

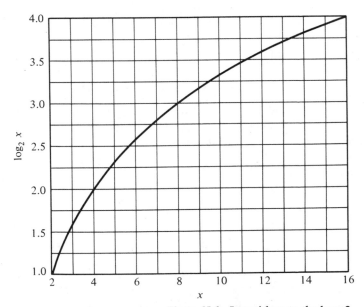

Figure 15.1 Logarithms to the base 2.

This is known as the *Hartley-Shannon law*. For its use we provide a curve of logarithms to the base 2 in Fig. 15.1.

The received signal may be partially masked by some noise, and the received message will be inaccurate. This means that the S/N ratio is low and the channel capacity is reduced.

The channel capacity for accurate reception of a signal can be increased by widening the bandwidth B, as by use of FM or by use of a pulse-code system. The time T taken to transmit a message can also be increased, possibly by sending a message twice or repeating parts of it as is done in some data transmission systems. An increase in channel capacity can be accomplished by raising the transmitter power, giving a larger S/N ratio at the receiver. This method is expensive, however, because S/N appears in the logarithm term and an increase of power by about 10 is needed to make a useful improvement. A cheaper means of improving S/N is often found by use of a more effective antenna system, using directional gain for the signal and having some directional discrimination against the noise sources.

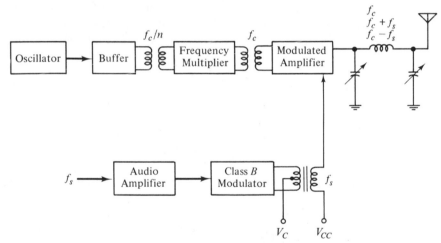

Figure 15.2 Block diagram for an AM transmitter.

15.4 An AM Transmitter

A block diagram of an AM transmitter is shown in Fig. 15.2. A submultiple of the output frequency is usually generated by the oscillator since lower-frequency crystals are more stable. A buffer amplifier is used to isolate the oscillator circuit from variations in the remainder of the transmitter, improving the oscillator frequency stability. The oscillator frequency is then multiplied by frequency doublers or triplers. These are overbiased amplifiers with loads tuned to twice or three times the input frequency. In addition to frequency multiplication, these amplifiers increase the power level sufficiently to drive the modulated amplifier.

An audio amplifier is driven by the microphone signal at frequencies f_s and raises the power level sufficiently to drive the Class *B* modulator to full output. The output power for the sidebands is transferred through the modulation transformer to the modulated amplifier. This transformer has a turns ratio that will produce the desired load for the Class *B* amplifier and transfer the power to the secondary load represented by the resistance of the modulated amplifier V_{CC}/I_C. Power modulation is obtained and the modulated signals are coupled to the antenna through the π network, which is designed to transform the large load value of the amplifier to the 50 or 75 Ω that the antenna represents. At the same time the π network acts as a filter and removes extraneous harmonics from the amplifier output.

An AM transmitter has simple circuits and is easy to adjust for modulation without distortion. Amplitude-modulated signals use a carrier that transmits no useful information, however, and can be considered to be a waste of power; AM signals also require a bandwidth that is twice the modulating signal band. More importantly, perhaps, AM suffers from *heterodyne interference*, created when another carrier is within a few kilohertz so that their difference or *beat* frequency produces a steady audio whistle in the output of the receiver.

The use of AM is declining because of heterodyne interference, its bandwidth requirements, sensitivity to noise, and its power requirement.

15.5 The Superheterodyne Receiver

Reception of AM signals is done with the *superheterodyne receiver*, shown in block form in Fig. 15.3(a).

The incoming signal has a carrier frequency f_c, with sidebands at $f_c + f_s$ and $f_c - f_s$, where f_s is the audio or information signal. This AM signal is translated to a new frequency band, with carrier frequency f_k and sidebands at $f_k \pm f_s$, by mixing with the first oscillator. The oscillator frequency is f_x and that frequency is adjusted so that $f_x = f_c + f_k$ for all incoming signals to which the receiver is tuned; that is, f_x is made to vary in step with f_c. Amplification follows at f_k, the *intermediate frequency* (I.F.), in an amplifier with fixed tuning at f_k. Since the intermediate frequency is usually lower than the frequency of the input signal, the tuned circuits in the I.F. amplifier can have more selectivity (in terms of bandwidth in hertz) than can be obtained with simple parallel-resonant circuits at the incoming radio frequency. The I.F. circuits employ double-tuned transformers that give a flat top and steep

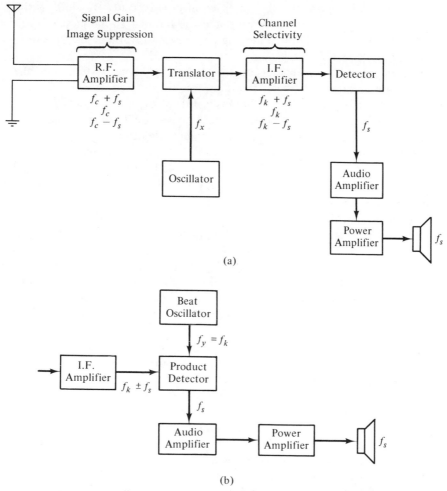

Figure 15.3 Block diagram of a superheterodyne receiver for AM signals; (b) demodulation of SSB-AM signals.

skirts to the overall response curve. This flat response is made just wide enough to include all the sideband frequencies present in $f_k + f_s$ and $f_k - f_s$.

After I.F. amplification the signal goes to the diode detector and the original modulation frequency f_s is derived. Audio frequency and power amplification follow before the signal reaches the loudspeaker.

There will always be unintentional positive feedback, introduced by common couplings in power supplies, by fields between connecting wires, or by stray capacitances. This positive feedback, although small, will limit the amount of gain possible at one frequency because when A becomes large, the product $A\beta$ will approach -1 and instability will occur. By shifting the frequency band it is possible to approach the limiting stable gain in each frequency band in succession. Thus high gain is an advantage of the superheterodyne receiver.

Because of the requirement of the superheterodyne system that the intermediate frequency f_k be a constant for all received frequencies, the oscillator frequency f_x must continuously differ from the frequency of the carrier f_c by the amount of the I.F., i.e., f_k. For any oscillator frequency, two signals can give an output to the intermediate frequency amplifier. These signals differ by $2f_k$, one being below f_x at $f_x - f_k = f_{c_1}$ and one above f_x at $f_x + f_k = f_{c_2}$. If the desired signal is at $f_{c_1} = 545$ kHz and the I.F. is $f_k = 455$ kHz, then the oscillator must be tuned to $f_x = f_{c_1} + f_k = 1000$ kHz. A second signal at $f_x + f_k = 1000 + 455 = 1455$ kHz $= f_{c_2}$ can also be received simultaneously and would be called the *image frequency*.

By using sufficiently selective radio-frequency circuits in the R.F. amplifier, the strength of the undesired image signals can be lowered and it is possible to reduce greatly or to eliminate image responses. At signal frequencies of 20 MHz or more, however, an I.F. of 455 kHz places the image at only 20.910 MHz. The radio-frequency tuned circuits cannot give sufficient attenuation for a frequency so close to the desired carrier and images will be received with the desired signals.

To eliminate images in such high-frequency receivers, a higher I.F. is used, possibly 5.5 MHz, so that for the 20-MHz signal the image frequency will be at 31 MHz and sufficiently separated from the desired signal to be rejected by the input tuned circuits. A second translation from 5.5 MHz to 455 kHz can be carried out so that the skirt selectivity of the 455-kHz I.F. amplifier is also obtained for the 20-MHz received signal.

Thus the selectivity curve of the receiver is really that of the lower-frequency I.F. amplifier. This makes the receiver selectivity curve independent of the frequency range for which the receiver is designed. This is a second advantage of the superheterodyne form of receiver circuit.

Single-sideband signals can be received in the same circuit if a second oscillator is added to reinsert the carrier frequency as shown in Fig. 15.3(b). If the I.F. is 455 kHz, then f_y of the second oscillator should also be 455 kHz. A product detector is used and the low or difference output frequency chosen, giving an output frequency of $455 \pm f_s - 455 = f_s$. The plus or minus sign is used, depending on whether the upper or lower sideband is being received. The output of the detector is f_s, the original modulating signal. The oscillator frequency f_y must be stable within 10 or 20 Hz if intelligible speech is to be recovered.

A receiver for the broadcast band is designed to tune over the frequency range from 600 to 1600 kHz. This is a frequency range of $1600/600 = 2.67 : 1$. The resonant frequency of a tuned circuit varies with $1/\sqrt{C}$ for a constant inductance L. Most variable air capacitors have a maximum capacitance about 10 times their minimum capacitance so that $\sqrt{10} = 3.16 : 1$. This is about the usual tuning range for a variable capacitor. The broadcast band, requiring a tuning range of 2.67 : 1, can be satisfactorily covered with such a capacitor. A small variable capacitance is used in parallel to adjust the minimum capacitance and to obtain the exact ratio of maximum to minimum capacitance needed for the tuning range.

For a superheterodyne receiver with the I.F at 455 kHz, the oscillator frequency must range between

Signal Frequency	Oscillator Frequency
600 kHz	145 or 1055 kHz
1600 kHz	1145 or 2055 kHz

The required tuning ratio for the oscillator tuning capacitor would be either

$$\frac{1145}{145} = 7.9 : 1 \quad \text{or} \quad \frac{2055}{1055} = 1.94 : 1$$

The usual air variable capacitor has a range from 35 to 350 pF, or 10 to 1. Such a variable capacitor could not tune over the wide-frequency range from 145 to 1145 kHz. Therefore the oscillator is designed to tune from 1055 to 2055 kHz and the oscillator frequency is placed *above* the signal frequency, or $f_x = f_c + f_k$. Because the oscillator tuning range will be only 1.94 : 1 compared to 2.67 : 1 for the radio-frequency signal circuits, the oscillator tuning capacitor usually has a lower maximum capacitance.

Since the signal frequency is the *difference* between the oscillator frequency and the I.F., $f_c = f_x - f_k$, the image frequency occurs at a signal frequency that is the *sum* of the oscillator frequency and the I.F., $f_c = f_x + f_k$, and we have

Signal Frequency	Oscillator Frequency	Image Frequency
600 kHz	1055 kHz	1510 kHz
1600 kHz	2055 kHz	2510 kHz

This places the range of possible image frequencies between 1510 and 2510 kHz, almost entirely outside the broadcast band and in a region where strong signals are not prevalent. Therefore the reception of an image signal is not usual, although a strong broadcast station near 1500 kHz may appear at tuned frequencies near 600 kHz.

15.6 The SSB Transmitter

A transmitter for SSB signals is more complex than the AM unit and one is shown in block form in Fig. 15.4. A radio frequency of perhaps 5 MHz, f_a, and the audio signal f_s are introduced to a balanced modulator and sideband pairs obtained without a carrier. A filter selects the upper or lower sideband to be transmitted and in a second translator this sideband signal is added to a carrier f_b such that $f_a + f_s + f_b = f_c + f_s$, or $f_a - f_s + f_b = f_c - f_s$ for the lower sideband. This gives a single-sideband signal at the frequency f_c at which it will be transmitted. The Class B linear amplifier provides a power gain for this varying amplitude signal. The output of the amplifier is coupled to the antenna through the π network for harmonic filtering of the Class B output.

A frequency of about 5 MHz is chosen for the first oscillator so that the sideband filter can adequately separate the upper and lower sidebands, which differ by only 300 or 400 Hz. Such filters are usually composed of quartz piezoelectric crystals with high-Q values for sharp cutoff.

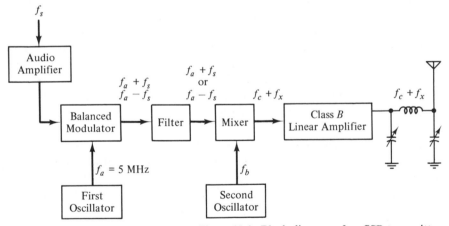

Figure 15.4 Block diagram of an SSB transmitter.

Single-sideband signals are being generally used for point-to-point telephone service, especially for *frequency multiplex* as in the radio relay service. Frequency multiplex is illustrated in Fig. 15.5, where eight separate speech channels are used to prepare eight lower sideband signals with carriers spaced every 5 kHz from 15 to 50 kHz. After being "stacked up" in frequency as shown, the composite signal, with frequencies from 12 to 49.7 kHz, is used as a modulating signal for a single carrier at 4000 MHz to yield a signal capable of transmission by radio relay.

At the receiver the composite signal is detected back to the original range of 12 to 49.7 kHz and the eight signals separated by filters. The proper

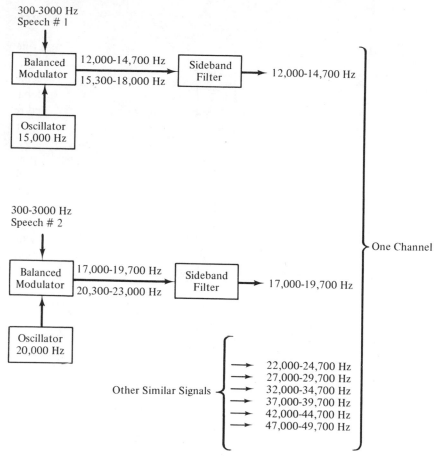

Figure 15.5 Frequency-division multiplexing of eight voice channels.

carriers for each channel are added in separate product detectors and the original speech signals recovered.

15.7 An SSB Transceiver

The block diagram of Fig. 15.6 illustrates a combined transmitter-receiver or *transceiver*, a type now commonly employed for SSB at high frequencies. The same frequency is used for both transmitting and receiving and the transmitting path of signals is indicated by dashed lines in the diagram.

Let us assume that we tune to an SSB signal of upper sideband, derived from a carrier at 7.2 MHz. The band of frequencies near 7.2 MHz is amplified and mixed with a fixed oscillator operating at 11 MHz. The difference frequencies are in a band near 4 MHz, including the desired signal, now at

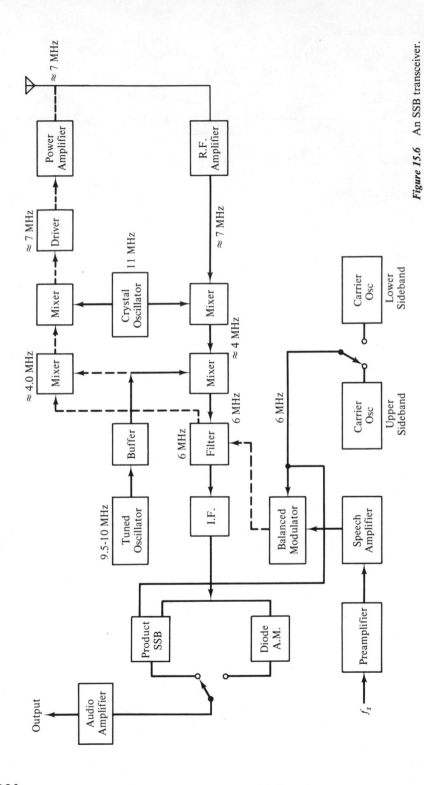

Figure 15.6 An SSB transceiver.

$11 - 7.2 = 3.8$ MHz. This band is mixed with the output of a variable-frequency oscillator, tunable over a 500-kHz range from 9.5 to 10 MHz. When the oscillator is set at 9.8 MHz, our desired signal will produce difference frequencies close to $9.8 - 3.8 = 6.0$ MHz. Only our sideband signal can pass through the filter with a narrow passband of about 2.5 kHz at 6.0 MHz, as shown in Fig. 15.7. All other signals in the original narrow band selected by the radio-frequency tuner are removed here by the filter.

An I.F. amplifier provides gain for the 6.0-MHz signal, which is then applied to the product detector. Here a 6.0-MHz carrier is reinserted and the speech frequencies f_s are obtained as output.

In the transmit mode the speech signal f_s passes through the speech amplifier to a balanced modulator, where a 6-MHz signal is supplied from either of two carrier oscillators. These differ slightly in frequency so that the selected sideband will be centered in the passband of the same filter used for receiving at 6.0 MHz + 2.5 kHz. The signal at 6 MHz is mixed with the variable-frequency oscillator output at 9.8 MHz, giving a 3.8-MHz output.

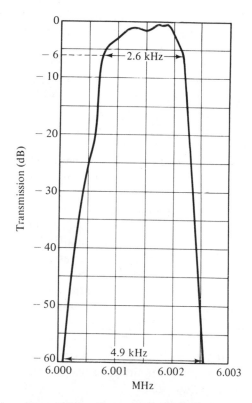

Figure 15.7 Selectivity characteristics of 6-MHz sideband filter; 60 dB/6 dB shape factor $= 4.9/2.6 = 1.9$.

A second mixing with the 11-MHz oscillator frequency gives a difference frequency at 7.2 MHz, actually the same frequency as originally received.

Power amplification with a driver and power amplifier operating in Class B follows.

Advantages of SSB transceiver operation include

1. Only a single channel is used for transmission and reception.
2. SSB gives a narrow bandwidth equal to that of the modulating signal.
3. The transceive mode employs less expensive equipment, especially the filter.
4. The power required is only that of one sideband of an AM signal.

The complexity of the equipment is the offset to these advantages.

15.8 AM versus FM

For an AM signal the bandwidth is fixed. The rate of transmittal of information can be improved by increasing the transmitter power to raise the S/N ratio but this is expensive. With an FM signal, the bandwidth can be increased arbitrarily by increasing m_f, assuming that frequencies are available. We can then receive a useful signal with a reduced S/N ratio.

For instance, the total signal power in a 100 per cent modulated AM wave might be 150 W, while the power in an FM wave carrying the same information need be only that of one AM sideband or 25 W. The bandwidths might be 10 and 100 kHz, respectively. It is found that the information capacity or the usability of the FM channel is nine times greater than that of the AM channel; in fact, the FM system is superior as long as the $2f_d$ value is greater than 10 kHz, the bandwidth of the AM signal.

Since the power in an FM wave is constant, there need be no provision for the large instantaneous power and voltage peaks of an AM transmitter. The FM equipment can be smaller and cheaper.

Noise signals caused by atmospheric static or man-made electrical discharges create reception problems for radio signals. In AM these noise signals may override the wanted signal and they cannot be separated. The noise can be reduced by narrowing the reception bandwidth but the minimum bandwidth of AM signals is determined by the modulating frequencies of the signal.

If a receiver is designed for reception of wide-band FM signals and made insensitive to amplitude-varying noise by use of a limiter, the noise can be largely eliminated.

Interference between two signals on the same frequency is a major problem in radio communication. In AM, if the two signals have carriers differing only a few kilohertz, the difference appears as *heterodyne* interference, or a steady whistle in the receiver output. This is a frequent occurrence in the crowded AM bands.

When this type of interference occurs between the higher-order sidebands of two AM broadcast stations with carrier frequencies spaced 10 kHz apart, the resultant noise is called "monkey chatter." Filters are often added to AM receivers to eliminate this form of interference.

When two FM signals are on the same center frequency, we encounter the *capture effect*, by which the stronger signal captures the receiver and blocks reception of any signal significantly weaker. Wider frequency deviation makes this capture effect greater. E.H. Armstrong was the first to demonstrate the value of FM in combating interference.

Thus FM has a number of advantages over AM, particularly where noisy channels must be used with weak signals as in the mobile service and where power input must be limited.

15.9 FM Systems

The FM transmitter diagrammed in Fig. 15.8 employs a variable-capacitance diode to shift the frequency of an oscillator in accordance with the modulation signal. The varying oscillator frequency may be generated with a center frequency of f_a at which a suitable frequency deviation can be obtained. That varying frequency is translated by the output of a second oscillator at f_b, such that $f_a + f_b = f_o$, where f_o is the assigned output frequency of the transmitter. The side frequencies are then clustered around f_o in the same manner that they were originally generated around f_a. The power amplifier provides the antenna power at f_o and side frequencies.

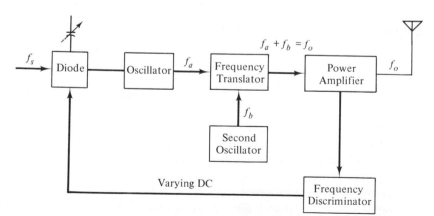

Figure 15.8 A reactance-modulated FM transmitter.

The output is also applied to the tuned circuit of the frequency discriminator of Sec. 14.13, which yields a voltage as a measure of the average drift of frequency from the assigned channel at f_o. If the frequency is not at normal, a correction voltage from the discriminator is applied to the frequency-controlling diode, to change its capacitance and to restore the oscillator frequency to its assigned channel.

The transmitter is relatively simple because of the constant power requirements of an FM signal. Circuit bandwidths can be adjusted by variation of circuit Q to obtain sufficient bandwidth for transmission of the important side frequencies.

The FM receiver of Fig. 15.9 follows the general design of the superheterodyne up to the detector. At that point a limiter-discriminator circuit is inserted to change the frequency variations to amplitude variations. The audio amplifier following the discriminator is standard.

To provide for adequately large frequency deviations to reduce interference, FM signals are used in the frequency bands above about 40 MHz. An I.F. of 10.7 MHz has become standard so that image interference will not be serious. The I.F. of 10.7 MHz also permits the design of double-tuned transformers with adequate bandwidth to handle the FM deviations. With 200-kHz radio-frequency bandwidths, some loading of the tuned circuits with parallel resistances may be necessary to obtain the needed bandwidths.

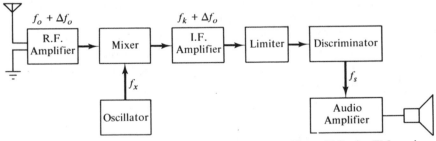

Figure 15.9 An FM receiver.

15.10 A Radar System

The radar system of Fig. 15.10 illustrates the manner in which electronic systems are designed by the assembly of relatively simple component circuits and functions into an ultimate complex objective.

A typical radar system employs a pulse sequence as in Fig. 15.11(a), transmitting a short very powerful pulse at a rate of perhaps 400 per second. This radio signal travels to a distant target and a very small portion of the energy is reflected back to the receiver. The time taken for the round trip is

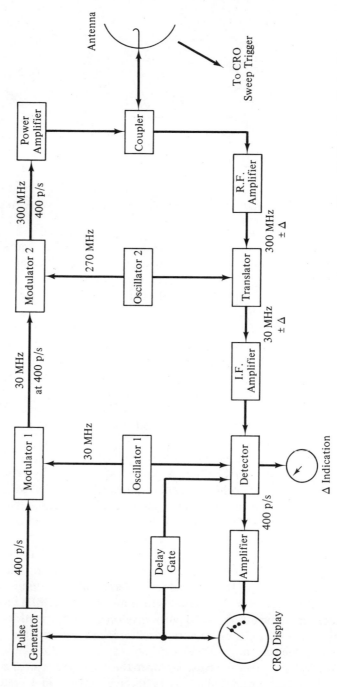

Figure 15.10 A simplified radar system.

369

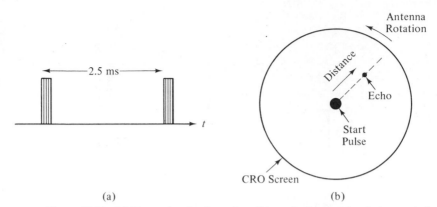

Figure 15.11 (a) Transmitted radar pulses; (b) received radar signal as presented on the CRO plot.

a measure of distance to the target and is plotted as radial distance on a cathode-ray oscilloscope screen with a linear radial sweep, rotating in step with the rotating antenna. The returned pulse is used to brighten the scope display at its instant of arrival, presenting a visual dot at the target position, as in Fig. 15.11(b).

The transmitter starts with a pulse generator, at 400 Hz, followed by a limiter producing square waves. These are used to produce short pulses at 400 times per second. An oscillator at 30 MHz drives modulator 1, which is turned on by the pulses as in Fig. 15.11(a), and results in pulses of 30 Mhz signal with pulse lengths of a fraction of a microsecond. By mixing with another oscillator at 270 MHz, the final pulses are obtained at 300 MHz. These are amplified to hundreds of kilowatts and radiated in a very narrow beam by a sharply directive antenna.

While the pulse is transmitted, the receiver input is blocked to avoid overload. The receiver input is opened after the pulse is sent to await the returned signal or echo. The echo pulse will be at 300 MHz $\pm\Delta$, where Δ represents a shift in frequency if the target is in motion. This is the *Doppler effect* and $\Delta = 2vf/c$, where v is the target velocity and c is 300×10^6, the velocity of radio waves in meters per second.

The received echo is amplified at 300 MHz and then mixed with the 270-MHz oscillator frequency in the first translator. The signal is there converted to 30 MHz $\pm \Delta$; here wide-band amplification is used to preserve the pulse waveform. The signal is then mixed with 30 MHz from oscillator 1. Retained in the output is the pulse envelope at 400 repetitions per second and this is passed to the cathode-ray oscilloscope to control the brightness of the spot on the screen.

The delay gate can be used to open the detector circuit at a given time after the pulse is transmitted so as to receive only a single returned pulse. In automobile speed measurement on the highway, this gating allows the observer to concentrate on the speed of only one target.

The basic circuit form of the receiver is that of the superheterodyne. To the equipment above must be added numerous power supplies and control equipment for feeding the antenna position into the oscilloscope. While rather easy to describe, a radar is, in fact, a complex system.

15.11 Frequency Classification for Radio Signals

Portions of the broad radio spectrum have varying characteristics and in describing radio waves and the equipment suited to them, we have the broad classification of Table 15.1.

TABLE 15.1 **Radio-Frequency Classification**

Band Classification	Frequency Range	Use
Low frequency (LF)	30–300 kHz	Marine point-to-point; navigation systems
Medium frequency (MF)	300 kHz–3 MHz	Commercial broadcast
High frequency (HF)	3–30 MHz	Moderate and long-distance communications
Very-high frequency (VHF)	30–300 MHz	Television, FM, aircraft navigation
Ultra-high frequency (UHF)	300 MHz–3 GHz	Television, radar
Extreme-high frequency (EHF)	3–300 GHz	Radar, space communications, radio relay

REVIEW QUESTIONS

15.1 What is meant by the signal-to-noise ratio at the receiver input?

15.2 Why do receivers for signals from space use input circuits cooled to liquid-air temperatures?

15.3 How does noise vary with bandwidth received?

15.4 What is meant by the *noise figure* of a receiver?

15.5 What is thermal noise?

15.6 What is a *bit* of information?

15.7 In computing information content of a signal, why is the logarithm taken to the base 2?

15.8 You are receiving an important telephone conversation and ask the caller to repeat a word. What have you done to the channel capacity?

15.9 What are the three factors available for trade-off in increasing the bit capacity of a communications channel?

15.10 Why is it uneconomic to increase power to improve the channel capacity?

15.11 What is the cause of heterodyne interference in AM reception?

15.12 What determines the bandwidth of an AM signal?

15.13 What situation leads to a steady whistle in the output of an AM receiver?

15.14 List the major circuit functions employed in a superheterodyne receiver.

15.15 Name two major advantages of the superheterodyne receiver.

15.16 What is meant by I.F.?

15.17 What is the reason that we can obtain very high signal gain in a superheterodyne receiver?

15.18 What is an image frequency?

15.19 A signal at 970 kHz is being received with an oscillator at 1430 kHz in a superheterodyne receiver. What is the I.F.? What will be the frequency that might be received as an image signal?

15.20 Name two advantages in the use of SSB over double-sideband AM.

15.21 Why must the carrier reinsertion oscillator of an SSB superheterodyne receiver be very stable? What are its stability limits in frequency?

15.22 What is a transceiver?

15.23 Can you see an advantage for the transceiver form of SSB transmitter and receiver?

15.24 Name several advantages of FM over AM.

15.25 What is the capture effect in FM reception?

15.26 From the Hartley-Shannon law, explain why increasing the bandwidth improves FM reception.

15.27 What causes the Doppler effect in a received radar signal?

15.28 The time from transmitted pulse to echo reception in a radar set is 30.5 μs. How far away is the target?

15.29 What frequency range is covered by VHF signals?

15.30 What is a major service employing VHF frequencies?

15.31 What frequency range is covered by UHF frequencies?

PROBLEMS

15.1 A teletypewriter uses a code with five possible positions for holes in a paper tape per letter symbol. Letters and spaces are sent at the rate of 360 per minute. What is the information rate in bits per second?

15.2 A radio receiver has an input resistance of 70 Ω. A signal of 4.7 μV is applied along with a steady noise voltage of 0.97 μV. What is the signal-to-noise ratio at the input?

15.3 The receiver of Problem 15.2 has a noise figure of 3.7 dB. What is the signal-to-noise ratio in dB at the receiver output?

15.4 How many bits can be transmitted per second through the channel represented in Problem 15.2 with a 3-dB bandwidth of 10^5 Hz?

15.5 The signal received in Problem 15.2 is reduced to 1.7 μV across the 70-Ω input circuit. With a 3-dB bandwidth of 10^5 Hz and circuit temperature of 300°C absolute, what is the signal-to-noise ratio at the input?

15.6 A wirephoto picture size is 12.5 cm $\times$ 18 cm and it is scanned at a rate of 40 lines per centimeter, with an equal resolution or number of dots along the lines. We assume that the eye can see 10 different levels or density gradations in the picture. Determine the S/N ratio in decibels required for the channel, of bandwidth 1200 Hz, if the picture is scanned in 3 minutes.

15.7 A superheterodyne receiver is tuned to a signal at 1.85 MHz. A signal at 2.76 MHz is found to interfere. What is the reason? What is the I.F. of the receiver? What is the receiver oscillator frequency?

15.8 A superheterodyne receiver tuned to a signal at 21.7 MHz has its oscillator at 25.2 MHz. What is the I.F. of the receiver? What is the possible image frequency?

16
Digital Circuits

Digital signals, using pulses, have become an increasingly important part of electronics in recent years, with the widening application of data processing and the digital computer. Beneath the complexity of these devices we find relatively simple and inexpensive circuits. This simplicity results from the fact that the circuits employ only two current levels or states, ON and OFF, in a *binary* or two-level code.

For a transistor to discriminate among 10 levels of current for decimal numbers would place an almost impossible requirement on transistor accuracy. But electronic devices are well suited to a binary code of on and off pulse signals, which require only that the transistors operate at saturation (ON) or cutoff (OFF). The signal-present state is usually designated 1 and the off state as 0; thus binary or two-level numbers are represented by chains of on-off pulses or 1's and 0's.

There are two basic types of circuits used in handling binary signals. They are logic on-off *gates* and *multivibrator* switches. We shall show how these are combined to process digital signals.

16.1 Binary Numbers

The decimal number system, with a base or *radix* of 10, has been a part of our lives since early childhood. With the decimal system we employ 10 marks or symbols, which we designate as 0, 1, 2, 3, . . . , 9.

If we explore the meaning of *position* in decimal numbers, we shall see that a numbering system with a base other than 10 is quite possible; we might use 2, 3, 8, or 9 as examples. Each position or place in a decimal number carries with it a weighting factor expressed in powers of 10:

Power of 10:	10^6	10^5	10^4	10^3	10^2	10^1	10^0	Decimal point
Weight of position:	1,000,000	100,000	10,000	1000	100	10	1	.

We use a zero to indicate that a particular weighted value is absent; we use our numbers, 1, 2, 3, . . . , 9 to tell us how many of a particular power of 10 are present at a position.

For instance, 521 in the decimal system employs the first three positions to the left of the *decimal point* and may be expressed

$$521 = (5 \times 10^2) + (2 \times 10^1) + (1 \times 10^0)$$
$$= \quad 500 \quad + \quad 20 \quad + \quad 1$$

The position of a digit is weighted by $10^2 = 100$, $10^1 = 10$, $10^0 = 1$, respectively.

In the *binary system* of numbers, the idea of position is the same as in the decimal system but powers of 2 are used for the weights with each position. Only two symbols are used, 1 and 0. The weighting factors are

Power of 2	2^8	2^7	2^6	2^5	2^4	2^3	2^2	2^1	2^0	Binary point
Weight of position (decimal value)	256	128	64	32	16	8	4	2	1	.

The decimal number 325 is expressed as 101000101 in the binary system. This means

$$(1 \times 2^8) + (0 \times 2^7) + (1 \times 2^6) + (0 \times 2^5) + (0 \times 2^4)$$
$$+ (0 \times 2^3) + (1 \times 2^2) + (0 \times 2^1) + (1 \times 2^0)$$

In decimal values we have $256 + 64 + 4 + 1$, which totals 325. The previously used decimal 521 would be written 1000001001.

In the decimal system, a decimal fraction such as 0.812 is expressed as

$$0.812 = (8 \times 10^{-1}) + (1 \times 10^{-2}) + (2 \times 10^{-3})$$

Similarly the binary fraction 0.1101 means

$$0.1101 = (1 \times 2^{-1}) + (1 \times 2^{-2}) + (0 \times 2^{-3}) + (1 \times 2^{-4})$$

Giving this the weight in decimal numbers

$$= \tfrac{1}{2} + \tfrac{1}{4} + 0 + \tfrac{1}{16}$$
$$= 0.500 + 0.250 + 0.062 = 0.812 \text{ (decimal)}$$

A simple method for conversion of a decimal number to binary representation employs repeated division of the number by 2. The remainder of 1 or 0 after each division becomes a digit of the binary number. For the decimal number 232 we have

	Remainder	
$232 \div 2 = 116$	0	Least significant digit
$116 \div 2 = 58$	0	
$58 \div 2 = 29$	0	
$29 \div 2 = 14$	1	
$14 \div 2 = 7$	0	
$7 \div 2 = 3$	1	
$3 \div 2 = 1$	1	
$1 \div 2 = 0$	1	Most significant digit

The last 1 obtained is the largest digit in the binary number. For decimal 232 we have 11101000 as the binary expression.

Fractional decimal numbers may be converted to binary form by successive multiplications by 2. For each step that results in a 1 to the left of the decimal point, record a binary 1 and carry on with the fractional portion of the decimal number. With a 0 to the left of the decimal point, record a binary 0 and carry on. For instance, to convert decimal 0.9375 to binary form, we operate as follows:

	Binary	
$0.9375 \times 2 = 1.8750$	1	Most significant digit
$0.8750 \times 2 = 1.7500$	1	
$0.7500 \times 2 = 1.5000$	1	
$0.5000 \times 2 = 1.0000$	1	
$0.0000 \times 2 = 0.0000$	0	Least significant digit

The binary equivalent of decimal 0.9375 is expressed as 0.11110. The largest digit is the first binary number obtained and it is placed to the right of the *binary point*.

The requirement for only two symbols in binary simplifies our electronic circuits but we offset this simplicity with the necessity for handling many more digits in the binary representation. For instance, decimal 10 is 1010 in binary.

A digit in binary is referred to as a *bit*, from the initial and final letters of binary digit.

16.2 Binary Arithmetic

Addition and subtraction in binary numbers is easier than the procedures used in the decimal system. Four rules apply for addition and these can be stated:

$$0 + 0 = 0$$
$$0 + 1 = 1$$
$$1 + 0 = 1$$
$$1 + 1 = 0 \text{ with a forward carry of 1}$$

A carry is handled similarly to decimal system procedure. Binary addition can be demonstrated in the following:

Binary		Decimal
11011		27
1011		11
10000	Sum	
1 11	Carry	
00110	Sum	
1	Carry	
100110		38

Subtraction in digital computers is almost universally handled by changing the negative number to its complement and adding. Negative numbers are usually manipulated in complement form.

With decimal numbers we form the 9's complement by subtracting each digit of the negative number from 9. We then add 1 to form the 10's complement. Suppose we wish to subtract 548 from 2012. The 9's complement of 548 is 451 and $451 + 1 = 452$, which is the 10's complement of 548. To subtract, we add the 10's complement and drop 1 from the leftmost digit:

Negative Numbers	Complementary Subtraction
2012	2012
− 548	+ 452
1464	2464
	−1
	1464

In binary numbers we employ a similar process. To form a 1's complement, we subtract each digit of the binary number from 1. For instance, the

1's complement of 10011 is 01100; to change this to the 2's complement, we add 1 so that 01101 is the 2's complement. To subtract 10011 from 11001, as an example, we add the 2's complement:

$$
\begin{array}{r}
11001 \\
\text{2's complement} \quad \underline{01101} \\
100110 \\
\underline{-1} \\
00110
\end{array}
$$

after subtracting 1 from the leftmost digit.

Since $11001 =$ decimal 25 and $10011 =$ decimal 19, the result should be $25 - 19 = 6$; in binary, $00110 =$ decimal 6.

In the decimal multiplication process, we add the multiplicand to itself the number of times specified by the least significant term in the multiplier, shifting one place to the left and repeating for each term of the multiplier. Binary multiplication follows the same procedure but is simplified because each term needs to be set down only once to represent multiplication by binary 1.

Thus 10101×1011 (decimal 21×11) leads to

$$
\begin{array}{ll}
10101 & \\
\underline{1011} & \\
10101 & \text{Multiply by 1} \\
10101 & \text{Shift, multiply by 1} \\
00000 & \text{Shift, multiply by 0} \\
\underline{10101} & \text{Shift, multiply by 1} \\
11100111 & \text{Sum} = \text{decimal 231}
\end{array}
$$

Binary division is carried out by successive subtractions or additions using the 2's complement in each case plus shifts to the right.

Thus, with the use of complements, all four of our arithmetic operations are reduced to the process of addition, plus right or left shifts. This greatly simplifies the arithmetic unit of a computer.

16.3 Binary Codes

To convert from the decimal system of the business community to the binary system in the digital computer has called for the development of computer codes. Foremost among these is the *binary-coded-decimal* (BCD) representation. The BCD code uses four binary bits to represent each of the 10 decimal digits. This form is also called the 8421 code, a name derived from the weight of each the four binary digits. The code groups are simply derived from binary representations as shown in Table 16.1.

TABLE 16.1 The BCD Code

Decimal	Binary	Decimal	Binary
0	0000	5	0101
1	0001	6	0110
2	0010	7	0111
3	0011	8	1000
4	0100	9	1001

The decimal number 3582 would be represented in four code groups as

$$0011 \quad 0101 \quad 1000 \quad 0010$$

Binary-coded-decimal form is frequently used at computer input and output. Switching circuits are available to convert such groups of bits to full binary number representation or to decode BCD groups into decimal numbers. Such circuits are usual at the output of most digital-reading instruments, such as digital voltmeters.

The number of bits employed in a code increases as does the number of symbols represented; that is, three bits are sufficient to transmit the eight decimal numbers $0, \ldots, 7$, while four bits are needed to transmit decimal numbers $0, \ldots, 15$. To transmit the English alphabet of 26 letters requires that $2^n \geq 26$ bits be used; with $2^5 = 32$, a five-bit code allows 32 different characters to be represented and the letters, punctuation marks, and start-stop signals can be transmitted. This is the alphanumeric code used for teletypewriters.

There is also in common use a code with seven bits, known as ASCII (American Standard Code for Information Interchange). With $2^7 = 128$ combinations of bits, the numerals, capital letters, and lowercase letters are available. Actually there are 64 characters and 64 operational controls. An example of the latter is the group 0001010, which calls for a line feed or the movement of the printing position to the next line. The computer can readily differentiate between a control function with the first two bits as zeros and an alphanumeric character when the first two bits are not both zeros.

Binary digits are sometimes handled in groups called *bytes*. The number of bits in a byte is not standardized but in one system a byte consists of eight binary bits, which might include two decimal digits in BCD code or one alphanumeric character in an eight-bit code.

16.4 Other Number Systems

Another number system employed in computers uses the base 8 and is called the *octal system*. By processing base-8 numbers in the binary-coded form, we need employ only the 1, 0 representation or the on-off switching of binary numbers. Therefore we can use the base of 8 without requiring our electronic devices to recognize more than the usual on-off states.

The number positions are given weights of powers of 8 and the symbols employed are 0, 1, 2, . . . , 7. The number 352 in octal means

$$\text{octal } 352 = (3 \times 8^2) + (5 \times 8^1) + (2 \times 8^0)$$
$$= (3 \times 64) + (5 \times 8) + (2 \times 1)$$
$$= 192 + \quad\quad 40 \quad + 2 = 234 \text{ decimal}$$

To transmit octal 352 in binary-code groups we use

$$0011 \quad 0101 \quad 0010$$

Since 7 is the largest number to be represented in octal, we could reduce this to

$$011 \quad 101 \quad 010$$

Note that the binary system representation for decimal 234 is

$$\overline{011}\ \overline{101}\ \overline{010}$$

so that if a binary number is segmented into groups of three bits, we have the binary code for octal numbers. This easy conversion from binary numbers to

TABLE 16.2 **Number Systems**

Decimal	Binary	Octal	Hexadecimal
0	0	0	0
1	1	1	1
2	10	2	2
3	11	3	3
4	100	4	4
5	101	5	5
6	110	6	6
7	111	7	7
8	1000	10	8
9	1001	11	9
10	1010	12	A
11	1011	13	B
12	1100	14	C
13	1101	15	D
14	1110	16	E
15	1111	17	F
16	10000	20	10
17	10001	21	11
18	10010	22	12
19	10011	23	13
20	10100	24	14

octal binary code is one reason for the employment of octal numbers within a given computer.

The representation of decimal 8 and 9 requires four bits in BCD code. Without adding any more equipment, we can expand the BCD representation to a base-16 system, or a *hexadecimal* system, with 16 symbols employed as 0, 1, 2, . . . , 9, *A*, *B*, *C*, *D*, *E*, and *F*. Decimal 10 is represented as *A*.

Table 16.2 shows the relationship of the several number systems discussed.

Subscripts are used to identify the number system employed. For example, 352 as an octal number would be written as 352_8; 011 in binary representation would be 011_2; and decimal 654 would appear as 654_{10}.

16.5 Series and Parallel Processing of Bits

The internal language of the digital computer is in the form of binary numbers and codes. The binary signals are handled as timed chains of pulses. A *clock* or timer generates the basic frequency, as shown in Fig. 16.1(a), and binary switches operating in on and off states code this waveform to represent binary data in on (1) and off (0) pulses, as shown in Fig. 16.1(b). The pulse chains may actually be positive and zero, zero and negative, or positive and negative in amplitude and there are many codes in which the data may be transmitted such as BCD or the seven-bit ASCII code.

There are two basic methods of handling bits in data processing. When transmitted over a wire line or a radio channel, the bits are sent in *series* and operated upon in sequence; the pulses in Fig. 16.1(b) may be visualized as sliding off the page to the right, with the first pulse being the most significant bit of the number. In *parallel* operation, the bits in the figure may be thought of as sliding down the page simultaneously in four separate channels; the weight ascribed to each bit is determined by the channel in which the pulse appears.

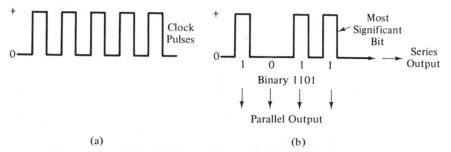

(a) (b)

Figure 16.1 (a) Clock or timing pulses; (b) binary code 1101 = decimal 13.

For a signal of 40 bits, the time for serial processing is 40 times that for the processing of 1 bit. With parallel operation the 40 bits are simultaneously processed and the time is approximately that for operating on 1 bit. Thus parallel operation would be 40 times as fast but requires 40 times as much equipment.

For reasons of speed, most digital computation is done with parallel processing. Registers or storage elements are used to accumulate n bits as they are received in series and the storage is periodically emptied or *dumped* into n parallel channels.

16.6 Logic Operations in Addition

It has been demonstrated that the basic arithmetical operation in digital computation is that of addition. To further analyze the addition process, we again carry out an addition with binary numbers A and B:

A	1011	
B	0111	
	1100	Sum
	11	Carry, C_1
	1010	Sum
	1	Carry, C_2
	10010	Sum

The requirements for the addition operation are summarized in the table below, in which columns A, B, and C_1 provide for all possible combinations of 0 or 1 input signals for these variables. The columns S and C_2 are the required results, reasoned from the addition example above and confirmed by the addition rules of Sec. 16.2.

A	B	C_1	S	C_2
0	0	0	0	0
0	0	1	1	0
0	1	0	1	0
0	1	1	0	1
1	0	0	1	0
1	0	1	0	1
1	1	0	0	1
1	1	1	1	1

The addition will be carried out in parallel channels so that one adding circuit will handle one bit from A and one bit from B. The process is not numerical in nature but requires circuits that provide 1- or 0-level output signals in response to several 1- or 0-level command signals at the inputs. Specifically, digital operations utilize circuits that recognize the presence of a pulse from one circuit AND the presence of a pulse from a second circuit, as well as when a pulse from one circuit OR a pulse from a second circuit is present. Also there must be recognition of a carry pulse C_1, coming from a preceding bit circuit, and provision must be made to transmit a forward carry C_2 to the next most significant bit channel.

The table shows that the sum S equals 1 when

$$A = 1 \quad \text{OR} \quad B = 1 \quad \text{OR } C_1 = 1, \quad \text{other variables} = 0$$
$$\text{OR} \tag{16.1}$$
$$A \text{ AND } B \text{ AND } C_1 = 1$$

These statements are reasoned from lines 2, 3, 5, and 8 of the table.

We also see that there must be a forward carry output C_2, or $C_2 = 1$, when

$$A \text{ AND } B = 1, \qquad C_1 = 0$$
$$\text{OR}$$
$$A \text{ AND } C_1 = 1, \qquad B = 0$$
$$\text{OR} \tag{16.2}$$
$$B \text{ AND } C_1 = 1, \qquad A = 0$$
$$\text{OR}$$
$$A \text{ AND } B \text{ AND } C_1 = 1$$

These statements result from lines 7, 6, 4, and 8 of the table.

To carry out the addition operation we need circuits that yield proper 1- or 0-level results from OR and AND combinations of two or three pulse inputs. We shall later find that a NOT or *inverter* operation is also needed. An inverter gives a 0-level output for a 1-level input, or vice versa.

Such circuits that recognize the presence of a group of 1- or 0-level signals at the input and give 1- or 0-level output dependent on OR, AND, NOT combinations of these input orders are called *logic circuits*. They act as off-on switches.

16.7 Logic Switches

Logic circuits are often called *gates* because they are open (OFF) or closed (ON) as called for by the specified combinations of pulse inputs.

A logic AND gate must provide a "logic-1" output only if *all* inputs are at the 1 level. The AND operation is illustrated in Fig. 16.2(a) as a series connection of switches. Only if all the switches are closed or at the 1 condition will an output voltage appear at F; that is, we write in logic algebra

$$ABC = F \tag{16.3}$$

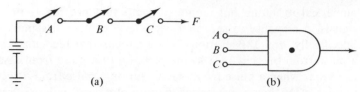

Figure 16.2 (a) Switches in an AND circuit; (b) AND logic-circuit symbol.

and read this as *A* AND *B* AND *C* equals *F*. The variable *F* is equal to 1 if the switches are all closed; $F = 0$ if any switch is open.

A table of switch outputs for an AND circuit with *A* and *B* inputs is

A	B	F	
0	0	0	
0	1	0	AND
1	0	0	
1	1	1	

A circuit symbol for an AND circuit is drawn in Fig. 16.2(b).

A logic OR gate must provide an output at "logic-1" level if *any* one or more of its inputs is at the 1 condition. The OR operation is illustrated in Fig. 16.3(a) as a parallel connection of switches. If switches *A* OR *B* OR *C*, OR any combination, is closed, the circuit is complete and $F = 1$; that is, in logic algebra

$$A + B + C = F \tag{16.4}$$

which is read as "*A* OR *B* OR *C* equals *F*."

A table of switching outputs for an OR circuit is

A	B	F	
0	0	0	
0	1	1	OR
1	0	1	
1	1	1	

A circuit symbol for an OR circuit is shown in Fig. 16.3(b).

The concepts of multiplication or addition should not be associated with the indicated symbols in Eq. 16.3 and 16.4; the equations are intended to convey ideas of switch connection.

A logic NOT circuit supplies the inverse of any operation and the inverse of switch *A* is written $\bar{A}$ (*A* NOT). If $A = 1$, then $\bar{A} = 0$, and vice versa. A common-emitter connected transistor often supplies the NOT operation with its inversion of signal.

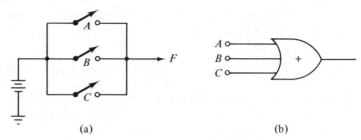

Figure 16.3 (a) Switches in an OR circuit; (b) OR logic-circuit symbol.

By combining the AND and NOT operations, we can form a NAND (negative AND) gate. A table of switching operations is

A	B	F	
0	0	1	
0	1	1	NAND
1	0	1	
1	1	0	

A circuit symbol for a NAND operation is shown in Fig. 16.4.

Similarly, if we combine OR and NOT gates, the result is a NOR (negative OR) operation. An operations table for the NOR gate is

A	B	F	
0	0	1	
0	1	0	NOR
1	0	0	
1	1	0	

A circuit symbol for a NOR gate is drawn in Fig. 16.4.

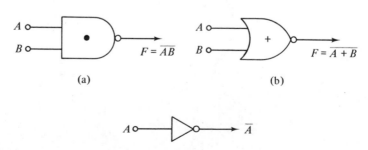

Figure 16.4 (a) NAND symbol; (b) NOR symbol; (c) NOT, or inversion, symbol.

Note that the results of NAND and NOR operations are exactly the opposite of AND and OR, respectively.

16.8 Logic Voltage Levels

In order to explain the operation of actual logic-switching circuits, it is necessary to choose voltages to represent logic-1 and logic-0 conditions in the gate circuits. Usually the choices are made to give

1. *Positive logic*, in which we make the 1-logic level more positive than the 0-logic level. Frequently the assigned voltages are $+5$ V for the 1 level and 0 V or ground for the 0-logic level.
2. *Negative logic*, in which we make the logic-1 level more negative than logic 0, and voltages of -5 V and 0 V are often assigned.

Our descriptions of gate circuit operation will be based on the selection of positive logic.

16.9 Diode Logic (DL) Gates

Many different circuits have been designed to develop the gate functions described in Sec. 16.7. With diodes employed as switches, *diode logic* (DL) gates will first be discussed because of their simplicity. An AND circuit is shown in Fig. 16.5(a); it is drawn with three inputs but the actual number may vary.

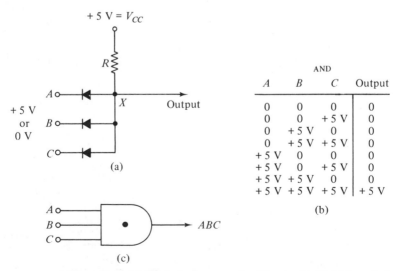

A	B	C	Output
0	0	0	0
0	0	+ 5 V	0
0	+ 5 V	0	0
0	+ 5 V	+ 5 V	0
+ 5 V	0	0	0
+ 5 V	0	+ 5 V	0
+ 5 V	+ 5 V	0	0
+ 5 V	+ 5 V	+ 5 V	+ 5 V

Figure 16.5 (a) Diode logic AND gate; (b) output table; (c) symbol.

With positive logic voltage of +5 V for the 1-logic level and ground for 0 level, the diodes are connected so that

1. If *all* the inputs are at logic 1 (+5 V), the diodes have no voltage across them and are open; the output voltage goes to +5 V through the load resistor *R*.

2. If *any* diode input is at logic 0 (0 V), its diode is forward-biased by V_{CC} and clamps the output voltage to zero or ground. This represents a logic-0 output.

These are the actions of an AND circuit, as predicted by the table in Fig. 16.5(b). The symbol for an AND circuit, is presented in Fig. 16.5(c).

Since the output voltage goes to +5 V through *R* and the diodes are then open, the output impedance is *R* at logic-1 output. For logic-0 output, the output impedance is that of the sources supplying the conducting diode or diodes. The capacitance of the circuit load must be charged through these resistances and the rise time due to *R* will usually be longer than the time of fall of the output signal, when the discharge is through the low diode resistance.

A diode logic OR circuit is diagrammed in Fig. 16.6(a). For a positive logic signal the circuit will act so that

1. If *any* diode input is at logic 1 (+5 V), that diode will be conducting and will place +5 V (logic 1) on the output. Two or more +5-V inputs will produce the same effect as for one +5-V input.

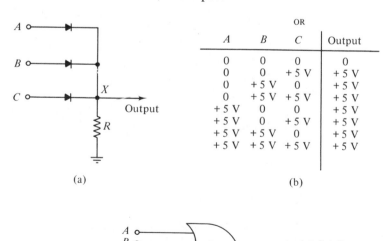

	OR			
A	*B*	*C*	Output
---	---	---	---
0	0	0	0
0	0	+ 5 V	+ 5 V
0	+ 5 V	0	+ 5 V
0	+ 5 V	+ 5 V	+ 5 V
+ 5 V	0	0	+ 5 V
+ 5 V	0	+ 5 V	+ 5 V
+ 5 V	+ 5 V	0	+ 5 V
+ 5 V	+ 5 V	+ 5 V	+ 5 V

(a)

(b)

(c)

Figure 16.6 (a) Diode logic OR gate; (b) output table; (c) symbol.

2. If *all* the inputs are at 0 V or ground (level 0), all diodes are open and the output is connected to ground through R and the output is at logic 0.

These are the actions of an OR circuit, as shown by the table in Fig. 16.6(b). The circuit symbol for an OR circuit is drawn in Fig. 16.6(c).

In computing systems a circuit may be supplied by many inputs and must absorb the input currents. The number of inputs of similar circuits that can be connected without disturbance of the voltage levels is called the *fan-in* property of the circuit. The ability of a logic circuit to supply output current for driving other similar gates is indicated by the *fan-out* property or rating. A fan-out rating of 8 would indicate that a logic circuit could provide the current to drive eight circuits with similar input-current requirements.

Due to the variable diode voltage drops, it is not possible to maintain the value of $+5$ V for the logic-1 level and this variability is a disadvantage of diode logic circuits. This variation is especially serious when many diode gates are used in series since the voltage losses become cumulative. The difference between the 1 and 0 levels may become insufficient for accurate operation of the circuits. Section 16.10 supplies a remedy for this situation.

Reasoning as applied above will show that if negative logic signals are employed with these AND and OR circuits, the operating functions reverse; that is, a positive logic OR circuit becomes a negative logic AND circuit and a positive logic AND circuit becomes a negative logic OR circuit. Because of the ready availability of more satisfactory logic circuits, diode logic is now rarely employed.

16.10 The NOT or Inversion Operation

The NOT function of inversion can be performed by a transistor in the C-E circuit in Fig. 16.7(a). The transistor is driven from saturation for an output at 0 level to cutoff for an output at the logic-1 level, as shown by the output characteristic in Fig. 16.7(b).

To ensure saturation, the base current should exceed the amount required at the intersection of the load line and the saturation line in Fig. 16.8; that is, the base current for saturation should be

$$I_B > \frac{I_{C(\text{sat})}}{h_{FE}} \qquad (16.5)$$

The collector current at saturation will be

$$I_{C(\text{sat})} = \frac{V_{CC} - V_{CE(\text{sat})}}{R} \cong \frac{V_{CC}}{R}$$

since $V_{CE(\text{sat})} \ll V_{CC}$. Then, to ensure saturation, we should have

$$I_B > \frac{V_{CC}}{h_{FE}R} \qquad (16.6)$$

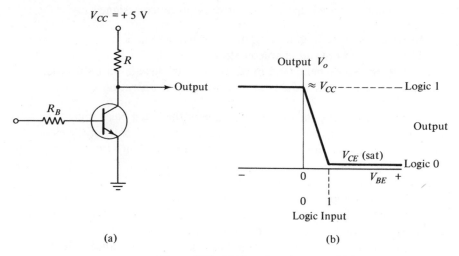

Figure 16.7 (a) Transistor inverter; (b) input-output curve.

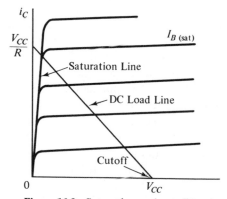

Figure 16.8 Saturation and cutoff levels.

This will ensure a logic-0 output for a logic-1 input, as desired for the operation of inversion.

The output will swing from $V_{CE(sat)} \cong 0$ V for logic-0 output to V_{CC} for logic-1 output.

16.11 Integrated Logic Circuits

Manufacturers produce many integrated circuit forms of the basic logic gates for AND, OR, NOT, NAND, and NOR operations. All will perform their stated functions and selection of a particular type depends on the application, with choice based upon the relative importance assigned to

1. Speed of switching, usually stated in nanoseconds.
2. Fan-in and fan-out characteristics.
3. Power requirements and rated dissipation; usually in milliwatts.
4. Cost, which usually increases with speed of operation.

Each circuit is designed to perform one of the listed basic logic functions and an overall system represents the interconnection of many of the elemental circuits. Frequently four or six similar and independent circuits are placed on one silicon chip and sold as *quad* or *hex* combinations to save space in the logic-circuit assembly.

In general a manufacturer makes available a complete family of circuits, with compatible input and output voltage levels. The individual circuit forms are simple, as will be shown. The internal circuit designs of typical units will be discussed here only to indicate the reasons for performance differences of the circuit types.

16.12 Diode-Transistor Logic (DTL)

In order to improve the fan-out capabilities of diode logic circuits and to maintain a standard logic-1 level at the output, a transistor amplifier may be added to the output of the diode logic circuits and this leads to a circuit family known as *diode-transistor logic* (DTL).

When the transistor is connected as an emitter follower, the output current is increased without loading the diodes. The transistor is more usually connected in the C-E circuit, however, and it then performs the NOT function. Since an AND circuit followed by an inverter performs the NAND operation

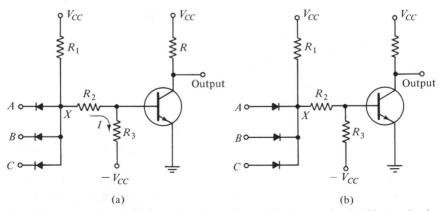

Figure 16.9 (a) Diode-transistor NAND circuit; (b) NOR circuit.

and an OR circuit followed by an inverter yields a NOR function, NAND and NOR circuits are available as illustrated in Fig. 16.9. These circuits with transistors are much less sensitive to output loading changes and have higher fan-out ratings than the diode logic circuits. Normal logic-level voltages are also restored at each gate.

Resistors R_1, R_2, and R_3 adjust the potentials between $+V_{CC}$ and $-V_{CC}$ so that in Fig. 16.9(a), when point X is at ground, the base of the transistor is placed well below cutoff and the circuit is immune to random noise pulses. With X at ground, current I is

$$I = \frac{V_{CC}}{R_2 + R_3}$$

and

$$
\begin{aligned}
V_{BE} &= R_3 I - V_{CC} \\
&= -\frac{R_2}{R_2 + R_3} V_{CC}
\end{aligned}
\tag{16.7}
$$

Thus when X is at ground, the base of the transistor is negative and the *npn* transistor is cut off. The output rises to $+V_{CC}$ or the logic-1 level. When all inputs are at the 1 level, X is positive and the base of the transistor is driven to saturation. The output goes to $V_o \cong 0.1$ V, which is the logic-0 level, and the circuit is of NAND form.

Another integrated circuit for DTL NAND operation is shown in Fig. 16.10. Transistor Q_1 is driven by the diode output and supplies the saturation bias

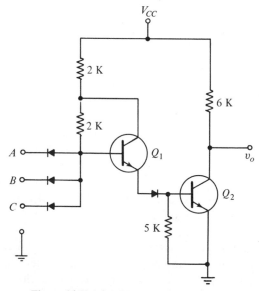

Figure 16.10 A DTL NAND integrated circuit.

current for Q_2; Q_1 is not driven to saturation. The speed is reasonably fast and the fan-out characteristics are good.

A general disadvantage of DTL circuits is the necessity for two power supplies.

16.13 Resistance-Transistor Logic (RTL)

The *resistance-transistor logic* circuit family replaces the individual diodes of the DTL circuits with the base-emitter junctions of transistors; a NOR circuit form is shown in Fig. 16.11 as an example. The use of transistors instead of diodes does not increase the circuit complexity or cost when integrated circuit production methods are used.

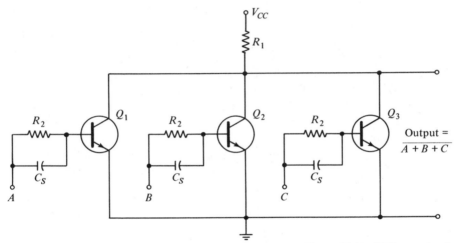

Figure 16.11 *RTL-*NOR *circuit.*

Introduction of a sufficiently positive pulse (logic 1) to any input will drive the transistor to saturation. If two or more of the inputs are driven at the same time, the output voltage across the parallel transistors remains at the saturation level. The operation is that of OR and this is inverted in the respective transistors. The result is that of a NOR circuit, with logic-0 output for any logic-1 input to A OR B OR C.

Resistors R_2 are present to prevent the emitter-base junctions from short-circuiting the pulse sources. The input capacitance of the transistor must be charged through resistor R_2 and this introduces a delay. To allow fast pulses to charge the internal capacitance more rapidly, *speedup* capacitance C_s is added to bypass R_2 for each transistor. Because of the time needed to discharge the transistor capacitances after saturation, however, the circuit is somewhat slower in operation than is the DTL form. The necessity for any one transistor to be able to lower the output from V_{CC} to $V_{CE(\text{sat})}$ means that

V_{CC} must be small and the discrimination between logic-1 and logic-0 levels at the output is limited.

16.14 Transistor-Transistor Logic (TTL)

The *transistor-transistor logic* circuit family is illustrated by a TTL NAND gate in Fig. 16.12, adapted to integrated circuit manufacture. This circuit can be likened to the DTL circuit, except that the switching diodes of the DTL form are replaced by separate emitters on the base electrode of a common transistor. If all the input emitters are given a logic-1 pulse, Q_1 will be turned off and the input to Q_2 will rise, driving Q_2 into saturation with an output at $V_{CE(\text{sat})}$ or the logic level of 0.

Speed of switching is increased by use of the common-base form of circuit, increasing the high-frequency capabilities of the input transistor. The base of Q_2 is supplied through the low-resistance collector circuit of Q_1 and the stored charge in the input of Q_2 can be rapidly removed upon switching. The TTL family of circuits operates with switching speeds of 10 to 20 ns (nanoseconds); operating clock frequencies can be as high as 30 MHz.

The gate output is either at logic 1, with Q_4 in saturation and Q_3 cut off, or at logic 0, with Q_4 cut off and Q_3 in saturation. Because of the manner in which the output circuit is usually drawn, it has been given the name of "totem pole" circuit. It is assumed that when Q_3 turns on, Q_4 turns off simultaneously, and vice versa. Both transistors are on for part of the transition, however, and the output circuit then shunts current to ground. This current is limited by the low value of R. The result is that the circuit draws a spike of

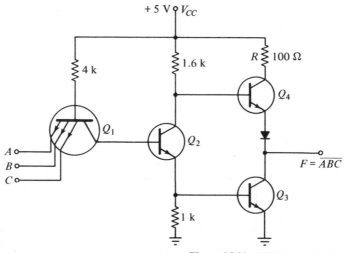

Figure 16.12 TTL-NAND circuit.

current (a *glitch*) in each transition and this abrupt current drain may cause a spiked voltage drop on the V_{CC} line, which is transmitted as electrical noise to other circuits supplied by that line. These noise pulses can cause random switching of circuits in which the voltage difference between logic 1 and logic 0 is not great. If many gates switch simultaneously, a large peak current may be required from the power supply.

In addition, the transition pulse of current represents power consumption and the power taken by TTL circuits increases with circuit operating rate, or the rate of the system clock. Under quiescent conditions the power may be only 6 mW per gate but at a clock rate of 20 MHz the power consumption may increase to 20 mW per gate.

A great variety of related and compatible circuit functions is available for TTL system design.

16.15 Emitter-Coupled Logic (ECL) Circuits

When a transistor is in the saturation state, it has a base current higher than necessary and excess charge is stored in the base region and in the collector-base junction region. A transistor cannot change to the cutoff state until this charge is removed and the charge transfer slows transistor switching. *Emitter-coupled logic* (ECL) circuits obtain greater speed of switching by operation of the transistors in a nonsaturated condition, through limiting the lowest collector voltages to values above $V_{CE(\text{sat})}$. An ECL circuit for the OR, NOR function is shown as an example in Fig. 16.13.

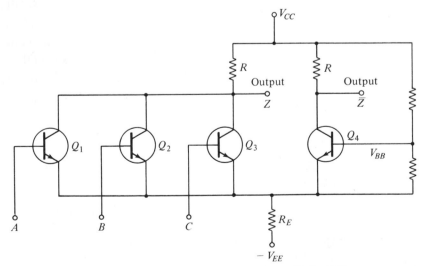

Figure 16.13 ECL-OR, NOR gate.

With fixed bias, transistor Q_4 supplies a constant current to the emitter resistor R_E; this current is less than the saturation collector current. A logic-1 input voltage to Q_1 will turn on that transistor and drop the output voltage at Z. As the switch transistor Q_1 increases its current, the increased bias across R_E causes the current in Q_4 to fall and the drop across R_E is maintained constant, but with current from Q_1 replacing the current from Q_4. Reduced current in Q_4 raises the output voltage at $\bar{Z}$ and with outputs at Z and $\bar{Z}$ the circuit provides its own inverted output.

The circuit is very fast, operating to switching speeds of 70 MHz; gives simultaneous OR and NOR outputs; and has high fan-in and fan-out capability. A complete logic family is available.

16.16 CMOS Logic Circuits

Because of major differences in the design and operation, the *complementary-metal-oxide-semiconductor* (CMOS) series of logic gates will be given more extensive treatment.

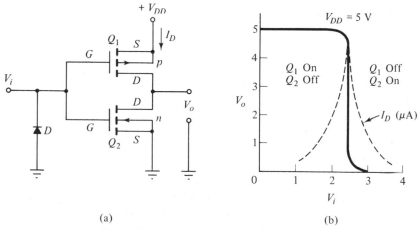

(a)   (b)

Figure 16.14 (a) Basic CMOS inverter; (b) transfer characteristic.

The CMOS switch employs FET enhancement-mode transistors in both *p*-channel and *n*-channel forms, justifying the complementary name. The basic switch appears as an inverter in Fig. 16.14(a). This circuit shows *p*- and *n*-channel units connected in series across a power supply and with a common gate input. These devices are diffused on a monolithic silicon chip, by processes that will be described in Sec. 16.18.

In the *n*-channel transistor Q_2, the majority carriers are electrons. A positive gate-to-source voltage greater than a threshold value V_T will increase the channel current. For a gate voltage at ground or source potential, the channel is cut off.

For the p-channel unit Q_1, the majority carriers are holes. A gate voltage negative to the source S increases the channel current; for the gate at source potential the channel is cut off.

With the gates connected together, a positive input V_i, equal to V_{DD}, will turn on the n-channel Q_2 and turn off the p-channel unit Q_1. With the gates at ground, the n-channel gate is at source potential and less than V_T; Q_2 is cut off; the gate of Q_1 is then negative to its source S and Q_1 conducts heavily. The effect of a common input voltage to both gates is complementary and results in a push-pull action of the two transistor elements.

Figure 16.14(b) shows the input-output transfer curve when the power supply is $V_{DD} = +5$ V. When V_i is near $+5$ V, the gate of Q_2 is at $+5$ V to its source at ground and Q_2 is on. The gate-to-source voltage of Q_1 is then zero and less than the threshold voltage, however, so that Q_1 is in an off condition. The device is therefore operating in the right half of the transfer diagram; it appears to have a resistance of a few hundred ohms through Q_2 to ground. The resistance of the off transistor will be in excess of 1000 MΩ. The current through both devices in series will be less than 5 nA, resulting in very low power dissipation.

We now place the gates near 0 V or ground. The conduction conditions reverse, Q_2 is off and Q_1 is on, and the output voltage reaches $+V_{DD} = +5$ V. Operation is in the left half of the transfer diagram. The quiescent current through both devices is again of 5-nA order. During the switching transition a small current passes as shown by the dashed curve on the transfer characteristic.

With positive logic-1 input, the output voltage is at 0 V and in the logic-0 condition; with logic-0 input, the output is at $+5$ V and in the logic-1 condition. The device is therefore an inverter or NOT logic circuit.

The switching time will approximate 25 to 200 nS. Because the circuit switches near one-half of V_{DD}, its immunity to random circuit noise is good. Because of the very thin oxide layer that serves as gate insulation, the diode D is shunted between gate and ground to reduce negative overvoltages that might break down the oxide insulation.

16.17 NAND and NOR Circuits with CMOS Logic

By adding devices in parallel or series, the NOR and NAND circuits in Fig. 16.15 can be obtained with CMOS logic. In Fig. 16.15(a), if we make either input A or B positive, we turn on Q_3 or Q_4; but making A or B positive also turns off either Q_1 or Q_2 and thus we have $V_o = 0$, or logic 0. We find the same result for both A and B positive.

If we place A and B at ground potential or logic 0, transistors Q_1 and Q_2

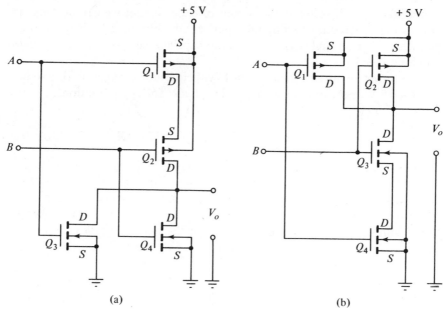

Figure 16.15 Positive logic CMOS: (a) NOR gate; (b) NAND gate.

are turned on and Q_3 and Q_4 are turned off; this is the condition for $V_o =$ +5 V, or logic-1 level.

The operation table is

A	B	V_o
0	0	1
0	1	0
1	0	0
1	1	0

and we can identify this circuit as giving a NOR output.

The circuit in Fig. 16.15(b) is an upside down version of that in Fig. 16.15(a). If A and B àre positive or at logic 1, Q_3 and Q_4 are turned on and $V_o = 0$, or logic 0. If either A or B, or both, are at ground or logic 0, then Q_1 or Q_2 are on but Q_3 or Q_4 will be off and so the output becomes $V_o =$ +5 V, which is logic 1. The operation table is

A	B	V_o
0	0	1
0	1	1
1	0	1
1	1	0

and the circuit is identified as giving NAND performance.

By adding an inverter at the output of either circuit we can develop OR and AND logic circuits. Additional inputs can be obtained by placing more transistors in series and parallel; the input capacitance increases and reduces switching speed.

The power requirements are very low, being in the range of 20 μW per gate for 15-V operation down to less than 1 μW for 5-V operation.

16.18 CMOS Manufacture

The construction of a monolithic CMOS integrated circuit is illustrated by the cross section in Fig. 16.16(b). Manufacture starts with an *n*-silicon substrate wafer having a thickness of about 0.006 in. (0.015 mm). A silicon dioxide layer is grown over the substrate by heating the silicon and a hole for the deep *p* diffusion at the left is etched through the oxide. Acceptor impurities are diffused into this large *p* region to the desired depth. The original oxide layer is removed and a new layer grown overall.

Holes are etched open for the heavily doped *p*+ diffusions at the right; these become the drain and source diffusions for the *p*-channel device. After the *p*+ regions are in place, the remaining oxide is removed and a new SiO$_2$ layer grown; holes are etched open for the heavily doped *n*+ diffusions at the left. These areas serve as the drain and source electrodes for the *n*-channel device.

Oxide is again removed, regrown, and holes etched for the gate areas. A very thin oxide layer, about 10^{-4} mm thick, is grown over the gate regions and etched away from the metal contact areas elsewhere. Metalization is then added over the gate dielectric areas and for connection to the elements as shown.

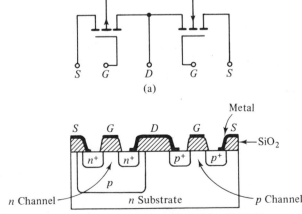

Figure 16.16 Circuit and monolithic construction of the CMOS inverter.

16.19 The Adder Circuit as a Logic System

To show how these logic circuit elements are connected to produce a computer subsystem, let us return to the addition process, analyzed in Sec. 16.6. If we identify logic-0 values by the complementary symbol $\bar{X}$ (X not), we can expand the logic statements of Eq. 16.1, showing that $S = 1$ when

$$A \text{ AND } \bar{B} \text{ AND } \bar{C}_1 = 1$$

OR

$$\bar{A} \text{ AND } B \text{ AND } \bar{C}_1 = 1$$

OR

$$\bar{A} \text{ AND } \bar{B} \text{ AND } C_1 = 1$$

OR

$$A \text{ AND } B \text{ AND } C_1 = 1$$

These statements can be made operative by use of inverters to obtain the complementary values, followed by four AND circuits of three inputs each, connected to a four-input OR circuit. The complete logic system for this portion of the addition operation is shown in the left portion of Fig. 16.17 and yields the sum bit S.

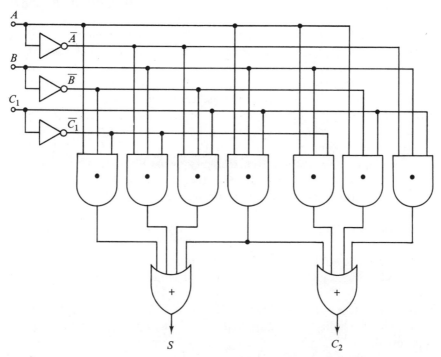

Figure 16.17 Logical adder for one bit.

The second statement in Eq. 16.2, for the forward carry $C_2 = 1$, results in an expansion as

$$A \text{ AND } B \text{ AND } \bar{C}_1 = 1$$
OR
$$A \text{ AND } \bar{B} \text{ AND } C_1 = 1$$
OR
$$\bar{A} \text{ AND } B \text{ AND } C_1 = 1$$
OR
$$A \text{ AND } B \text{ AND } C_1 = 1$$

These logic statements can be realized by three more AND circuits feeding into a four-input OR circuit. Since the last logic statement is identical to the last statement of the first set, the output of that AND circuit is used and one AND circuit is saved. The result appears in the right portion of Fig. 16.17 and the circuit yields the carry bit C_2.

The complete circuit must be duplicated for each bit of the parallel-processed binary number, with appropriate interconnections for the carry bits C_1 and C_2. Such complete adders are available on single integrated chips.

16.20 Multivibrators

The name *multivibrator* designates a group of circuits widely applied for switching as shift registers or temporary memories and as square-wave timing oscillators or clocks. The circuits are basically closed-loop feedback circuits operating with positive feedback. Because of the cumulative effects produced by this feedback, the circuits drive themselves to either of two limit or latch-up conditions, in which the transistors are at either cutoff or saturation. There are two output terminals A and B and when in a latched condition these terminals have potentials that are logic opposites as $A = $ logic 1, $B = $ logic 0, for example. Switching can be made to occur between the two latched limits and the logic levels interchange so that $A = $ logic 0, $B = $ logic 1.

Three general types of multivibrator are

1. The bistable or flip-flop, which can be triggered from one stable latched condition to the other by an external signal.
2. The monostable multivibrator, which can be switched from one stable state to the other; it then returns to the first latch-up condition after a time delay.
3. The astable or free-running multivibrator, which continuously switches between its two limits without application of an external signal; it is a square-wave oscillator.

The flip-flop is of great importance in digital operations; the monostable is useful as a delay and timing circuit; and the free-running type is used for the timing oscillator or clock of a computer system, as well as a frequency source.

16.21 The Bistable Multivibrator or Flip-Flop

The circuit in Fig. 16.18(a) illustrates the operating principle of the important *bistable multivibrator* or *flip-flop*. The circuit employs two inverter amplifiers in a positive feedback loop, with the output of inverter 1 fed to the input of inverter 2 and the output of inverter 2 returned to augment the input of inverter 1. The two stable or latched conditions occur with inverter 1 in saturation and inverter 2 in cutoff and the reverse with inverter 2 in saturation and inverter 1 at cutoff. The outputs are complementary and if point A is high at logic 1, then B is low at logic-level 0. If an external triggering signal is applied, the circuit may be made to switch and latch in its reverse state with A at low potential and logic-level 0 and B high at logic-level 1.

For further understanding, a form of discrete circuit is drawn in Fig. 16.18(b). Assume the circuit is in a latched state with Q_1 at cutoff and Q_2 in saturation. The voltage at A is equal to V_{CC}, while that at B approximates zero. Cross coupling from the low voltage at B through R_2 and R_3 and aided

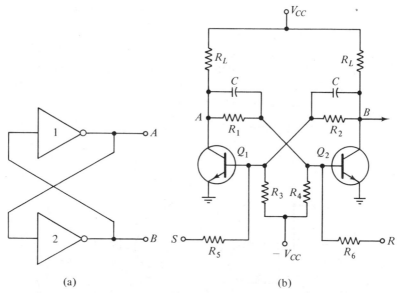

(a)   (b)

Figure 16.18 (a) Bistable inverter loop; (b) circuit of bistable multivibrator.

by the bias $-V_{CC}$ maintains the base of Q_1 at cutoff voltage. Cross coupling from A through R_1 and R_4 places a positive potential on the base of Q_2. Thus Q_1 is held at cutoff and Q_2 in saturation and the circuit is actually in a stable or latched state.

A negative trigger pulse may then be applied to R, the *reset* terminal, and to the base of Q_2, driving that transistor toward cutoff. The voltage at B rises as the Q_2 collector current drops and by coupling through R_2 the potential rise at B is transmitted to the base of Q_1, turning that transistor on. The rise in its collector current then reduces the voltage at A; this fall of voltage is transmitted through R_1 to the base of Q_2 and drives that transistor further toward cutoff. This action raises the voltage at B further and the cumulative action proceeds around the loop until the second latched state is reached, with Q_1 in saturation and Q_2 at cutoff. The voltage at A is then near zero or at logic 0 and the voltage at B is V_{CC} and at logic 1. A negative pulse applied to S, the *set* terminal, will cause the circuit to return to its original state, with A at logic 1 and B at logic 0.

The time required to transfer conduction from one state to the other is known as the *switching time*. Fast switching is possible if the transistors are chosen with high values of f_T, indicating low values of input capacitance. The time constants associated with the speedup capacitors C should be small with respect to the time between switching pulses to allow the speedup capacitors to recharge.

16.22 The RS Flip-Flop

Figure 16.18(b) shows the R and S trigger terminals of a *reset-set* flip-flop. In Fig. 16.19(a) we have a circuit composed of a cross-coupled pair of NAND gates for the multivibrator switch, with input logic supplied by two additional NAND gates. This circuit will act as does an RS flip-flop, although the actual circuit may be much more complex.

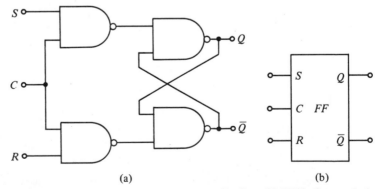

(a)   (b)

Figure 16.19   (a) NAND gate simulation of an RS flip-flop; (b) RS flip-flop symbol.

The circuit will trigger on the positive-going edge of the clock pulse inserted at C if a positive 1 exists at the S or R inputs. A logic-1 input to S with the input to R at 0 will cause Q to go to 1 and $\bar{Q}$ to 0; this is the *set* state. If R is given a logic-1 signal when S is at 0, the output condition becomes $Q = 0$ and $\bar{Q} = 1$; this is known as the *reset* state.

Simultaneous 0 inputs to S and R will not give outputs from the NAND logic circuits when a clock pulse arrives and the flip-flop is left in its previous condition. Simultaneous 1 inputs to R and S will cause circuit confusion, however, since the circuit will attempt to reach logic 0 on both outputs. This is an input condition that must be avoided.

The circuit operations are predicted by

S	R	Q	$\bar{Q}$
0	0	Unchanged	
0	1	0	1
1	0	1	0
1	1	?	?

In effect the RS flip-flop indicates and remembers the last input signal received. The switching action is synchronized with the arrival of a positive-going clock pulse. Input pulses need be only long enough to initiate the switching action programmed by the R or S signal and the input pulse can then be removed.

16.23 The T or Toggling Flip-Flop

The T multivibrator in Fig. 16.20 will switch between alternate states on successive negative trigger pulses. The two diodes D_1 and D_2 are *steering diodes*, which direct the trigger pulses to the base of whichever transistor is on. A reset input may also be added to the base of a transistor to set the initial condition of the circuit before toggling starts.

With Q_1 off and Q_2 on, a -5-V negative trigger pulse is applied and takes point X down to $+5$ V. Diode D_2 is reverse-biased between C at $+5$ V and B at 0 V. Diode D_1 is found to be forward-biased and transmits the negative trigger pulse past A to the base of Q_2, turning it off. Transistor Q_1 is turned on in normal fashion.

The next negative pulse finds D_1 reverse-biased due to the change in voltages at A and B and, with D_2 conducting, the trigger pulse is directed to the base of Q_1 to turn it off. The circuit then toggles or reverses its outputs.

Toggle flip-flops are used in counters to accumulate pulse counts. Each toggle circuit is known as a *scale-of-two* circuit since Q_2 switches on once for

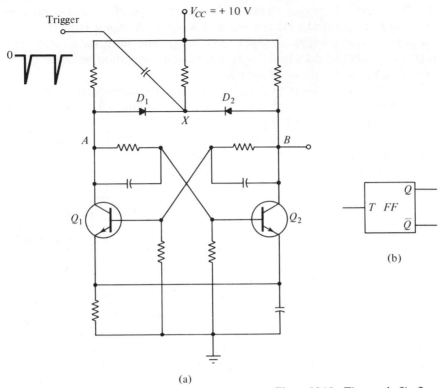

(a)

Figure 16.20 The toggle flip-flop.

two input pulses. In switching on, Q_2 generates a negative-going pulse at B that can be transmitted to a second toggle circuit as a triggering pulse. For two toggle circuits in cascade a pulse is obtained from the last circuit for every $2^2 = 4$ input pulses. This is the action of a *scale-of-four* circuit, as shown in

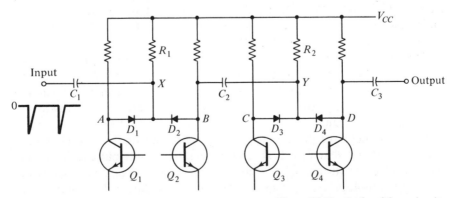

Figure 16.21 Scale-of-four circuit.

Fig. 16.21. The count can be extended by use of additional toggles in cascade, n circuits providing for a count of 2^n.

If positive logic signals are used, then the negative-going input triggers are obtained when the trigger pulse goes from $+V$ to 0 V and this is known as *trailing-edge triggering*.

16.24 The JK Flip-Flop

The *JK* flip-flop is the most complex of these circuits. It is similar to the *RS* circuit except that its operation is predictable when both trigger inputs are at the logic-1 level. In that condition the circuit always changes state and complements its previous outputs.

A circuit that will provide *JK* operation is drawn in Fig. 16.22; actual circuits are more complicated but being available in integrated circuit packages makes this internal complexity of little importance to the user.

Since each AND gate has one input fed back from an output, one of the AND gates has a logic-1 input in addition to the logic 1 supplied by the clock at *T*. We also have the *J* and *K* inputs to the respective gates and four possibilities exist; these are listed in the operation table:

J	K	Q	$\bar{Q}$
0	0	No change	
0	1	0	1
1	0	1	0
1	1	Toggles	

Signals may be applied and held at *J* and *K* but switching occurs only at the time of the positive-going rise of the clock pulse applied at *T*.

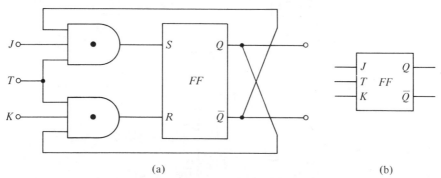

(a) (b)

Figure 16.22 (a) Equivalent circuit for a *JK* flip-flop; (b) symbol for a *JK* circuit.

As seen from the table, if both J and K are at logic 0 the AND gates are inoperative and no action occurs when the clock pulse arrives. If both J and K inputs are at logic 1, the action of the clock pulse is as a toggle and the output becomes the complement of the preceding condition. The circuit then operates as a T flip-flop.

For $J = 1$, $K = 0$ or $J = 0$, $K = 1$, the outputs follow the inputs and the circuit operates as an RS flip-flop.

Figure 16.23 illustrates a JK flip-flop as designed with CMOS switches. The transistors designated Q_1 and Q_2 constitute the basic NOR circuit of the RS form of flip-flop, with Q_3, Q_4 providing the JK features and Q_5 supplying the clock toggle action. The complexity is not a matter that need concern us when the circuit is designed as an integrated package.

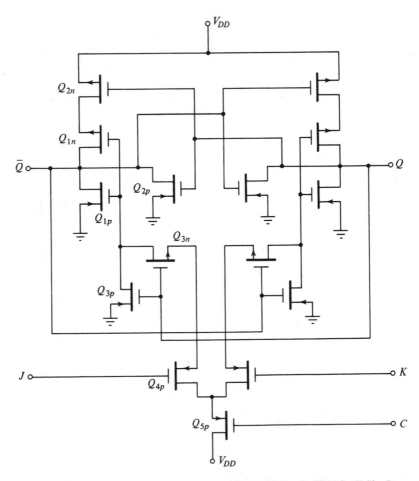

Figure 16.23   A CMOS JK flip-flop.

16.25 The One-Shot or Monostable Multivibrator

The basic circuit of the *monostable* or *one-shot multivibrator* is derived from the bistable form of circuit by replacement of one of the cross-coupling resistors with a capacitor *C*. The result is indicated in Fig. 16.24 where we have two inverters in a closed loop, with the positive feedback path between the output of inverter 2 and the input to inverter 1 consisting of the ac path provided by capacitor *C*.

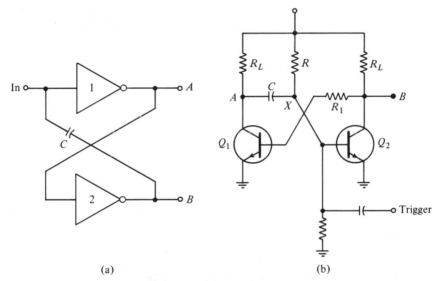

(a)   (b)

Figure 16.24 (a) Cross-coupled inverters acting as a monostable multivibrator; (b) a monostable circuit.

The normal stable condition finds inverter 1 at cutoff. If a positive pulse is given to inverter 1, driving it into saturation, its inverted output will drive inverter 2 to cutoff. The signal is coupled back to the input of 1 and will hold inverter 1 in saturation as long as the charge remains on *C*. This is a *quasi-stable state* for the circuit and after the charge drains off *C*, the circuit switches back to its original stable state in which inverter 1 is cut off.

The time duration *T* in which the circuit is in the quasi-stable condition is determined by *C* and its associated resistances and the delay time in that state can be varied by means of a variable resistance. The one-shot circuit is frequently used to provide a delayed pulse or to vary the length of a received pulse. It is then known as a "pulse stretcher."

A simplified circuit for a monostable multivibrator is drawn in Fig. 16.24(b), with one cross-coupling resistance replaced with capacitor *C* as the feedback element. Normally Q_2 is in saturation by reason of the positive bias supplied through R and Q_1 is cut off because the voltage at B and the base of Q_1 is essentially zero.

A negative trigger pulse will turn Q_2 to the off condition. As its collector current drops, the voltage at B and the base of Q_1 rises to V_{CC}, turning Q_1 on. As the Q_1 current rises, the voltage at A falls from $+V_{CC}$ to zero and this negative step potential is transmitted through C to the base of Q_2, placing it in the off condition. This change in base voltage is sketched in Fig. 16.25(b).

Transistor Q_2 is held off until capacitor C can recharge and the time delay is indicated as T seconds. When the base voltage of Q_2 rises to zero, Q_2 turns on and Q_1 goes to cutoff and the original stable condition is restored.

The time of delay is determined from the RC time constant as

$$T \cong 0.7RC \quad \text{(s)} \tag{16.8}$$

The outputs at A and B are complementary.

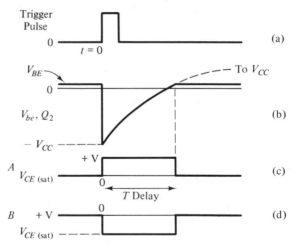

Figure 16.25 Circuit waveforms for a monostable multivibrator.

16.26 The Astable Multivibrator

The third version of the multivibrator circuit is the *astable* or *free-running form*, which acts as a square-wave oscillator. The circuit is used to provide the square-wave clock signals for digital processing.

The astable circuit is derived from the bistable form by replacing both feedback resistors with capacitors, as indicated in Fig. 16.26(a). With feedback supplied by capacitors, there is no stable state for the circuit and it oscillates with a square-wave output at B and a complemented square-wave output at A, as shown in Fig. 16.27.

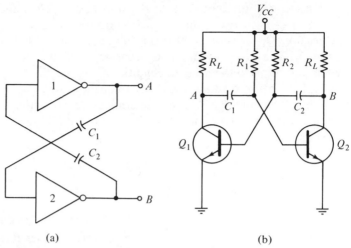

(a)   (b)

Figure 16.26 (a) Astable action by feedback with inverters; (b) stable multivibrator circuit.

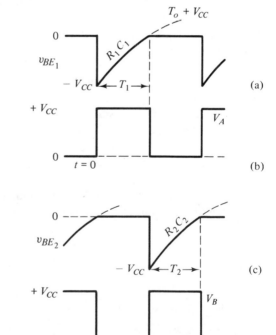

Figure 16.27 Waveforms of the astable multivibrator.

409

Consider Q_1 just turned off at $t = 0$ in Fig. 16.27(a) and (b). The base voltage of Q_1 will rise from $-V_{CC}$ toward zero as C_1 charges. When v_{BE_1} reaches 0 V, Q_1 turns on and turns off Q_2, as shown in Fig. 16.27(c) and (d) at $t = T_1$. The cycle repeats as C_2 charges, after which Q_1 and Q_2 again switch, returning to the original condition.

The oscillation continues and the frequency is determined by the time constants R_1C_1 and R_2C_2. These determine the respective time delays and since these need not be identical, the frequency of the oscillation can be stated as

$$f = \frac{1}{T_1 + T_2} = \frac{1}{0.7(R_1C_1 + R_2C_2)} \quad \text{(Hz)} \qquad \textbf{(16.9)}$$

16.27 Synchronization of the Free-Running Multivibrator

In the freely oscillating form, the rectangular output waves of the multivibrator circuit include many harmonic frequencies. This was the reason for the multivibrator name for these circuits; however, the stability of the frequency generated is not great. The circuit can be *synchronized* to a standard frequency or pulse chain by applying pulses to the base of the off transistor so that these pulses add to the rising base-voltage wave, as in Fig. 16.28. The added pulses cause triggering of the circuit at a time determined by the pulses rather than by the rising charging curve.

The synchronization need not be at the pulse rate but may take the form of frequency division as shown, in which the multivibrator triggers for every n pulses. The figure shows a countdown ratio of 8. The harmonics in the square output waves then provide standard and accurate frequencies covering a wide frequency range.

The horizontal- and vertical-sweep oscillators of TV receivers are synchronized in this manner to pulses received with the picture signals.

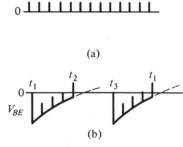

Figure 16.28   (a) Synchronizing pulses; (b) countdown by a factor of 8.

16.28 The Schmitt Trigger Circuit

The *Schmitt trigger* shown in Fig. 16.29 replaces one of the cross-coupled feedback paths of the multivibrator with feedback coupling across a common-emitter resistor R_E. The circuit switches at two input voltage levels, one when the input voltage is rising toward the trigger level and the other when the input voltage is falling from above the trigger level. The difference in these levels is called the *hysteresis* of the circuit.

The triggering voltage level is set by V_E, the voltage across the emitter resistor R_E and due to the current of Q_2. For input voltages less than V_E the input transistor is cut off by this emitter bias, Q_2 being in conduction with a low voltage at B. As v_i approaches V_E, the collector current of Q_1 starts to rise, lowering the voltage at A as well as the base voltage of Q_2. As Q_1 turns further on, Q_2 cuts off because of the feedback through R_1 from A and the voltage at B rises to V_{CC}. A reverse action follows when the input voltage falls from a value above the original V_E.

The change of output voltage at B versus the input v_i is plotted in Fig. 16.29(b) and shows the hysteresis in switching voltage that exists when the input voltage falls from above the V_E value. By always approaching the switching level from below V_E, the operation will accurately occur at the same input voltage.

Thus a switched voltage is available at B when $v_i = V_E$, and this voltage change can be used to indicate equality of the voltages for voltage comparison

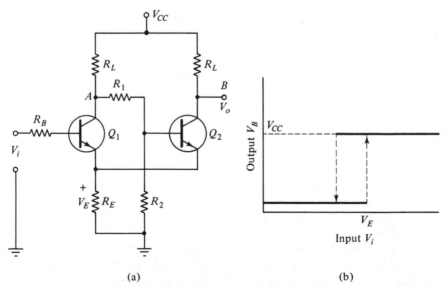

(a)   (b)

Figure 16.29   (a) The Schmitt trigger; (b) input-output relations.

circuits. The circuit also will provide squared output waves for sinusoidal or other pulse forms. In particular, it is used to reshape pulses that have been distorted in transmission, as shown in Fig. 16.30.

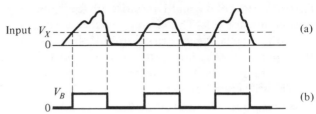

Figure 16.30   Reshaping of pulses by the Schmitt trigger.

16.29 The Shift Register

One of the most common applications of the flip-flop is in the *shift register*. This is a form of temporary memory in which data pulses, 1 and 0 levels, can be transferred serially from one flip-flop to the next flip-flop. Readout of the condition of all flip-flops can be made on call, with the data flowing into parallel channels for further processing in parallel form.

In the circuit of Fig. 16.31(a), serial data is continuously supplied from a telephone line or other data channel into the first flip-flop FF_1. The input pulse is complemented by an inverter and the pulse and its complement are fed to two AND circuits along with the clock pulses. A 1 input pulse present at the time of the first clock pulse goes into the AND circuit along with the clock pulse and the upper AND circuit output feeds a logic 1 to the set input of FF_1. If a logic 0 had come from the line, it would have been inverted and supplied as a 1 to the lower AND circuit along with the clock pulse and a 1 output would then have gone to the R input, resetting FF_1 with a 0 at Q.

On the next clock pulse the conditions at the outputs, Q and $\bar{Q}$, will be transferred through the second set of AND circuits to S and R of FF_2; new data will be supplied to the inputs of FF_1. On the third clock pulse the condition of FF_2 is transferred to FF_3, the state of FF_1 is transferred to FF_2, and the new input pulse enters FF_1. The input data pattern flows through the register with identical waveforms, as shown in Fig. 16.31(b).

The first stage samples the input data at the instant of the trailing edge of a clock pulse; even though the data input changes state at random times, the data at the output of each flip-flop changes only in synchronism with the clock.

If four-bit numbers are used, then gates placed at the outputs 1, 2, 3, and 4 can be activated once every four clock pulses and the bit values of the four-bit number then present in the register will be read into four parallel channels, for translation of serial input data for parallel processing and further

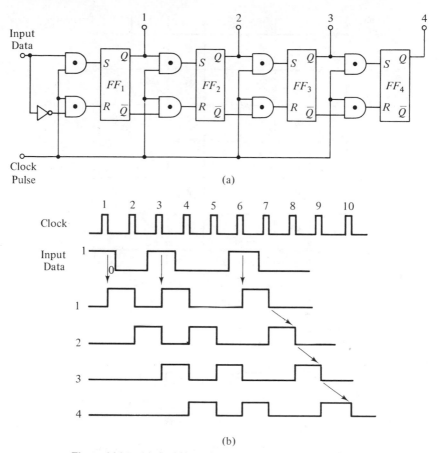

Figure 16.31 (a) A shift register; (b) flow of signals through the register.

computation. The addition of more flip-flops to the register would increase the number of bit positions that could be handled from the input data.

The shift register can also be used as a delay element. Input pulses may be fed serially to the input of a register of appropriate length and the same data taken out of a later flip-flop, delayed by the time of one clock pulse per stage included in the register.

16.30 Decoding Matrices

A decoder matrix is usually a diode logic network, actuated by a shift register as a temporary storage element, and used to decode binary signals to decimal or other code. The application in Fig. 16.32 is intended to translate three-bit BCD numbers into decimal outputs 0, . . . , 7; another flip-flop in the register and more diode logic switches would allow extension to decimal 15.

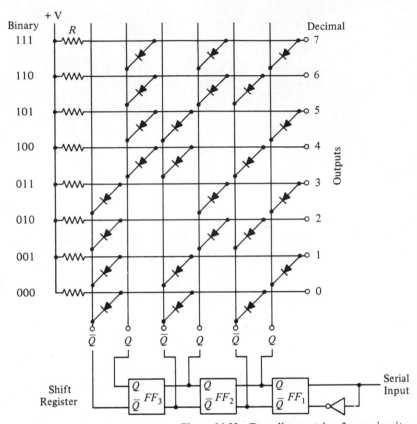

Figure 16.32 Decoding matrix of AND circuits.

The outputs of the flip-flops in the shift register are applied to lines connecting to diodes, each vertical bus constituting a four-input positive logic AND circuit. With $+V$ connected to the horizontal lines, any diode connected to a $\bar{Q} = 0$ flip-flop state will conduct and ground that particular horizontal bus. For any state of the flip-flops, only one bus will not be at zero potential and that bus will place $+V$ at the terminal corresponding to the decimal equivalent of the binary number to be indicated.

A similar diode matrix can be designed for any input code by noting that a 1 in the binary number calls for a diode at a Q connection; a 0 in the binary code requires that a diode be connected at $\bar{Q}$.

16.31 Decimal Counting

A chain of four flip-flops will count to $2^4 = 16$ pulses before the last flip-flop returns to its initial state. When using decimal numbers, it is more convenient if we can count by decades and modifications of the basic scaler circuits are available, using feedback, to cause them to count by 10's.

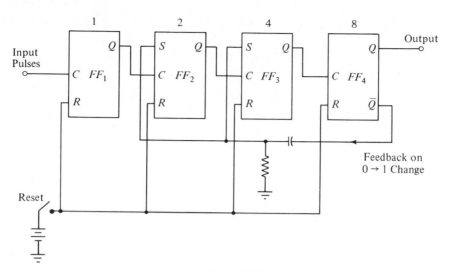

Figure 16.33 A divide-by-10 feedback counter.

One such circuit is shown in Fig. 16.33. Initially a reset pulse is applied to place all the flip-flops in the normal condition. Counting is normal up to the eighth pulse, and the conditions of the flip-flops are indicated in Table 16.3. At the eighth pulse, $\bar{Q}$ of FF_4 goes from 0 to 1 and this transition provides a pulse that is fed back to FF_2 and FF_3 to advance them by one pulse each. Including the weight of 2 for FF_2 and 4 for FF_3, the counter is therefore advanced by a total count of $2 + 4 = 6$. With the count of 8 already present at the output of FF_4 and an artificial advance of 6, the count now appears to be $8 + 6 = 14$; two more input pulses advance the output to 16, giving a 0

TABLE 16.3

Pulse No.	Q_1	$\bar{Q}_1$	Q_2	$\bar{Q}_2$	Q_3	$\bar{Q}_3$	Q_4	$\bar{Q}_4$
0	1	0	1	0	1	0	1	0
1	0	1	1	0	1	0	1	0
2	1	0	0	1	1	0	1	0
3	0	1	0	1	1	0	1	0
4	1	0	1	0	0	1	1	0
5	0	1	1	0	0	1	1	0
6	1	0	0	1	0	1	1	0
7	0	1	0	1	0	1	1	0
8	1	0	1	0	1	0	0	1
8*	1	0	0	1	0	1	0	1
9	0	1	0	1	0	1	0	1
10	1	0	1	0	1	0	1	0

*By feedback.

to 1 transition at Q of FF_4, an output pulse, and resetting all flip-flops to the initial zero count condition.

Actually the output pulse is transmitted after

$$N = 2^4 - 6 = 10 \text{ pulses}$$

and so we have a *divide-by-10 counter*. Additional circuits can be added in cascade for counting of further decades.

The condition of the individual flip-flops is indicated in Table 16.3.

At the tenth pulse the flip-flop conditions correspond to a zero count and a 0 to 1 transition pulse is transmitted from Q_4.

Such decade counters are available as complete integrated circuits. Many variations of the basic scheme are employed as elements in frequency counters.

REVIEW QUESTIONS

16.1 Explain the meaning of zero in a decimal number.

16.2 Why is a binary code well adapted for use with transistors?

16.3 What is the weight assigned to each position in the binary number 0101? What is the decimal value of that number?

16.4 Convert decimal 812 to its binary equivalent.

16.5 Convert decimal 0.764 to its binary equivalent.

16.6 What is the meaning given to the word *bit*?

16.7 In decimal numbers, what is the 10's complement of 371?

16.8 In binary numbers, what is the 2's complement of 100110?

16.9 Write the BCD number for decimal 1028.

16.10 What is the decimal equivalent of the BCD number written as

0111   0101   1000

16.11 How many bits are needed to transmit the decimal number 137 as a binary number?

16.12 What is a byte?

16.13 Write decimal 317 in octal code.

16.14 What is meant by parallel processing of digital signals?

16.15 A signal contains 32 bits per byte. If the time to process one bit is 1.1 μs, how long does it take to process one byte in series and in parallel operation?

16.16 Describe the operation of a two-input AND circuit.

16.17 Describe the operation of a two-input OR circuit.

16.18 In logic algebra, what is the meaning of the symbol $\cdot$? Of the symbol $+$?

16.19 What is a NAND circuit? A NOR circuit?

16.20 Define positive logic; negative logic.

16.21 Two voltage levels of $+5$ V and -5 V are available in some equipment. How would you assign these in
(a) Positive logic?
(b) Negative logic?

16.22 What is meant by the fan-in of a circuit?

16.23 What is the meaning of fan-out?

16.24 What is a logic gate?

16.25 Name one disadvantage of DL circuits.

16.26 What happens to AND and OR circuits when the logic voltages are reversed?

16.27 What is a NOT operation?

16.28 Name at least four factors to be considered in the selection of a particular form of logic circuit.

16.29 Name one advantage of DTL over DL.

16.30 Name one advantage of TTL circuits over DTL; also name one disadvantage.

16.31 Name one operating advantage of the ECL form of circuit.

16.32 To what design factor does the ECL family owe its switching speed?

16.33 Describe a CMOS inverter switch.

16.34 Why is the power consumption of a CMOS switch so small?

16.35 What is meant by a flip-flop?

16.36 What is a monostable multivibrator?

16.37 What is an astable multivibrator?

16.38 What is meant by multivibrator-switching time?

16.39 Why are speedup capacitors used?

16.40 What are the output conditions represented by a trigger pulse of a flip-flop to the set terminal? To the reset terminal?

16.41 Describe the action of an *RS* flip-flop.

16.42 How does a *JK* flip-flop differ from an *RS* flip-flop?

16.43 What is the function of a *T* flip-flop?

16.44 What is the cause of the quasi-stable state in the one-shot multivibrator?

16.45 What is the purpose of synchronizing a free-running multivibrator to a standard-frequency oscillator?

16.46 How would you use a Schmitt trigger?

16.47 What is the purpose of a shift register?

PROBLEMS

16.1 Decode the following binary numbers to decimal form:

10110	01010	00111
11010	10101	01011
01110	11000	11101

16.2 Write the following decimal numbers as binary numbers:

$$104, \quad 547, \quad 123, \quad 362, \quad 445, \quad 176$$

16.3 Translate the following decimal numbers to base-8, or octal, numbers:

$$123, \quad 387, \quad 462, \quad 97$$

16.4 Translate the following decimal numbers to base-3 numbers:

$$57, \quad 96, \quad 81, \quad 104$$

16.5 Perform the following operations in binary arithmetic, with $+$ meaning addition and $\times$ implying multiplication:

$$110110 + 011011 = (a); \qquad 110010 \times 10110 = (c)$$
$$101111 + 100101 = (b); \qquad 101010 \times 11100 = (d)$$

16.6 The following numbers are in binary notation:

$$01011101 \quad 01101100 \quad 0.1101$$
$$10110010 \quad 11100011 \quad 0.0111$$
$$11011011 \quad 10101100 \quad 0.1011$$

(a) Determine the decimal equivalent values.
(b) Determine the equivalents in octal notation.

16.7 For the indicated operations in Fig. 16.34(a), write the output expression in words as $F = A$ OR B. . . .

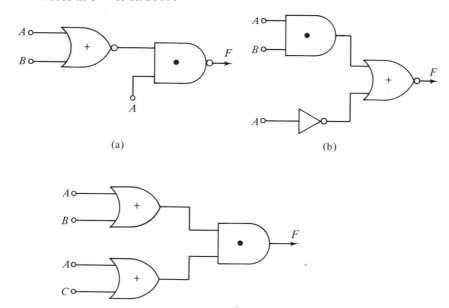

(a)

(b)

(c)

Figure 16.34

16.8 Implement the operations of Fig. 16.34(a) with diode-transistor logic circuits and positive logic, and draw the circuit.

16.9 Derive the operation table for the circuit in Fig. 16.34(a) by writing the table step-by-step.

16.10 Repeat Problem 16.7 for the circuit in Fig. 16.34(b).

16.11 Draw the circuit, using DTL gates, to carry out the operation in Fig. 16.34(b).

16.12 For $A = 0$, $B = 1$, find the output state F of the circuit in Fig. 16.34(b).

16.13 The waveforms in Fig. 16.35(a) are applied to a NOR gate. Draw the output waveform.

16.14 The waveforms in Fig. 16.35(a) are applied to an AND gate. Draw the output waveform.

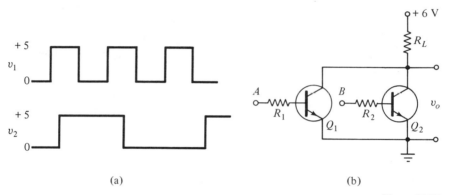

(a)   (b)

Figure 16.35

16.15 Both inputs to the *RTL* NOR gate in Fig. 16.35(b) are at $+6$ V. If $R_1 = R_2 = 10,000\ \Omega$, $R_L = 1000\ \Omega$, what are the collector currents of Q_1 and Q_2 if $h_{FE} = 50$? Consider the R_L drop.

16.16 In the circuit in Fig. 16.35(b), each transistor has a reverse saturation current of 20 μA in the off state. How many similar inputs could be added before the off-state output voltage drops to 5.5 V? Use the circuit resistances of the previous problem.

16.17 Consider the NOR gate in Fig. 16.35(b). How much current can be supplied to a load at v_o if v_o is not to be less than 5 V for logic 1, the gate transistors being in the off state.

16.18 For the *RTL* NOR gate of Problem 16.17, if $v_i = +5$ V, what is the input current? Will that input current saturate Q_1 if $h_{FE} = 30$, $V_{CC} = 5$ V?

16.19 Write the operation table for the circuit in Fig. 16.34(c). What is the output state when $A = 1$, $B = 0$, $C = 1$?

16.20 Determine if the two circuits in Fig. 16.36 are equivalent in operation by comparing their operation tables.

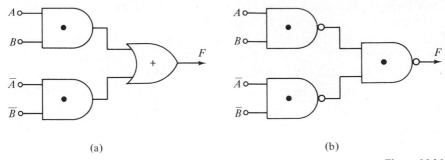

(a)   (b)

Figure 16.36

16.21 Figure 16.37(a) shows a CMOS bistable multivibrator. What are the stable operating conditions for the transistors of this circuit?

16.22 **(a)** Explain the operation of the circuit in Fig. 16.37(b).
    (b) What is the purpose of Q_1 and Q_2? Note that the circuit does not employ complementary symmetry.

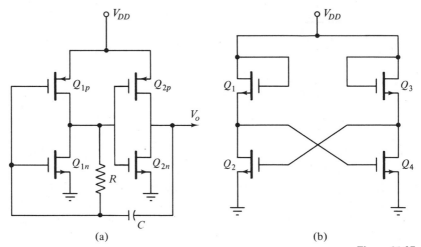

(a)   (b)

Figure 16.37

16.23 Determine the output frequency from an astable multivibrator having timing circuit component values of $R_1 = 2.7 \text{ k}\Omega$, $C_1 = 1000 \text{ pF}$ and $R_2 = 5.2 \text{ k}\Omega$, $C_2 = 1500 \text{ pF}$.

16.24 The *RTL* bistable circuit in Fig. 16.38(a) is to be changed to monostable form. Draw the circuit and give possible values for R and C to give an off time of 10 μs for Q_2.

16.25 Show whether the inverter in Fig. 16.38(b) is driven into saturation if $h_{FE} = 20$, with logic levels of $+6$ and 0 V applied at A.

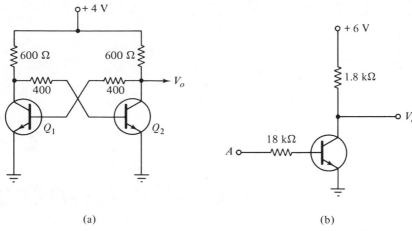

Figure 16.38

16.26 If a second trigger pulse is applied to a one-shot multivibrator while it is in the quasi-stable state, what will be the effect on operation?

16.27 The excess-3 code is sometimes used because no digit is represented by a complete null or zero signal:

0	0011	5	1000
1	0100	6	1001
2	0101	7	1010
3	0110	8	1011
4	0111	9	1100

Design a diode-switching matrix to translate excess-3 coded numbers to their equivalent decimal values.

17
Power Control

The basic design of a junction transistor is not well suited for switching in high-power applications. For efficiency with high currents, the emitter and collector regions should be of low resistance or of low resistivity material; for high h_{FE}, the base should be very thin. A thin base between a low-resistance emitter and collector is not suited to operation at the high voltages needed in industrial switching. By use of four material layers and three junctions in series, however, the *silicon-control rectifier* (SCR) family of devices has been developed to control high currents at high voltages. This family includes various related designs known as the SCR, thyristor, triac, and the gate-controlled switch.

17.1 The Silicon-Control Rectifier (SCR)

In contrast to the continuous control of current in the transistor, the *silicon-control rectifier* action is that of a trigger able to switch a current on. The current can only be stopped by reducing its magnitude below a certain holding value I_h. The four-layer *pnpn* construction of an SCR is shown in Fig. 17.1(a).

We can consider the unit to consist of two transistors in series, as diagrammed in Fig. 17.1(b). Transistor Q_1 is composed of layers n_2, p_1, n_1 and is of *npn* characteristics and transistor Q_2 employs layers p_2, n_2, p_1 and is of *pnp* characteristics. A *gate electrode* is connected to p_1. With the gate grounded or $V_G = 0$, transistor Q_1 is cut off and its collector current $I_{C_1} = I_{CO}$ and consists of the leakage current only. This is the base current to Q_2 and it is too small to turn that transistor on. The device is open-circuited for $V_G = 0$.

When we apply a sufficiently large positive voltage to the gate, we turn Q_1 on. The collector current is $I_{C_1} = I_{B_2}$ and so Q_2 turns on; but the collector current of Q_2 is $I_{C_2} = I_{B_1}$ and so Q_1 turns on further, increasing $I_{C_1} = I_{B_2}$. The action around the Q_1, Q_2 loop is cumulative and the device rapidly switches to a conducting condition between anode and cathode. With both bases driven to saturation the forward resistance between anode and cathode is very small. The time required for the cumulative turn-on action is approximately 0.1 to 1 μs.

Except for the initial triggering action, device conduction does not depend on the gate current. Therefore turn-off does not occur when the gate signal is removed since the two base currents are now internally driven. To stop conduction the anode voltage must be removed or the anode driven to a negative potential, reducing the current below I_h. Turn-off time is typically 5 to 30 μs.

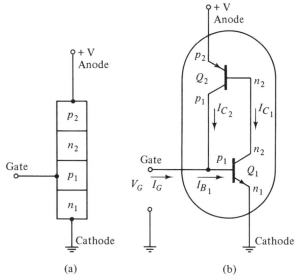

Figure 17.1 (a) A four-layer switch; (b) two-transistor simulation of four-layer action.

17.2 SCR Characteristics

The SCR is normally used with an ac voltage applied and conduction occurs on the half cycle in which the anode is positive. The characteristic for one value of gate current I_G is shown in Fig. 17.2(a), with the circuit symbol of the SCR presented in Fig. 17.2(b).

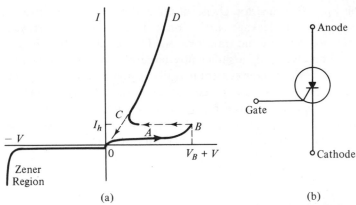

(a)   (b)

Figure 17.2 (a) Voltage-current relation for the SCR; (b) symbol.

For a particular value of gate current the SCR will trigger when the voltage of the wave reaches V_B. For larger values of I_G the triggering voltage will be lower. As conduction starts at point B on the voltage-current curve, the SCR voltage abruptly falls to C and then rises to its peak value along the CD portion of the curve. In the on condition the SCR acts as a forward-biased diode and the curve between C and D is that of a junction diode. A great many combinations of gate voltage and gate current will trigger the SCR but gate voltages greater then 3 V are usually required.

After the peak of the ac current wave is passed on the CD curve, the current value falls along the curve; when it reaches I_h, the *holding current*, the conduction ceases and the current goes to zero. With ac applied, conduction ceases when the ac sine wave approaches zero.

Applications with ac involve the use of a pulse of gate current i_G at the time it is desired that conduction start in the circuit in Fig. 17.3(a). This will be at some angle θ_1 in the wave of ac anode voltage illustrated in Fig. 17.3(b). The average or dc value of the rectified current varies with the triggering angle θ_1 as shown in Fig. 17.4. For a half-wave-controlled rectifier circuit with a

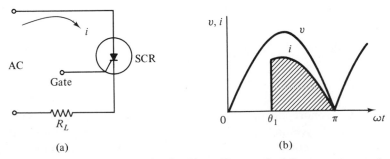

(a)   (b)

Figure 17.3 (a) SCR control circuit; (b) rectifier waveform for current control.

resistance load, the output voltage is expressed by

$$\frac{V_{dc}}{V_m} = \frac{1}{2\pi}(1 + \cos \theta_1) \tag{17.1}$$

For the full-wave rectifier circuit the result is twice as great, or

$$\frac{V_{dc}}{V_m} = \frac{1}{\pi}(1 + \cos \theta_1) \tag{17.2}$$

Mounting, cooling, and temperature limitations for the SCR follow those previously discussed for the power transistor. Devices available employ gate currents of 50 mA to control operating currents of 50 A or more.

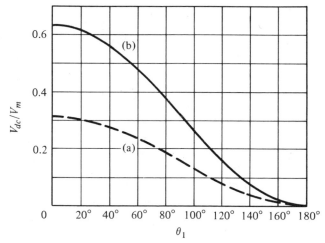

Figure 17.4 Control of average rectifier current: (a) half wave, resistance load; (b) full wave, resistance load.

17.3 The Unijunction Transistor (UJT)

Another trigger device is made with a bar of n silicon having a p junction of aluminum-doped material added below the midpoint as shown in Fig. 17.5(a). With only one junction, the device has become known as a *unijunction transistor* (UJT). The circuit symbol appears in Fig. 17.5(b). The emitter arrow points in the direction of forward current.

The ends of the bar act as base contacts B_1 and B_2. Base B_2 is maintained positive at V_{BB} and the bar appears as a resistance of 4000 to 10,000 Ω. The current through the bar creates a voltage drop along the bar. At the location of the p emitter the bar is positive to B_1 and ground by some fraction of V_{BB}, called ηV_{BB}. The lowercase Greek letter η (eta) is called the *intrinsic standoff ratio*.

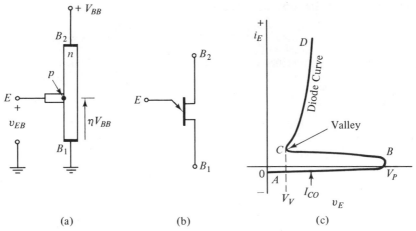

Figure 17.5 (a) Unijunction transistor construction; (b) circuit symbol; (c) performance curve.

If the voltage v_{EB} applied to the emitter is less than ηV_{BB}, then the emitter is negative to the n bar and the junction is reverse-biased. If the voltage v_{EB} is made more positive than ηV_{BB}, the emitter becomes forward-biased and holes move from the p emitter into the n bar and toward B_1. The presence of holes in the bar calls for electrons to enter the region from B_1 and the increased density of mobile charges lowers the resistivity of the bar between E and B_1. The voltage between E and B_1 then drops, allowing more holes to enter the bar and results in more electrons entering, rapidly reducing the voltage v_{EB} from V_p to that of a normal forward-biased diode, as illustrated in Fig. 17.5(c).

This action can also be explained by use of Fig. 17.5(c) in which the initial current at A is only the reverse value I_{CO}. The voltage can rise to the *peak voltage* V_p where $v_{EB} = \eta V_{BB} = V_p$, and emitter forward conduction begins. With the entrance of holes into the bar, the bar resistance falls abruptly and the voltage drops to C on the forward-bias diode curve. The *valley voltage* V_v is the lowest voltage value between emitter and base B_1. We thus have a device that will rapidly switch from V_p to the diode voltage approximating V_v.

The value of the peak voltage V_p can be predicted for a given V_{BB} by use of

$$V_p \cong \eta V_{BB} + 0.7 \text{ V} \tag{17.3}$$

where the term 0.7 V is an approximation to the inherent forward drop of a silicon junction.

The device yields an equivalent circuit of the form shown in Fig. 17.6. The diode is that of the junction; the resistances R_1 and R_2 are those of the

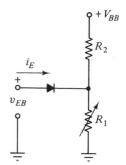

Figure 17.6　Equivalent circuit for the UJT.

two parts of the bar. The value of η can be found from the resistances of the bar as

$$\eta = \frac{R_{10}}{R_{10} + R_2} \qquad (17.4)$$

where R_{10} is the value of R_1 at $i_E = 0$. The resistance $R_{10} \cong 5000\ \Omega$ and may drop to 50 Ω when switched by i_E, for a typical UJT.

17.4 Triggering the SCR

The circuit in Fig. 17.7(a) will provide a variable-phase voltage for triggering an SCR at a desired angle θ_1 in the ac cycle. Either R or C may be varied to shift the phase of the control voltage V_c, derived between points B and E of the phase shift bridge. Voltage drops IR and $jIX_C = jI/(2\pi fC)$ must add to V_{AD}; this addition must be carried out at right angles because of the j coefficient of the reactance term. Then, by geometry, point E must always lie on the semicircular locus with a constant diameter equal to V_{AD}, as shown in Fig. 17.7(b). As either R or C is varied, the point E moves

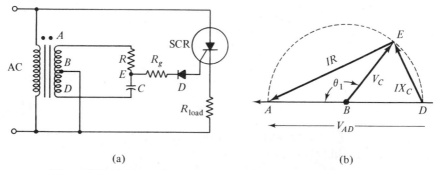

(a)　　　　　　　　　　　　　　　　　　(b)

Figure 17.7　(a) Phase shift control of an SCR; (b) phase shift analysis by circle diagram.

around the circle and voltage V_c lags the SCR anode voltage by angle θ_1; the magnitude of V_c remains constant at one-half of V_{AD}.

The lag angle θ_1 may be found from the circuit constants and the geometry of the triangle as

$$\tan \frac{\theta_1}{2} = \frac{R}{1/(2\pi f C)} = 2\pi f CR \tag{17.5}$$

If R is replaced with an inductor L and C with a resistor, then

$$\tan \frac{\theta_1}{2} = \frac{2\pi f L}{R} \tag{17.6}$$

If the range of variation of these components is sufficiently great, the angle θ_1 may be varied nearly from $0°$ to $180°$, resulting in a variation of the average value of SCR current as indicated in Fig. 17.4(a).

Diode D is present to prevent gate current on the negative half cycle and resistor R_g serves to limit the forward gate current to a safe value.

A common application of the UJT is in triggering an SCR, as shown in the typical circuit in Fig. 17.8. Diode D is incorporated to synchronize the gate signal with the positive anode of the SCR. With terminal A and the SCR anode positive, the diode blocks current and transistor Q is off. Capacitor C is then able to charge through control resistor R_x, introducing a time delay in each positive half cycle before the capacitor voltage builds up to V_p and triggers the UJT. This buildup of voltage is shown from t_0 to t_1 in Fig. 17.8(b) and is determined by the time constant CR_x. When the UJT triggers, its emitter voltage falls and C is discharged through the UJT and R_2. The sharp pulse of discharge current produces a peaked gate voltage for the SCR, as shown between t_1 and t_2 in Fig. 17.8(c), and the SCR turns on. The magnitude of the triggering current for the SCR is controlled by resistor R_2.

The circuit provides a triggering delay in each cycle and also provides a pulse waveform for precise triggering of the SCR. The time of triggering or the angle θ_1 is controllable by resistor R_x.

On the negative half cycle at A, diode D supplies voltage to Q and this transistor conducts and acts as a short circuit across C. The charging of C then starts precisely at the beginning of each positive half cycle at A.

Resistor R_x must vary within limits so that the peak point triggering current I_p can be obtained from the $+22$ V source. Thus R_x must be limited as

$$\frac{V - V_p}{I_p} > R_x \tag{17.7}$$

to turn on the UJT. At the valley point the emitter current must fall below I_v to turn off the UJT; that is,

$$\frac{V - V_v}{I_v} < R_x \tag{17.8}$$

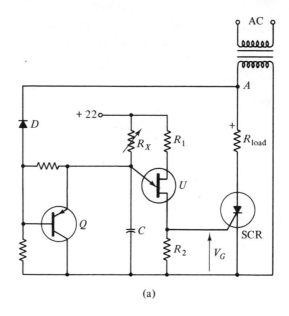

(a)

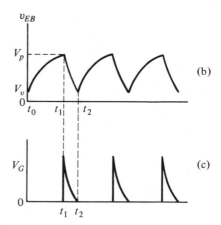

(b)

(c)

Figure 17.8 (a) Phase control with a unijunction trigger; (b) and (c) triggering waveforms.

The steady UJT current must not produce a voltage equal to the gate voltage across R_2; this places a limit on R_2.

Example: A typical UJT is rated

$$P_d = 300 \text{ mW} \qquad V_{BB} = 35 \text{ V}$$
$$I_{E(\text{peak})} = 2 \text{ A} \qquad R_{BB} = 7000 \text{ }\Omega$$
$$I_{E(\text{ave})} = 50 \text{ mA} \qquad \eta = 0.6$$

Typical circuit values may be chosen as $V_{BB} = 30$ V, $V_v = 1$ V, $I_p = 10$ μA, and $I_v = 10$ mA. Using Eq. 17.3,

$$V_p = \eta V_{BB} + 0.7$$
$$= 0.6 \times 30 + 0.7 = 18.7 \text{ V}$$

With Eq. 17.7

$$\frac{V - V_p}{I_p} = \frac{30 - 18.7}{10 \times 10^{-6}} = \frac{11.3}{10^{-5}} = 1.13 \times 10^6 \text{ } \Omega > R_x$$

With Eq. 17.8,

$$\frac{V - V_v}{I_v} = \frac{30 - 1}{10 \times 10^{-3}} = \frac{29}{10^{-2}} = 2900 \text{ } \Omega < R_x$$

Therefore we have available the range of 1.13 MΩ to 2900 Ω for the variation of R_x, in controlling θ_1 for triggering the SCR.

17.5 The Controlled Polyphase Rectifier

Most industrial power is supplied by three-phase circuits. When dc power is needed, it is convenient to employ a polyphase rectifier, one form of which is shown in Fig. 17.9. This is a bridge form of circuit, in which two rectifier elements conduct in series at a given time. Control of the starting time of current conduction in SCR_1, SCR_2, SCR_3 is furnished by a three-phase adaptation of the unijunction transistor circuit shown in Fig. 17.8. Each diode conducts at the time its corresponding SCR turns on.

More complex circuits are used for providing larger amounts of power.

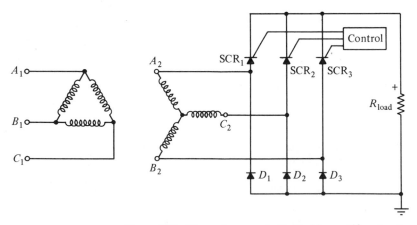

Figure 17.9 Three-phase controlled bridge rectifier circuit.

17.6 Shunt-Wound DC Motor Control

The speed-control circuit for a dc shunt-wound motor illustrates the application of some of the principles previously discussed. A simplified circuit is shown in Fig. 17.10, which operates by controlling the triggering time of an SCR so as to vary the average voltage applied to the motor armature.

The UJT circuit is similar to that of Fig. 17.8, with diode D supplying a positive voltage to the UJT only on the half cycle in which the SCR anode is also positive. Current from D charges C through R_x and when the capacitor voltage reaches V_p of the UJT, the latter triggers and discharges C through R_G. This sharp pulse of current produces a pulse of gate voltage for the SCR and it is triggered on. The current of the SCR is the armature current of the motor, occurring in partial half-wave pulses as shown in Fig. 17.10(b). Adjustment of R_x controls the triggering angle θ_1 and thereby the average current to the motor. The result is variation of motor speed.

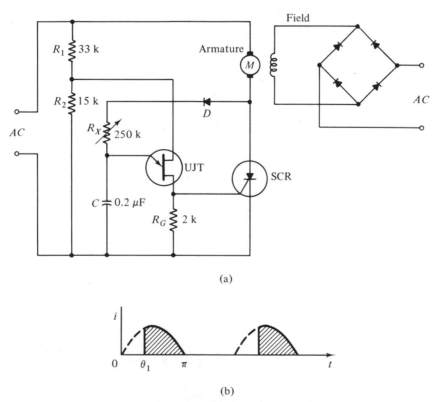

(a)

(b)

Figure 17.10 DC motor control circuit; (b) waveform of current.

If the motor is heavily loaded, it tends to slow down, the voltage across the armature will be reduced, and a larger voltage will appear at diode D and across the control circuit. This larger voltage decreases the time required to charge C and trigger the SCR and yields an earlier θ_1 and a longer current pulse to the armature. This longer pulse tends to speed up the motor and compensates for motor loading. Speed regulation of about 10 per cent is possible.

The motor field current is supplied by a separate bridge rectifier.

17.7 Comments

We now have a family of power control switches. Since all units of the family are basically similar to the SCR, we have discussed applications only in terms of that unit. A listing of the devices seems worthwhile, however, and is presented here:

SCR: Behaves as a rectifier diode, blocking current in the forward direction until a current pulse of sufficient magnitude is applied to the gate electrode.

Thyristor: A general name for the entire family of power control switches but most usually applied to the device we here call the SCR.

Triac: A two-way conduction version of the SCR, with current triggering by a gate pulse. The action of the gate controls the time of triggering in both current directions.

Diac: A two-electrode, three-layer device acting as two inverse-parallel diodes. It conducts in either direction after the applied voltage exceeds a value called the *breakover voltage*.

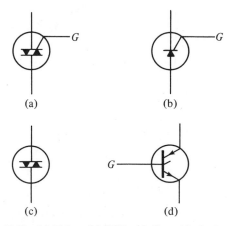

Figure 17.11 (a) Triac; (b) SCR; (c) diac; (d) GTO.

GTO: A gate-turn-off switch; a one-way device with turn-on characteristics as in the SCR. By driving a sufficient negative current into the gate, the GTO can also be turned off.

Circuit symbols for these devices are shown in Fig. 17.11. For further information, reference should be made to the manuals of several manufacturers, as

SCR Manual, G.E. Semiconductor Products Dept., Electronics Park, Syracuse, N.Y.

RCA Solid-State Power Circuits, SP-52, RCA, Somerville, N.J. 08876.

Westinghouse SCR Designer's Handbook, Westinghouse Corp., Semiconductor Division, Youngwood, Pa.

REVIEW QUESTIONS

17.1 What is an SCR?

17.2 Explain the process of cumulative current buildup when an SCR is gated to turn on.

17.3 What is the typical turn-on time of an SCR?

17.4 What turn-off time is expected for an SCR?

17.5 Explain the SCR action in the different current regions of Fig. 17.2.

17.6 What is meant by the *holding current*?

17.7 How do you turn off an SCR?

17.8 What is a UJT?

17.9 What is the intrinsic standoff ratio?

17.10 How is η (eta) measured?

17.11 Explain the process that occurs in a UJT when it triggers.

17.12 What is the peak voltage in a UJT?

17.13 What is the valley voltage?

17.14 If C is fixed and R is variable, what are the limits on R if θ_1 is to vary between $0°$ and $180°$ in a phase shift bridge?

17.15 What is the function of R_g in the gate circuit of an SCR?

17.16 What factors determine the time between voltage application and triggering of the UJT in the circuit of Fig. 17.8?

17.17 What is the purpose of transistor Q in the circuit of Fig. 17.8?

17.18 Why do we not have to use two sets of SCR elements in the polyphase bridge rectifier of Fig. 17.9?

17.19 Explain the method by which speed is controlled in the motor circuit of Fig. 17.10.

PROBLEMS

17.1 The ac supply to an SCR half-wave rectifier is 240 V rms. When the load resistance is 1000 Ω, what are the dc load currents when the SCR is triggered at 0°, 40°, 90°, and 135°?

17.2 A 10-Ω load is connected to a 120-V rms supply line through an SCR. The average load current is to be varied between 3.0 and 0.5 A. What range of triggering angles is needed?

17.3 A 10-Ω load is supplied with an average current from a 120–0–120 V center-tapped transformer through a full-wave SCR rectifier circuit. The load power is to be varied between 100 and 40 W.
(a) What are the load voltages?
(b) What is the range for the triggering angle?

17.4 A phase shift bridge at 60-Hz is to control an SCR from full on to 20 per cent of full-on current. If $C = 0.7\ \mu F$, find the range of variation of R.

17.5 A phase shift bridge at 60-Hz is to control an SCR from 10 per cent to full-on current. The value of $C = 0.15\ \mu F$. What is the range needed for R?

17.6 In Fig. 17.12, $R_1 = 5000\ \Omega$, $C_1 = 0.5\ \mu F$, and $R_L = 60\ \Omega$. What is the angle θ_1, and what is the average load current I_o for V = 60 V?

17.7 In Fig. 17.12, $R_1 = 2500\ \Omega$ and $R_L = 10\ \Omega$. What value of C_1 will cause the SCR to trigger at 50°?

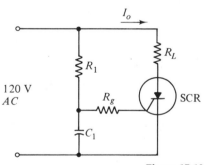

Figure 17.12

17.8 An SCR is used to control the power into a 1000-W, 57-Ω, heater element from a 540-V rms circuit. What triggering angles should be used for $\frac{1}{4}$, $\frac{1}{2}$, and $\frac{3}{4}$ rated heat?

17.9 A UJT has $R_{10} = 5000\ \Omega$ and $R_2 = 6000\ \Omega$. What is the value of η? With $V_{BB} = 25$ V, what is the expected peak voltage?

Index